THE
CRUSADER

THE

CRUSADER

RONALD REAGAN
AND THE FALL OF COMMUNISM

PAUL KENGOR, PhD

REGAN

An Imprint of HarperCollins*Publishers*

For editorial inquiries, please contact Regan, 10100 Santa Monica Blvd., 10th floor, Los Angeles, CA 90067.

Designed by Kris Tobiassen

ISBN 13: 978-0-06-113690-0

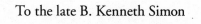
To the late B. Kenneth Simon

CONTENTS

"The U.S. President really does see himself as . . . some kind of latter-day crusader."

—YURI KORCHAGIN, "UNRESTRAINED,"
IZVESTIA, JUNE 11, 1982

"[T]he 'crusade' declared by U.S. President Reagan is not just talk. It is an action program aimed at . . . 'rolling back' communism. That is, a program of all-around struggle against world socialism."

—VADIM ZAGLADIN, "THE GREAT TRUTH
OF OUR TIME," *PRAVDA*, JULY 28, 1983

"Join me in a new effort, a new crusade."

—RONALD REAGAN, FEBRUARY 26, 1982

PREFACE

IT MAY SEEM ODD TO TITLE A BOOK ON RONALD REAGAN AND the Cold War, *The Crusader*. Some might think the choice sensational, chosen to overdramatize what Reagan believed, or even sarcastic. Yet, the label reflects Reagan's mindset and actions, and what the Soviets believed and said about him throughout the 1980s.

The title is instructive: In researching this book, I spent many days reading through Soviet media archives from the 1960s through the 1990s—articles from *Pravda*, *Izvestia*, and unmemorable publications. I vetted press releases from TASS, the official Soviet news agency, as well as transcripts of Moscow radio and TV broadcasts. I was surprised to encounter hundreds of examples of Soviet figures referring to Reagan's intentions vis-à-vis the USSR as a "crusade," dubbing him the "crusader," and calling his team "crusaders." It was not uncommon in Moscow in the 1980s to open the pages of *Pravda* or *Literaturnaya Gazeta* and be struck by headlines like "Twentieth Century 'Crusaders' " or "The Washington Crusaders: The 'Ideological War' Declared by Reagan Against Communism and Socialism."[1]

The Soviets frequently used these words to characterize Reagan's assault on their ideology and empire. It was standard fare for someone like commentator Genrikh Borovik to get on Moscow TV and explain that "a crusade against us was announced" by Reagan.[2] When the Soviets used the word, they often employed it within a religious context and with derision. Where did they come up with this colorful language? It was not devised by a Kremlin propagandist. Reagan himself had used it since at least 1950.

Many years later, in his emotional July 17, 1980 presidential nomination speech at the Republican convention, Reagan extemporaneously closed: "I'll confess that I've been a little afraid to suggest what I'm going to suggest—I'm more afraid not to: that we begin our crusade joined together in a moment of silent prayer."[3] As president, he frequently exhorted the faithful to join him in "a new effort, a new crusade," as he requested in a February 26, 1982 speech at the Conservative Political Action Conference.[4] As the Soviets knew, the word did not refer simply to Reagan's domestic thinking or political conservatism. He applied it to the great struggle, to the Cold War battle against Communism.

The president took the crusading message abroad, twice urging British audiences to join him. On June 8, 1982 in his Westminster Address in London, where he predicted Marxism-Leninism would be placed on the "ash heap of history," Reagan pressed for a "crusade for freedom"—a call that sent the Soviet media into a fury. Returning six years later in a June 1988 speech before London's Royal Institute of International Affairs, he offered a verse from the Bible (Isaiah 40:31) as the ingredient in his "formula for completing our crusade for freedom." "[T]hat crusade for freedom," he assured his distinguished audience, "that crusade for peace is well underway."[5]

Of course, the word *crusade* has a heavy religious connotation. In my work, *God and Ronald Reagan: A Spiritual Life*, I focused on the man's spiritual pilgrimage, a journey that culminated in his personal crusade to undermine the USSR—a peaceful crusade not marked by any hoped for or planned apocalyptic clash. He saw himself as an instrument of God, doing the Almighty's will according to what he called "God's Plan." Ronald Reagan believed America was chosen by God to confront the Soviet Communist empire and prevail. As the leader of the United States at a special moment in time, he sensed that God had ordained such a role upon, as he put it, his "team."

It was this religious dimension to Reagan's Cold War assault that enraged the Kremlin. The officially atheistic Soviet government had long pursued, in Mikhail Gorbachev's words, an open "war on religion."[6] For Soviet atheists, Reagan's talk of a crusade was too much to swallow; it became a source of highest condemnation and scorn. "[T]he present White House incumbent, invoking God, [has] declared the 'crusade' against socialism," observed Vitaliy Korionov in an angry *Pravda* analysis titled, "Production Line of Crimes and Hypocrisy."[7] Also in *Pravda*, propagandist Georgi Arbatov was unusually subdued when he raged at Reagan: "Frenzied calls are being made for crusades. . . . [This is] outright medievalism. And all this is covered up by hypocritical talk about faith and God, about morality, eternal good and eternal evil."[8]

To Moscow, Reagan's crusade was synonymous with his effort to undermine the Communist empire. The Kremlin understood his intentions completely. The objective of this book is to reveal Reagan's intent to undermine and, in the process, illuminate his personal role in this historic effort.

FEW EVENTS IN THE TWENTIETH CENTURY AND AMERICAN history as a whole were as consequential as the end of the Cold War. Vladimir Lenin's Bolsheviks seized power in 1917. The Cold War started in the 1940s. By 1989, the Soviet Communist empire was finished—a death that Ronald Reagan had hoped for long before his 1981–89 presidency. Historian John Lukacs declared that the twentieth century ended in 1989. And so it did.

A professor teaching a course on the twentieth century could tell much of it through Reagan's experiences—from the seven-year-old boy joining a flag-waving crowd welcoming home doughboys from WWI on the streets of Monmouth, Illinois in 1918, to the influenza epidemic that nearly took his mother in 1919, to the advent of radio, to the Great Depression, to the magic of Hollywood's golden age, to the New Deal and the rise of the federal government, to World War II, and on to the Cold War, Communism, the bomb, the Red Menace, and much, much more, all the way through his presidential races and the Cold War victory. One can trace the birth of the Bolshevik Revolution, six years after Reagan's birth, to its collapse when he was near eighty: the start and end of Soviet Communism were the bookends of Ronald Reagan's life.

Reagan not only hoped for Communism's demise; he often predicted it. More so, his administration went beyond hoping for the end, and beyond the commander-in-chief's forecast that Communism would end up on the ash heap of history. The Reagan administration went so far as to design and implement actions, policies, and even formal directives intended to reverse the Soviet empire and win the Cold War. This is not an exaggeration, not a sloppy statement, and the fact itself is historically critically important. And yet, despite all that has been written on Ronald Reagan, questions on the man's *personal role* in this Cold War strategy remain unanswered.[9]

Edmund Morris, who more than any outsider had the best opportunity to learn what Reagan knew and did, described in the prologue of *Dutch*, his official biography of Reagan, how he left the Oval Office one day after interviewing the president and puzzled for the "hundredth time"—"How much does Dutch really know?"[10] The *New York Times'* columnist Anthony Lewis, one of many on the political left not respectful of Reagan's mind, who usually

assumed the worst, said, "There's this mystery to Reagan that pervades everything, which is how much was he aware of what he was doing?"[11]

This book will show that the correct answer, as related to Reagan and the
Cold War specifically, was provided by Reagan foreign-policy adviser Richard
V. Allen, who said candidly that "the key factor in the winning side's team,"
meaning the United States in winning the Cold War, "was the president himself."[12] Allen's assessment, unlike that of Lewis, had the advantage of proximity. Unlike the pundits, Allen actually interacted with Reagan on a daily basis
for years, both before and during the White House period. That was even
truer for Judge William P. "Bill" Clark, former Reagan national security adviser, right-hand man, and close friend. Going further than Allen, Clark has
often said that Reagan "will be remembered by those of us who worked with
him in Sacramento and Washington as being far wiser than his Cabinet and
his staff combined."[13]

Historians can debate those claims to the letter. Not debatable were contemporary charges like this from Gary Paul Gates and CBS News' Bob
Schieffer in their book, *The Acting President*: "Ronald Reagan had very little
to do with his administration and the issues that came before it." The evidence overwhelmingly proves that that assertion was simply not accurate.[14]
Likewise inaccurate, and reflective of much partisan thinking at the time, was
a December 8, 1986 *Time* feature titled "How Reagan Stays Out of Touch."
It ran as a news article, not an opinion piece. The article declared:

> [Reagan's] briefing with his senior staff, which mainly concerns his daily
> schedule, lasts only about thirty minutes, and Reagan usually remains quiet,
> except for his trademark bantering. It is followed by a briefing from his
> national security council staff that is usually even shorter. When National
> Security Council staffers prepare Reagan for a full-fledged meeting of the
> NSC, the President typically does not ask any questions about the topic at
> hand; instead he inquires, "What do I have to say?" . . .
>
> Reagan's reading is not heavy. . . . Old friends and cronies have access
> to a special private White House post office box number and they can send
> him clippings that they think might strike his fancy. That box number is the
> source of many of Reagan's familiar "factoids," snippets clipped from ob-
> scure publications.
>
> Reagan is not notably curious. His aides say he rarely calls them with a
> question and that he knows in only a vague way what they actually do.[15] He
> does not sit down with his advisers to hammer out policy decisions. He is

happiest when his aides form a consensus, something they try awfully hard
to do. . . .

He sees himself less as an originator of policy than as the chief mar-
keter of it. . . .

[Reagan] can work only if he is supported by a competent and active
staff. During his first term, Chief of Staff James Baker protected Reagan
from his woollier notions and helped put many of his ideals into practice.
When Baker and Donald Regan pulled off their White House shuffle in
January 1985, with a typically detached Reagan looking on like a bemused
bystander, the new chief of staff proclaimed that his primary goal was to let
Reagan be Reagan.[16]

By that point in this astounding article, *Time* readers understood the danger of letting Reagan be Reagan: the president might not be able to find his
desk.

Today, many journalists, like many academics, have stepped back, allowed their emotions and partisan inclinations to cool, taken a careful look at
the record, and see Reagan much differently. "Ronald Reagan was not the
person that I thought he was when I was covering him in the early 1980s,"
says *Washington Post* reporter Don Oberdorfer, who has written a book on the
period.[17] His is far from an isolated judgment.

Many long-time Reagan critics, some of the harshest among them, seem
to have softened or outright changed. Historian Garry Wills now states:

Part of Reagan's legacy is what we do *not* see now. We see no Berlin Wall.
He said, "Tear down this wall," and it was done. We see no Iron Curtain. In
fact, we see no Soviet Union. He called it an Evil Empire, and it evaporated
overnight. . . .

Admittedly, Reagan did not accomplish all this by fiat. But it was more
than coincidence that the fall began on his watch.[18]

There is hyperbole here, particularly Wills' suggestion of immediate
causality. Yet, curtains and empires fell, as Reagan expected. It is striking that
a Reagan critic to Wills' degree, who seemed to have no respect for Reagan,[19]
now grants so much credit.

The biggest surprise, especially to Reagan conservatives, has been the
reevaluation of the fortieth president by academics. Reagan has received quite
fair, favorable reviews by top presidential scholars and historians—names like

Thomas Mann, Sidney Milkis, John W. Sloan, David Mervin, Stephen Skowronek, James T. Patterson, Robert Dallek, Matthew Dallek, and Alonzo Hamby, to cite a few.[20] George Mason University's Hugh Heclo maintains that Reagan was a "man of ideas" in the estimable company of Jackson, Madison, and Jefferson.[21] There are similar appraisals from giants like Harvard's dean of presidential scholars, the late Richard Neustadt, from popular historian and Pulitzer Prize–winner David McCullough, and from John Lewis Gaddis, the leading Cold War historian.[22] Reagan is even faring well in certain surveys of academics, nearly all of whom—like most to all of the aforementioned names—are politically liberal.[23]

There are many reasons for this rising Reagan.[24] Importantly, Ronald Reagan's Cold War success—now widely attributed to more than luck—heads the list. And yet, as the pages ahead show, there is much to learn about the man's Cold War achievement.

PART I

The Early Years

1.

Rock River Rescuer

STEPPING OUT OF HIS HOUSE THE MORNING OF AUGUST 2, 1928, Ronald "Dutch" Reagan was expecting another scorcher. As he walked across the street to the Graybills' to catch his ride to the river, he noticed that it was yet another muggy Thursday in Dixon, Illinois. It was a typical mid-summer afternoon in the Midwest, humid beyond any reasonable expectation, and with the advent of air-conditioning still years in the distance, the best form of escape could be found in a Ford automobile with windows open amid a breezy drive to the river at Lowell Park.

At Lowell, there were shady trees, cool water, and people, all kept under the watchful eye of seventeen-year-old Dutch Reagan. Already tall, he hovered above the swimmers in a ten-foot-high chair perched on the grassy banks, making himself a beacon for all to see. His height was emblematic of his swimming prowess, and a key factor in his swimming successes. At the YMCA in January, Reagan had sprinted to victory in the 110-yard freestyle by a half-length of the pool. When competing in the annual Water Carnival at the Rock River on Labor Day, he took first in the longest competition—the 220-yard River Swim.[1] He still holds the record for swimming fastest from the park entrance to the river's farthest bank and back. So adept were his swimming skills that he was allegedly approached by an Olympic scout who invited him to work out with the team preparing for the 1932 games—an offer Reagan said he refused because he could not give up his summer pay.[2]

On this August day, the river's rough waters and undertows were particularly active. Scattered throughout the choppy waters were hundreds of swimmers, and through the spectacles that rested atop his nose, Dutch gazed at the clusters of people, aware that he could not slack off for a moment. Reagan's regular pattern for patrolling the waters was tested on such a chaotic summer afternoon. According to the *Dixon Telegraph*, on a day like this Ronald Reagan often single-handedly watched over 1,000 bathers *at a time*, with *no* assistant.[3]

His most difficult concerns were toddlers who ventured too far out (there were legions of them) and adults cocky enough to think they could conquer the depths of the treacherous river. Toddlers that failed to listen were an easy nab for Reagan, who was vigilant in pulling them back right away before they disappeared into the murky water. Dutch always followed with a quick lesson to the child about wandering out.

Unfortunately, the adult swimmers were not as easy. They were bigger and stronger. If not secured in the right position, they tended to pull and grab, putting the lifeguard's own life in peril. A panicky six-foot-frame was the worst foe. Among them, the end of the summer brought brawny farm boys to the water, just finished with the annual harvest; they invariably underestimated the river, not giving it the respect Reagan learned to grant it. Once in the death grip of a current, they became exhausted, went vertical, and began struggling and clawing frantically. On more than one occasion, Dutch belted them with a right cross to the jaw in order to facilitate a safe rescue.[4] The unorthodox method was effective: Reagan never lost one.[5]

On occasion, there was another type of swimmer, a more unusual "rescue"—young girls who "accidentally" found themselves in peril to try and catch Dutch's eye.[6] "I had a friend who nearly drowned herself trying to get him to save her!" said one woman, recalling an occurrence that was not infrequent. "He was everyone's hero," said a Reagan schoolmate. "Every girl was in love with him. He was a handsome young man, built like Mr. Perfection, tanned to the hilt."[7]

On afternoons like this August 2, Reagan felt like the burning sun would never set. Mercifully, it finally obliged, quickly growing dark until the swimming section, which was surrounded by tall, full trees and lush, thick hills was covered in shadow. This meant that the area Dutch surveyed got darker quicker than the rest of flat, open Illinois.

With nighttime upon the beach, it was now officially after-hours. A party of four, two girls and two boys, were looking to have some fun. They giggled as they surreptitiously slipped into their bathing suits down shore. They entered the beach area from the side and quietly made their way into the deceptively gentle surf, in defiance of beach rules. Among them was Dixonite James Raider, who was not the proficient swimmer he figured.

It was 9:30 PM, the end of another very long day, and Dutch and Mr. Graybill were closing up the bathhouse when they heard splashing in deep water: James Raider had been sucked under. Another member of his group tried to save him but could not and was forced to abandon efforts when he, too, almost drowned in the swift current.

Dutch sprinted to the water and dove into the darkness. With only the stars to light the way, Reagan relied on himself, on his inner eye, the one that knew the way better than anyone else. There was a major struggle in the black water. Witnesses recall noisy splashing, some yelling, and arms flailing in the air. Suddenly, a mass of human appendages began moving in their direction. The lifeguard wrapped one arm under the victim's arms and dug water profusely with the other, kicking his feet under the current as rapidly as he could. Raider was brought ashore. Young Ronald dragged him onto the grass.

Artificial respiration was started. The party was no longer in a partying mood; the festive tone had been muted by a sense of horror. They watched, hoped, and probably prayed. Raider responded, and there was a collective sigh of relief. An exhausted Raider was transported to his home with an unexpected new lease on life. Ronald Reagan headed home as well. When his parents, Jack and Nelle, asked about his day, he might have shrugged that it was not especially unusual. It was, after all, the second near drowning in two weeks.

This early rescue gave Ronald Reagan one of his first tastes of notoriety: the front page of the Friday, August 3, 1928 *Dixon Evening Telegraph* carried a top-of-the-fold headline that read "James Raider Pulled From the Jaws of Death," about the rescue made the previous evening by "Life Guard Ronald 'Dutch' Reagan." It informed residents that Dutch had notched his twenty-fifth save.

This was Reagan's first page-one headline. He shared the top-of-the-fold with King George, who was reportedly enthusiastic about the Kellogg-Briand Pact to "outlaw war," and with the customary story on election fraud in Chicago. He was a news story for the first time—thousands more would follow. He had made the front page, as a hometown hero, early in life. And this would not be an isolated occasion for the Dixonite: The *Telegraph* continued to broadcast Reagan's heroics as he continued to make saves, each time updating Dutch's rescue tally to the wonderment and amusement of the locals.[8] Such episodes forged in the young man a supreme confidence that never left him, one that would forever affect Ronald Reagan's actions, from Dixon to Hollywood, and eventually all the way to Washington.

WHEN THE REAGANS ARRIVED IN DIXON IN 1920, THEY FELL IN love with Lowell Park. That first summer, Jack, Nelle, and sons Neil and Ronald drove to the park in the Model T, as the youngest Reagan, nine-year-old Dutch, dreamed about working there one day.

In the spring of 1926, as a sophomore in high school, Dutch's big chance arrived. There was a vacated lifeguard job at Lowell. The teenager applied through beach owner Ed Graybill, who had doubts about Reagan's age. Graybill first checked with Dutch's dad: "Give the kid a chance," Jack Reagan prodded. "He can do it." Graybill hired him, and never regretted his decision. His wife, Ruth, later remembered of Dutch:

> Even as a high school sophomore, he was a perfect employee. He was my boy. He was wonderful—dependable and polite. I never heard one complaint about him. He always knew his duties. We never had a basket of clothes left due to a body being at the bottom of the river. That was because we had a good lifeguard. . . . He was dependable and firm. . . . He would give out orders—"stay inside the lifeline"—and he meant it. When the beach was not busy, he taught kids to swim. And if he was in a jovial mood, he'd start walking like a chimp and give us a little entertainment.[9]

Prior to Reagan, there had been many baskets of clothes left behind, as several people had drowned in the river in the immediate summers before Reagan's arrival.[10] But once Dutch joined the ranks of the lifeguards, all that changed. This sense of purpose played a formative role in Reagan's life. Of all his jobs prior to adulthood, Reagan most enjoyed "my beloved lifeguarding," as he later referred to it—"maybe . . . the best job I ever had."[11] It was a job he did not consider work, despite toiling riverside seven days a week, ten to twelve hours per day, typically 10 AM to 9 PM, for a mere $15–16 a week plus so-called "meals," which usually meant something from the snack stand.[12] It was a grueling schedule, but despite its difficulties Reagan kept it up, continuing it for seven summers until he finished his education at Eureka College in 1932.

Local resident Burrel J. Reynolds remembered being watched by Dutch. "I was seven or eight and couldn't swim," said Reynolds sixty years later, "but

I'd go in deeper water. He'd blow a whistle—blow you in. He was always after me."[13] Not all were so appreciative. Most, in fact, never thanked Reagan. Some were too embarrassed, especially the guys, who usually only thanked him grudgingly at the urging of girlfriends. He came to learn that many people seemed to hate being saved. "[A]lmost every one of them," he claimed, "later sought me out and angrily denounced me for dragging them to shore. 'I would have been fine if you'd let me alone.' "[14]

But it was not his job to let them alone. For seven summers, Ronald was the rock at the Rock River, always watching, always there. During his seven-year tenure, Reagan saved the lives of seventy-seven people as a lifeguard, a number he held dear. "One of the proudest statistics of my life is 77," he said many decades later.[15]

Actually, Reagan may have short-changed himself. He netted a seventy-eighth save at the river a year after he left the job. He had returned to Lowell to visit the new lifeguard—a buddy. His friend asked him to watch the water for a minute while he used the bathroom. "Would you believe I had to go in and make a rescue while he was gone?" Reagan later told an inquirer.[16] And there were still others yet to come in the years ahead, as Reagan would make rescues at pools in Des Moines, Iowa and even in Sacramento, California, the latter as a 56-year-old governor.[17]

Saving a drowning victim is not an easy task under any circumstance, but it was especially difficult in the treacherous Rock River with its swirling currents and murky, dark water. Downstream from the river was a dam. When the sluices were opened, the normally slow current quickly picked up its tempo. Reagan remembered the "deeper thrust" this created in the water. The area surprised swimmers by sloping off just a few feet from the edge. Dutch learned that the shortest route from shore to shore was never a straight line but a swooping up-stream arc.[18] This swift-moving body of water is so thick, with currents so strong and dangerous, that swimming there *today* is banned and has been for years.

FORGING A TEFLON CONFIDENCE

The object here is not to transform Ronald Wilson Reagan into a political superhero, nor to focus on the positive to the complete exclusion of any negatives. The point is that these very real rescues, these feats of physical daring (of which this is a short list), impacted him greatly, forming an indelible mark on his psyche.[19] And they shaped him not just as a person but, one day, as a president of remarkable ambition.

Generally, these experiences taught Reagan quite a bit about life. A later close friend and National Security Advisor, Bill Clark, often listened to Reagan reminisce about those summer days and eventually came to believe that the lifeguarding instilled in the young man a basic respect for the dignity and sanctity of human life. This quality later manifested itself in President Reagan's opposition to abortion, abhorrence of the prospect of nuclear war, and empathy for the suffering citizens of the so-called Captive Nations behind the Iron Curtain.[20]

More fundamentally, the rescues did wonders for Reagan's self-esteem. Not many youngsters have saved a life, certainly not seventy-seven souls in such a risky fashion.[21] What better reinforcement to a man's self-worth than such incidents? How many people have directly saved another human being? How many individuals have done so well over seventy times, salvaging lives out of the grim pull of dark water? All of this happened in real life for Ronald Reagan, not in movies. Here were the seeds of his can-do attitude.

Of course, there were other influences. Downstream from the Rock River sat the First Christian Church on 123 South Hennepin Avenue, only nine blocks from Ronald Reagan's boyhood home. The faith that he acquired there, under the nurturing of Rev. Ben Cleaver and his mother Nelle—it was Nelle who was the most formative figure in his life—became a fundamental source of Reagan's confidence.[22] He was sure that God had a plan for his life and guided him daily along a preordained path that was just and right, and which made him confidently optimistic. Nelle taught him that "all things were part of God's Plan, even the most disheartening setbacks, and in the end, everything worked out for the best."[23] He never abandoned that belief.

It seems fitting that the church which influenced Reagan was perched alongside the Rock River, as the river would help shape him through its religious and personal significance. Indeed, he had two baptisms which took place on the banks of the river, one spiritual and one sacrificial. The water of the Rock River forever changed Reagan, and he followed its currents not only downstream to the First Christian Church but eventually to the oceans of the world.

These cold waters of Illinois forged a confidence in Reagan that ran far deeper than the Rock River. It was this confidence that transcended his actions, weaving its way into every aspect of his life. Its pervasive nature exists throughout his biography—from the Gipper all the way to the Oval Office. This confidence became the bedrock of Reagan's personality, an unappreciated intangible that helped him achieve his goals. His was not a mere confidence but an unshakable one.

Indeed, pundits in the 1980s called Reagan the "Teflon President" because, like a Teflon frying pan, nothing unseemly appeared to stick to him. He just shook or washed it off. He was seemingly invincible. "I'm amazed at this Teflon Presidency," CBS Evening News anchor Walter Cronkite awed. "Reagan is even more popular than [Franklin] Roosevelt, and I never thought I'd see anyone that well-liked. . . . Nobody hates Reagan. It's amazing."[24] Even the president's fiercest critics usually liked him as a person. And even the Soviets remarked upon his " 'Teflon' qualities," as they put it, and "protective 'Teflon' coating."[25] That Teflon quality can be ascribed to Reagan's confidence.

It took someone of this immense confidence to predict and believe that he could win the Cold War. This required not only uncommon fortitude in his own vision but also in the boldness to pursue the provocative policies that he felt were necessary to achieve that vision. As will be seen throughout the pages ahead, the steps Reagan authorized were extremely daring; and yet, he was unwavering in his belief that he could lead America to victory in the Cold War. In the face of boisterous criticism, he never shied away from his inner voice. It was the voice that led him to James Raider's rescue in the murky depths of the Rock River. It was the voice that led him to seek another save—to pursue Cold War victory when few put stock in the possibility, and amid naysayers all around him.[26]

2.

Reagan's Long March: The 1940s

IN 1946, RONALD REAGAN WAS ON LOCATION FILMING A MOVIE in a remote area in rural California. Now a seasoned actor and active member of the Screen Actors Guild (SAG), Reagan was almost ten years into his film career. He was told by a crew member that he had a telephone call waiting at a nearby gas station. Reagan hopped in a car and made his way to the station. At the time, he was preparing a report for SAG concluding (correctly) that the 1946 strike was jurisdictional rather than over wages and hours—a position unpopular with Communists who wanted to portray the strike as the fault of greedy studio heads.[1] Spearheading the strike was the Red-dominated Conference of Studio Unions (CSU), led by a Communist thug named Herb Sorrell, who Reagan charitably described as "a large and muscular man with a most aggressive attitude."[2]

Reagan arrived at the gas station and answered the phone. "I was told," he said later, "that if I made the report a squad was ready to take care of me and fix my face so that I would never be in pictures again." Specifically, the caller threatened to splash acid upon Reagan's unsuspecting million-dollar face—the source of his livelihood.[3]

Such fears were nothing new for Reagan: in another instance during this period, the bus he was scheduled to ride through studio picket lines was bombed and burned just before he boarded. As a result of such episodes, police began guarding Reagan's home and children, and he began packing a

Smith & Wesson revolver, which he took to bed each night.[4] For the first of innumerable times throughout his very public life, from Hollywood to Sacramento to Washington, Reagan started receiving death threats.[5] Reagan would always remain a nice guy, but now he was chastened—a warrior, a Cold Warrior. From then on, he was in the ring.

THE SCREEN ACTORS GUILD

When Ronald Reagan arrived in Hollywood in 1937, Communism was not the first thing on his mind. After his success at the lifeguard stand, he embarked on an equally successful career in radio, becoming a broadcaster of Chicago Cubs games from the 50,000-watt WHO, the NBC affiliate in Des Moines, Iowa. Here he was no longer calling out to absent-minded swimmers; instead, his voice could be heard in kitchen after kitchen, barn after barn, and car after car throughout middle America during the Great Depression. As the Depression came to a close, Dutch began looking for a new challenge, seeking to change mediums and parlay his radio experience into something much grander—a career in the movies. While many people labored for years to make this transition, Reagan's intrinsic confidence apparently told him that such a career shift would be neither impossible nor long in the making.

Indeed his instincts proved correct and between 1937 and 1964 he would take part in fifty-three films and work with some of the most legendary names in Hollywood—names like Barbara Stanwyck, Humphrey Bogart, Bette Davis, John Wayne, Lionel Barrymore, Ginger Rogers, Doris Day, and Eddie Arnold, among many others. In *Knute Rockne, All-American* (1940), he beat out Wayne and William Holden for the part of Gipp.[6] That same year, he made *Santa Fe Trail* with Errol Flynn, a picture now considered a classic, with a lineup that included Olivia de Havilland, a huge star coming off the previous year's epic, *Gone With the Wind*, for which she was nominated for an Academy Award. Contrary to the modern image of him as a "B" movie actor, at one point Reagan was one of the top five box office draws in all of Hollywood, receiving more fan mail than any actor at Warner Brothers except Flynn.[7] He became so successful that in October 1944 he made the cover of *Modern Screen* magazine.[8]

"I only recall respect for him as an actor," remembered actress Judith Anderson, in a typical assessment.[9] Film historian Robert Osborne, the face of Turner Classic Movies, says Reagan was "exceptionally likeable," had all the ingredients required of a leading man in the 1930s and 1940s, and had won "worldwide recognition."[10] This had the important, wider effect of expanding Reagan's self-confidence.

Although his star was eventually outshone by many of his industry counterparts, Reagan became revered throughout Hollywood more for his role in SAG than for any part that he played on camera. "Revered" is not a stretch: In one SAG event honoring him, a black-tie gathering at the Friars Club, a procession of dignitaries one by one praised Reagan for the "stature and dignity" he lent to the industry. Cecil B. DeMille reminded the group of the bad pre-Reagan days of Hollywood. Pat O'Brien saluted him as no less than another George Gipp himself. Really piling it on, the great Al Jolson sang "Sonny Boy" and wished and gushed that his own sonny boy, Al Jolson, Jr., could one day ascend "to be the kind of man Ronnie is."[11]

In the end, much of what Reagan eventually accomplished in Hollywood came from his organizing SAG, a natural extension of his deep interest in politics. Spurred in part by his role in short propaganda films that he made for the U.S. War Department during World War II, this interest grew into a passion during the years immediately following the war. By 1946 and 1947, Reagan began a career juggling act, bouncing back and forth between making pictures, political activism, and union work. His effectiveness as SAG president can be measured objectively: he was reelected to the post seven times beginning in March 1947, by huge margins. He was never rejected, and eventually granted a gold life membership card.[12]

While Reagan had many responsibilities as SAG president, movies and studio contracts were only half of the Hollywood coin for Reagan, as it was during this time that he was hammered into an iron-clad anti-Communist. Despite the guise of entertainment and glamour, these crucial years had an immeasurable impact on his view of international Communism and America's responsibility to confront it. His presidency with SAG planted seeds that would come to fruition four decades later. "I know it sounds kind of foolish maybe to link Hollywood, an experience there, to the world situation," he said from the White House in 1981, "and yet, the tactics seemed to be pretty much the same."[13] When aide Lyn Nofziger cautioned him about the Soviets at Reykjavik, he responded: "Don't worry. I still have the scars on my back from fighting the Communists in Hollywood."[14] He judged his Hollywood experience "hand-to-hand combat."[15]

One of his first scrapes came July 2, 1946, at his first council meeting of the Hollywood organization HICCASP, a group which had moved so far to the left, and included so many Communists, that it was frequently accused of being a Communist front. Though Reagan's involvement with HICCASP started as early as 1944, he later admitted that he was very "naïve" regarding its true colors when he first began his association. Those true colors were made remarkably clear during that July day in 1946. It was then that group member James Roosevelt, FDR's son and, like Reagan, a non-Communist liberal, suggested a group statement repudiating Communism. By Reagan's description, "a Kilkenny brawl" erupted. Musician Artie Shaw sprang to his feet, offering to recite the Soviet constitution by memory, which he claimed was "a lot more democratic" than the U.S. Constitution. A writer yelled that if there were ever a war between the United States and USSR, he would volunteer for the Soviets. When Reagan endorsed Roosevelt's proposal, he was showered with epithets, called a fascist, "capitalist scum," "witch-hunter," "red-baiter," an "enemy of the proletariat."[16]

This fracas was not unusual for Reagan, who found himself in a number of similar situations in 1946. Actor Sterling Hayden vividly remembered one such occasion, when he went head-to-head with Reagan. Hayden, who played the Air Force general who launched nuclear war in "Dr. Strangelove," had been a marine in World War II, for which he was decorated for his actions parachuting behind enemy lines in Yugoslavia.

He was also a Communist.

One evening in 1946, Reagan, along with fellow actor and anti-Communist William Holden, decided to crash a meeting convened by Hollywood Communists who were making a final push to organize the film industry. The Reds were counting on new recruit Hayden to lead the discussion, and this was to be a big moment for the organization.

The crowd of about seventy-five was "astonished and miffed" when Reagan and Holden walked through the door. Reagan sat with the rest of the gathered, listened politely, and then waited for the right moment to ask for the floor. He managed to keep his temper during a forty-minute presentation in which he was repeatedly interrupted, booed, and cursed. He would not be bullied, and held forth.

As Hayden himself later put it, Reagan coolly "showed up and took over and ground [Hayden] into a pulp; . . . he dominated the whole thing." As historians Ronald and Allis Radosh show, this seemingly somewhat minor incident was a seminal moment in helping to end the "golden era" of Hollywood Communists, who were unable to hijack the unions and were ultimately exposed as stooges to Moscow.[17]

THE "DPs"

While much of Reagan's early fights against Communism happened behind closed doors and between the lines of SAG papers, his first public confrontation with the USSR has been overlooked by historians. A bridge from his frequent attacks on Nazism during the war period to his subsequent assault on the next totalitarian threat, Bolshevism, this particular fight stemmed from Reagan's speeches as SAG president on behalf of the so-called "DPs," the Displaced Persons. A daily headline in 1947, the DPs initially were survivors of World War II fascism, mainly from Germany, Italy, and Austria, and were primarily Jews. Once the war ended, the list of DP-designated peoples widened to include 1.3 to 1.5 million individuals escaping Soviet-occupied areas in Eastern Europe, though they still included numerous Jews who longed for the creation of a homeland in Palestine (later called Israel). The Eastern Europeans, reported the *New York Times*, would "dare not go back . . . because they will not submit to the arbitrary governments which have been imposed on their homelands [by the USSR]."[18]

The DPs were held in camps, at a large cost to the United States.[19] Soviet officials, doing what they did best, outrageously claimed that the United States was holding the DPs as a source of semi-slave labor—a charge dismissed by Eleanor Roosevelt, who was intimately involved in the issue, as "utterly untrue."[20] Moscow demanded that the DPs be forcibly repatriated to areas of Eastern Europe now subject to Soviet control. Secretary of State George C. Marshall adamantly rejected the demand.[21]

A bill was introduced in Congress by Rep. William G. Stratton (R-IL) to permit entry of 400,000 DPs into the United States.[22] Though a drop in the bucket, the Stratton bill was a lifeboat to hundreds of thousands. And yet, it faced stiff opposition—an opposition Reagan resolved to resist.[23] Reagan agreed with U.N. official Herbert H. Lehman: "Apparently there are some people who would rather bury the Stratton bill in red tape and thus bury the DPs in a mass grave. They would be burying Protestants, Catholics, and Jews alike."[24] He also agreed with Earl Harrison, Commissioner of Immigration and Naturalization, who said that the DPs were fleeing totalitarianism and would become good Americans, "eager to support our government and institutions."[25]

On May 7, 1947, Reagan, through the New York-based Citizens Committee on Displaced Persons, released a statement urging passage of the

Stratton bill.[26] It was his first public campaign against Moscow, and for the first of many times to come, Reagan riled the Kremlin.

REAGAN AT HUAC

A few months later, on October 25, 1947, Reagan made a highly visible overture to the anti-Communist cause in testimony before the Congressional House Un-American Activities Committee (HUAC). He and a number of actors and producers were called as friendly witnesses to testify about Communist infiltration in the motion-picture industry.

While this period is today portrayed as a frightening example of Red-under-every-bed paranoia by grandstanding McCarthyites in the U.S. Congress, that caricature ignores the very real penetration that had taken place by Hollywood Communists. Only a year before the HUAC hearings, Communist screenwriter Dalton Trumbo had written that "every screenwriter worth his salt wages the battle in his own way—a kind of literary guerrilla warfare." The Hollywood enforcer for that warfare was an unpleasant individual named John Howard Lawson, a screenwriter known as Hollywood's "commissar." Lawson gave total subservience to the Marxist motherland. Befitting Moscow's representative in Tinsel Town, the Leninist Lawson also despised religion, desiring an orchestrated "campaign against religion, where the minister will be shown as the tool of his richest parishioner." In his aptly titled book *Film in the Battle of Ideas* (published by the Communist house Masses & Mainstream), Lawson provided this advice to his comrades: "As a writer do not try to write an entire Communist picture, [but] try to get five minutes of Communist doctrine, five minutes of the party line in every script that you write." Comrade Lawson also issued marching orders to fellow travelers among the acting class: "It is your duty to further the class struggle by your performance."[27]

Reagan was not exactly sympathetic to Lawson's edict, and neither was HUAC, which, despite its ruthless reputation, was slow to the dance. It was not until late 1947 that Congress began to take seriously the repercussions of the surreptitious advice of the likes of Lawson. Hollywood Communists wanted to use the film industry to spread their ideas. Congress decided it should have a dog in that fight—and a dogfight it became.[28]

By October 1947, Reagan's film career was ten years old. He had made dozens of movies and had been elected president of SAG at a very young age.

His youth was not lost on *Newsweek,* which, in a story on the hearings, said the pink-cheeked and sandy-haired Reagan looked so boyish that when he stood to speak the room was filled with "oh's and ah's," mainly from the contingent of star-struck girls who came to ogle him, Robert Taylor, Gary Cooper, Robert Montgomery, and some of the other leading men.[29]

By all accounts, from the left and the right, Reagan's testimony was much more mature than his looks. Indeed, it is ironic that a man known for raining hellfire upon "evil" Communists gave a decidedly sober, cautious statement, particularly in contrast to the scathing denunciation provided by fellow actor Adolphe Menjou.

Reagan was first questioned by Robert E. Stripling, HUAC's chief investigator, and then by chairman J. Parnell Thomas. Other committee members declined to pose questions, including Cold Warrior Richard M. Nixon. (Only a crazy person watching that hearing on that October 25, 1947 would have predicted that the two people sitting in the room who would later become president were Nixon and Reagan.) In his best moment, when his questioners allowed him a substantive, lengthy response, Reagan struck a reasonable balance between desiring to battle Communists and respecting a free, democratic society. Speaking of Hollywood Communists, he lectured:

> [W]e have exposed their lies when we came across them, we have opposed their propaganda, and I can certainly testify that in the case of the Screen Actors Guild we have been eminently successful in preventing them from, with their usual tactics, trying to run a majority of an organization with a well organized minority.
>
> So that fundamentally I would say in opposing those people that the best thing to do is to make democracy work. In the Screen Actors Guild we make it work by insuring everyone a vote and by keeping everyone informed. I believe that, as Thomas Jefferson put it, if all the American people know all of the facts they will never make a mistake.
>
> Whether the [Communist] party should be outlawed, I agree with the gentlemen that preceded me that that is a matter for the Government to decide. As a citizen I would hesitate, or not like, to see any political party outlawed on the basis of its political ideology. We have spent 170 years in this country on the basis that democracy is strong enough to stand up and fight against the inroads of any ideology. However, if it is proven that an organization is an agent of a power, a foreign power, or in any way not a legitimate

political party, and I think the Government is capable of proving that, if the proof is there, then that is another matter.

I do not know whether I have answered your question or not. I, like Mr. [Robert] Montgomery, would like at this moment to say I happen to be very proud of the industry in which I work; I happen to very proud of the way in which we conducted the fight. I do not believe the Communists have ever at any time been able to use the motion picture screen as a sounding board for their philosophy or ideology. I think that will continue as long as the people in Hollywood continue as they are, which is alert, conscious of it, and fighting. . . .

I abhor their [Communist] philosophy, but I detest more than that their tactics, which are those of the fifth column, and are dishonest, but at the same time I never as a citizen want to see our country become urged, by either fear or resentment of this group, that we ever compromise with any of our democratic principles through that fear or resentment. I still think that democracy can do it.[30]

It was a nuanced assessment, and one which featured the words "fight" or "fighting" three times. He condemned Communism, while still championing the virtues of an American society that respects civil liberties. Executing a balance between levelheadedness and a willingness to mix it up, he satisfied voices on both the right and left, making it difficult for anyone to disagree with his words. In fact, chairman Thomas immediately followed Reagan's closing line by conceding: "We agree with that. Thank you very much." At the other end of the spectrum, among liberals, James Loeb, the executive secretary of Americans for Democratic Action (ADA), dubbed Reagan's testimony "by all odds, the most honest and forthright from a decent liberal point of view," adding that his closing statement was "really magnificent." Loeb called him "the hero" of the Washington hearings, and held high hopes that Reagan's golden testimony would "win over the liberals of the stage and screen" to the ADA position, which opposed both "party-liners" and "witch hunts."[31]

To the left of the ADA, Reagan's testimony was so solid that he even managed to please the *Daily Worker*. The New York-based Communist Party organ called the HUAC session a "kangaroo court" but was impressed that Reagan "proved a reluctant witness, refusing to parrot back to . . . Thomas his suggestion that outlawing the Communist Party was desirable." The *Daily Worker* underscored Reagan's quote from Jefferson and remarks on making democracy work, while castigating Thomas.[32] It was perhaps the first mention of Reagan in the Communist press and it was positive, but it was a far cry from what was to come.

Reagan himself was satisfied with his HUAC appearance. He had confirmed to the congressmen that there were a number of Communists present in Hollywood, but he had stood by his principles and avoided outing any of them. Reagan did, however, oppose the presence of party members in SAG leadership positions, and his testimony only reaffirmed his thinking. After the long train ride home from Washington to Los Angeles (Reagan refused to fly in those days), he redoubled his efforts, considering a variety of new tactics for combating Communism in the years to come.

On November 10, 1947, the SAG board voted that no officer or board member could serve without first signing an affidavit swearing "that he is not a member of the Communist Party nor affiliated with such a party." Board meetings revolved around the Communist issue.[33] At Reagan's insistence the SAG board was cleaning house, and there would be no welcome mat for Red infiltration.

INTO THE FIFTIES

While his work in the forties built the backbone of his opposition to Hollywood Communism, in 1950 Reagan launched his first anti-Communist crusade—literally. That year saw the formation of the Crusade for Freedom, headed by the colorful, brash General Lucius Clay, who two years earlier had advised President Truman that U.S. tanks crash Joseph Stalin's blockade at Berlin. Long before Reagan resurrected the notion as president over thirty years later, the Crusade for Freedom called for "rollback"—for reversal of the Soviet empire. The eventual 1980s Crusader fully embraced the 1950 Crusade from the moment of its inception. He appeared at Crusade for Freedom rallies and raised money for the organization in its bid to free the "captive peoples." "More than any other performer," writes author Peter Schweizer, "Reagan embraced the cause and spread the word."[34]

In an attempt to bolster additional Hollywood support for the movement, Reagan passed a resolution in SAG to encourage other actors to get involved in the Crusade. For Reagan, the policy of liberating Eastern Europe from Soviet control was not simply just another political issue; it was an imperative. He used much of his industry clout to encourage support for the cause, taping a short film for civic groups, churches, and schools to bolster support at the grassroots level and beyond.

Shortly thereafter, Reagan enlisted in another crusade, when he began giving speeches in rallies before Dr. Fred Schwarz's Christian Anti-Communist
Crusade.[35] This crusading group, as its name suggested, devoted special attention to the institutionalized atheism and endemic religious hatred that inspired the USSR and Communist movement.

Generally speaking, Reagan was becoming a seasoned veteran of anti-
Communist campaigns. He continued crusading in Hollywood, always looking to protect the industry from subversion. "By the early 1950s," wrote author
John Meroney, referring to Reagan's efforts at SAG in particular, "the back of
the Communist Party in Hollywood" was "essentially broken."[36] No less than
Jack Warner said Reagan "turned out to be a tower of strength, not only for
the actors but the whole industry."[37] More than one Hollywood associate described him as a "fearless foe."[38]

An eye-opening testimony to this effectiveness was the April 10, 1951
HUAC statement by Sterling Hayden. Hayden admitted he had been a Communist but left the party after the direction he saw Communism take under
Stalin. He said that joining the party was the "stupidest, most ignorant thing
I have ever done" and explained how Hollywood Communists tried to organize all of the entertainment industry's labor unions under one giant union
controlled by the party. When asked by HUAC what stopped them, Hayden
said assuredly: "They ran into a one-man battalion named Ronald Reagan."[39]

A large part of Reagan's success in his fight stemmed from the fact that he
studied the tactics of the enemy. One such tactic was the "diamond formation" seating strategy employed by just four conspiratorial Communists to
manipulate a crowd during a speech. One of the four would sit in the back of
the room in the middle of the row, while another Communist would be to his
left near the wall in the middle of the room. Yet another would be just opposite to his right, and the fourth would be up front in the middle of the first
couple of rows. Each of the four heckled and asked abrasive questions from
their strategic vantage point, and this simple method quickly became a highly
valuable tool for rattling the speaker and suggesting a strong presence of dis-
senters.[40] This was down and dirty hardball politics. Perhaps, then, it was no
coincidence that at this time Reagan often invoked words like "victory" and
"fight" in his rhetoric.

As a further indication of how the confrontation pervaded his thoughts,
Reagan infused anti-Communist statements into SAG pronouncements that
had nothing to do with Marxism. An example was a declaration at his final
meeting as president of the guild on November 9, 1952. Under Reagan's

tutelage, SAG called upon motion picture producers and the entertainment industry in general to provide greater employment for African-American actors;
the statement was critical of recent "well-intentioned but ill-directed" efforts to
avoid casting black artists in stereotypical black roles, which, said SAG, had
caused "inadvertent harm" by creating a sudden dearth in available roles for
black actors. And yet, Reagan's SAG could not help but link its declaration to
the battle against Moscow. "In a critical world period, when the democratic
credo is under fire from our Communist foes," declared SAG, "it becomes increasingly important that the expanding role of our Negro citizens in the community of this nation be adequately portrayed in the entertainment art."[41]

WITH PROCLAMATIONS SUCH AS THIS, REAGAN HAD TAKEN HIS
fight to the front lines of his SAG presidency. Announcing his agenda to both
the entertainment industry and the nation, Reagan's name as an actor was becoming synonymous with his efforts to curb the spread of Communism at
home and abroad. By the time his tenure at SAG ended in 1952, Reagan had
overseen the union through some of its most dangerous moments, ensuring that
the integrity of the institution remained sound and that the Communists failed
to gain a foothold in America's most influential medium.

Having accomplished certain goals in Hollywood, it was now time for
him to shift his focus to a burgeoning new industry, a young but promising
field that gave him the opportunity to speak to groups that were previously
out of reach. With his film career floundering, his attention would turn to a
medium that gave him access to the hearts and minds of millions of Americans around the country, a medium that placed him straight into their living
rooms.

3.

The TV Crusade:
1950 to mid-1960s

BY THE EARLY 1950S, RONALD REAGAN'S POLITICAL INTERESTS were flourishing, but these interests were beginning to take their toll on his wife Jane Wyman. In her divorce papers, she cited his obsession with politics as a primary factor in their separation. It was a divorce that Reagan had not wanted and one that disturbed him greatly. To Reagan, marriage was an institution that should last forever and his inability to find permanence in his bond with Wyman became a source of major failure for him.

Further complicating matters was the fact that while politics were his passion, at this stage they did not pay him anything. Despite his considerable earnings of the last few years, he still needed to earn a salary, and after his time at SAG came to an end, acting opportunities became more difficult to come by. The loss of his movie success threw him for a loop, and by all indications, this was the most difficult period of the Crusader's life, a time when, for the first time, personal and professional failure seemed like a possibility.

While he was crushed by the breakup with Wyman, it was during this time that he met the woman that would forever change his life: Nancy Davis.

In 1952, the life of the forty-one-year old turned around when he married Nancy, who, more than anyone, supported him totally, including his political ambitions. She was head over heels for her husband. "Ronnie is my hero," she glowed. "My life began when I got married. My life began with

Ronnie."[1] She watched his back for him, distrusting seemingly nearly every-one he trusted, and, from his front, encouraged him. She was willing to follow him wherever it took to ensure that his career was once again fruitful.

Though his marital life was on the right track, he was still faced with the difficult reality of his career. In February 1954, unready to say goodbye to show business for good, Reagan turned to a stint in a Las Vegas vaudeville act, at the aptly named Last Frontier hotel, to provide him with the income he needed to keep his career afloat.[2] Despite this apparent setback, Reagan re-mained resolute, telling his new bride that God had something greater in store. For Reagan, who had always felt that Providence had a plan for him, this was merely a brief stop on an otherwise upward trajectory.

Once again Reagan's optimism was vindicated; he was not in Las Vegas for long. Soon after his move, CBS began considering Reagan as a host for its programming. While they originally wanted him to host the popular *Om-nibus*, which up until that point had been hosted by the future *Masterpiece Theatre* host, Alistair Cooke, CBS executives found use for Reagan as host of the new show *GE Theatre*. Reagan struck CBS as quintessentially American and ideally suited for this emcee type role. It was a role that Reagan would come to know well over the next several years and it was a position that he would inhabit from September 1954 to August 1962.

Reagan was already well known from his movie days. This new position provided him with unprecedented exposure, making him (according to sur-veys) one of the most recognized names in all of America—up there with President Eisenhower. By the mid-1950s, two-thirds of American homes al-ready owned at least one TV set, and millions of families sat perched in front of "the tube" for hours on end, terrifying the movie industry—which was los-ing clients at the theater. That was fine with Reagan, who was positioned per-fectly for the shift; he rode the wave. Moreover, there were very few channels on the dial in 1954. The typical American could not turn on the TV on a given evening without seeing Ronald Reagan's face in the living room.

Adding to Reagan's notoriety, *GE Theatre* was a smash. The show took off, eclipsing *I Love Lucy* only weeks into its debut, and attracting the very best actors: Ethel Barrymore, Joseph Cotten, Bette Davis, Jimmy Stewart, James Dean, Natalie Wood, Alan Ladd, Jack Benny, Lee Marvin, Charles Bronson, Angie Dickinson, Vincent Price, Walter Matthau, Charlton Heston, Donna Reed, Greer Garson, David Janssen, to name a few.[3] These were megastars at the height of their careers. Reagan often acted alongside them, since he not only hosted *GE Theatre* but starred in more of its 200 episodes than any actor. The show also attracted top musical talent, like Judy Garland, Ethel Merman, Harry Belafonte, and Fred Waring and his Pennsylvanians.[4]

According to Reagan, who was also supervisor of *GE Theatre* and involved in episode creation and development, it was the top show in the 9:00 PM slot for eight years.[5] If it was not consistently number one each week, it was always near the top, and all of this exposure combined to make Reagan so popular that his face graced the cover of *TV Guide* twice—the November 22, 1958 and May 27, 1961 issues.

GE'S ANTI-COMMUNIST SPOKESMAN

Of course, Reagan was not about to let this TV experience interfere with politics. Like his movie career, he insisted that there be time for both the camera and anti-Communism, and throughout the latter 1950s and early 1960s, Reagan traveled the country as a spokesman for General Electric—part of his duties as *GE Theatre* host. During these travels, he toured GE plants around the country, where he visited with executives and employees and gave lunchtime and dinner speeches. While these speeches usually contained innocent, humorous anecdotes about Hollywood, they often became quite ideological, as Reagan laid out a litany of attacks against big government at home and abroad—especially in Moscow.

Despite this exciting new forum for his political ideas, not all of his political remarks at this time were made on GE tours. For instance, in a June 1957 speech at Eureka College, his first of many at his alma mater, Reagan's battlefield rhetoric was on display, as was his sureness of an American responsibility to fight the bad guys in the red hats. Citing a possibly apocryphal story about the signing of the Declaration of Independence in Philadelphia, he quoted an unidentified man in the state house on that day. "Were my soul trembling on the verge of eternity, my hand freezing in death," said the stranger allegedly, "I would still implore you to remember this truth: God has given America to be free." Here, said Reagan, "was the first challenge to the people of this new land, the charging of this nation with a responsibility to all mankind. And down through the years with but few lapses the people of America have fulfilled their destiny."[6] Here again was a common Reagan refrain that God not only chose America to be free, but gave the nation that freedom with a larger responsibility to all mankind.

In that speech, he expressed aggressive words, telling his audience: "You are fighting for your lives. And you're fighting against the best organized and the most capable enemy of freedom and of right and decency that has ever been abroad in the world." He pointed to the 1930s, when he claimed that Communism came to Hollywood via a man he cryptically identified only as a technician who came to town "on direct orders from the Kremlin." "When he quietly left our town a few years later the cells had been formed and planted in virtually all of our organizations, our guilds and unions. The framework for the Communist front organization had been established."[7]

Although most of these anti-Communist statements were made as Reagan toured the country for GE, he used *GE Theatre* on several occasions to air his views. An example of this occurred during a February 3, 1957 broadcast, in which he played a boxing trainer. At the close of the broadcast, Reagan stepped back into his host duty to give his customary goodbye and plug for GE products. This time, however, he put in a word for Hungarian refugees, fresh off the disaster of October–November when Soviet tanks, under orders from Moscow, rolled in and killed tens of thousands of Hungarians, causing a large number to flee the country. "Ladies and gentlemen, about 160,000 Hungarian refugees have reached safety in Austria," reported Reagan to his huge audience. "More are expected to come. These people need food, clothes, medicine, and shelter. You can help." He told his fellow Americans to send donations to the Red Cross or to the church or synagogue of their choice.[8] Those Hungarians were Reagan's heroes: the Captive Peoples of the Communist bloc suffering the sword of Soviet repression, and this was perhaps the Great Communicator's first use of the TV bully pulpit on behalf of Eastern Europeans.

There were other *GE Theatre* occasions where Reagan assumed political roles, most notably in a December 13, 1959 episode called "The House of Truth," in which Reagan played a U.S. intelligence officer in an Asian village in which Communists burned down an American library. The officer not only helped reopen the library but countered the Communists. In yet another *GE Theatre* show, aired September 24, 1961, titled "The Iron Silence," Reagan played a Soviet Major named Vasily Kirov during the occupation of Budapest. At the end of the episode, Kirov releases two Hungarians in his custody, telling them, "I never knew what freedom was until I saw you lose yours."[9] Reagan liked the line so much that he shared it years later when making a point in one of his 1970s radio broadcasts.[10]

Chris Matthews, a former speechwriter for Democrats and now a popular pundit and TV host, recalls tuning in to *GE Theatre* one night as a kid and observing Reagan saying, "This is a program I care a lot about personally."[11]

Matthews was referring to a two-part finale titled, "My Dark Days," broadcast on March 18 and 25, 1962. Reagan starred as the husband of a housewife who was asked by a friend to attend a meeting of a liberal Los Angeles group called the Alien Protection Committee, which claimed that its purpose was to promote the advancement of foreign-born Americans. The housewife and the FBI suspected the group was a Communist front. She infiltrated the group and became an informant. The script was adapted by Richard Collins from Marion Miller's autobiography, *I Was a Spy: The Story of a Brave Housewife* (Bobbs-Merrill, 1960).[12]

Indeed this episode was deeply personal to Reagan, who, a few weeks after it aired, wrote a letter to two friends in which he spoke of the difficulties he faced trying to get the show produced. He complained that it was "near impossible" to "cram five years of espionage into thirty minutes." But that was the least of his problems: "I had to fight right down to the wire to make the Communists villains," said Reagan. "When I say 'fight' I really mean that." The problem, explained Reagan, was that there were liberals on the producing staff who believed that Communist infiltration was a fantasy "dreamed up" by "right-wingers," and as such they attempted to sabotage the show. Reagan was especially irate over the fact that two of the producers and one director tried to cut the scene in which a little girl said her prayers. "Finally in a near knock-down, drag-out," wrote Reagan, "they admitted their objection was because they were atheists." While Reagan was victorious in this battle, the company remained ignorant of his struggle to get the program produced.[13]

EARLY 1960S—"WE ARE IN A WAR" AND "WE ARE LOSING"

During these later GE years, SAG once again became a part of Reagan's life, as he returned to his role of union president in 1960 after an eight-year hiatus. Almost immediately after resuming this high-profile position, Reagan faced the difficult responsibility of negotiating a new general contract with the studios. Despite the complex nature of this hurdle, Reagan won. As a measure of his triumph, when he announced the strike-settlement package to a mass meeting of the membership on April 18, 1960, Reagan received a standing ovation and a landslide approval vote of 6,399 to 259.[14]

As SAG president, Reagan found new venues for voicing his anti-Communist cause, and while he had labored hard to stymie the spread of Communism in Hollywood, he felt that the battle was far from over. In 1960, he boycotted a 20th Century-Fox banquet for Soviet Premier Nikita

Khrushchev, and a few months later in May 1961 he asserted that Communists in Hollywood were "crawling out of the rocks." To Reagan, the Communist Party had "ordered once again" a massive infiltration of TV and motion pictures. "We in Hollywood broke their power once," Reagan rallied, "but it was only an isolated battle."[15] In the years since his absence from SAG, Reagan believed the Communists had once again gained a foothold in Hollywood and were again in the process of spreading their anti-democratic sentiment. "They are renewing in the spirit of Lenin's maxim of two steps forward and one backward," said the future president in July 1961. This particular Reagan clarion call was picked up by UPI in a story syndicated throughout the country.[16]

Trying to convey the gravity of what was happening, he contended in another July 1961 speech that the "ideological struggle with Russia" was the "number one problem in the world." In the same talk where he shared that thought, he raised the intensity, striking the notion of "war." Offering his view of how America should react to the hazards of Communism, Reagan criticized those who "subscribe to a theory that we are at peace, and we must make no overt move which might endanger that peace." He declared that "the inescapable truth" is that America is "at war," and "we are losing that war simply because we don't, or won't realize that we are in it."[17]

Some deemed this rhetoric too bellicose, but Reagan felt that such candor was necessary. *We may not, after all, think we're at war,* Reagan repeatedly told his *GE Theatre* soldiers, *but the Communists certainly feel they're at war with us.* He assailed Communism's aggressive and expansionary nature, and its expectation of inevitable conflict with the capitalist West:

Karl Marx established the cardinal principle that communism and capitalism cannot coexist in the world together. Our way of life, our system, must be totally destroyed; then the world Communist state will be erected on the ruins. In interpreting Marx, Lenin said, "It is inconceivable that the Soviet Republic should continue to exist for a long period side by side with imperialistic states. Ultimately, one or the other must conquer."

Last November, the Communist parties of 81 countries held a convention in Moscow; and on December 6, reaffirmed this principle of war to the death. In a 20,000-word manifesto, they called on Communists in countries where there were non-Communist governments to be traitors and work for the destruction of their own governments by subversion and treason.[18]

These were not the passive words of a cheerful Hollywood dinner speaker who came to chat about the Oscars. Moreover, much of this Reagan sentiment was candidly reiterated, reexplained, and redefended again and again. He harped on his points, and people that knew him, knew that this battle was at the heart of what he represented. Reagan was determined to make America see the very real threat that he believed Communism posed to the nation's way of life. America could not afford to be lulled into complacency by the fact that no one was dying in this war. This was a war that he wanted to win.

1962 TO 1965

Firmly established in his talks in 1961, this "war" rhetoric became a regular drumbeat by Reagan throughout 1962. He took that specific message from the heartland to the Badlands, from the plains to the mesas, writing and refining speeches that were designed to convince Americans of Communism's manifold dangers. In Bartlesville, Oklahoma in February, he reminded a crowd that, "We are in a war, whether we admit it or not."[19] In two other stops in the final week of February, including at a press conference before a speech in Dallas, he made remarks such as "the free world is at war" with Communism and "the war isn't over."[20] "The weapons in this war frequently are strange to us," said the Crusader, "such as subversion, propaganda, and deliberate infiltration of many institutions of our free society." Here, Reagan no doubt had Hollywood in mind. He went on: "The enemy has not resorted to the traditional instruments of war, partly because he has been doing so well without them." He admonished, as he would in the 1980s: "Communism is a single, world-wide force dedicated to the destruction of our free-enterprise system and the creation of a World Socialist State. Communism is not a political party, it is a quasi-military conspiracy against our government."[21]

A common Reagan speech in the early 1960s, which he gave innumerable times, was one he titled, "A Foot in the Door," which warned of Communists trying to surreptitiously subvert institutions like the motion-picture industry. In one version of the speech, he again used the Lenin-Marx quote on "one or the other must conquer." Also in this speech is a quote attributed to Lenin about Latin America falling into Communist hands like "overripe fruit," which Reagan cited for decades to come, including during his presidency, though it was likely not a direct Lenin quote, even if accurate in spirit.[22]

While much of this anti-Communist campaigning took place as Reagan was a Democrat, as the 1960s progressed he found himself growing more and more disillusioned with the direction of the party. In his estimation, the party had abandoned traditional ethics and values in favor of big government solutions to problems. He felt that liberal Democrats had created a permanent welfare state that (in his view) FDR would not have supported; such a system was fostering a "dependency class." Rather than scaling back that system, Reagan saw that the party was committed to a huge expansion of government under LBJ and his Great Society. Reagan feared a "creeping socialism" under a party that he suspected was increasingly naïve to the dangers of Communism, and in 1962, he switched to the Republican Party.

The GOP was happy to have him. Then, on October 27, 1964, he was asked to speak on behalf of Senator Barry Goldwater (R-AZ), the Republican presidential nominee. In this speech, which would come to be known as the "Time for Choosing" speech, Reagan declared, "We are at war with the most dangerous enemy that has ever faced mankind in his long climb from the swamp to the stars, and it has been said that if we lose that war, and in so doing, lose this way of freedom of ours, history will record with the greatest astonishment that those who had the most to lose did the least to prevent its happening."[23] He vowed to do something; the will was there; the specific policies and plans would need to wait for a while.

This speech for Goldwater became an instant landmark and a focal point for Reagan's political support. Republicans everywhere immediately urged him to run for office, something that took him by surprise but invigorated him nonetheless. "I had never given a thought to public office," he later claimed, recalling the reaction to the speech. "I was happy to be in show business. They kept after [us] till Nancy and I were having trouble sleeping. I thought they were crazy."[24]

This shift in horizons could not have come at a better time. By 1964, the GE tours had ended, but Reagan continued to search out more forums for warning the world about Marxism. To that end, he opted for what he knew best—a camera and the spoken word—producing two documentaries. One was a two-part 1963 film titled "The Truth About Communism," which he hosted and narrated, focusing on the Hitler-Stalin Pact, the Comintern, the massacre in Poland's Katyn Forest, Moscow's desire for world revolution, the words of Marx in the *Communist Manifesto*, and much more.[25] The second documentary, released in 1965 and conarrated with friend and actor Robert Taylor, was called "Let the World Go Forth."[26] Both were intended to "wake up" the world to the threat emanating from Moscow. Not only did these documentaries embody Reagan's anti-Communist ideology, but they also represented the marriage of his two passions: film and politics.

But by the mid-1960s, it was becoming increasingly clear that while one passion was coming to a close, the other was just beginning to take off. Reagan's speech for Barry Goldwater thrusted his views into the national arena in a once unthinkable manner, and now it was time for him to stop commenting on politics and start doing them.

Hollywood had given Reagan the notoriety and platform to make these robust, hard-hitting statements, statements that would come to define him both politically and personally. While his showbusiness background would be the object of ridicule during his eventual presidential campaigns, it was this background that provided him with the experience he needed to begin his political career. Indeed, the irony was that despite the claim that his Hollywood background deemed him unworthy of Oval Office stature, it was this very thing that gave him the name recognition, confidence, and even some of the leadership experience that he needed to be president. Lew Uhler, who as a member of the Counter-Intelligence Corps reported on Communist infiltration in the Los Angeles area in the 1950s, and later worked for Reagan in Sacramento, shrewdly observed that only an individual who came through the crucible of combating Hollywood Communists on a daily basis—and in the manner in which Reagan clashed with and defeated them—could in turn appreciate the strengths and weaknesses of Communists generally.[27] That experience would be indispensable when Reagan got to Washington and formally commenced his Cold War crusade.

Reagan had fought the nation's chief enemy on the home front for over twenty years, but now recognized that it was finally time to take his views to a new stage, a stage where he could impact Communism with more than mere words, a stage that would eventually help make him president.

4.

Cold War Governor:
Late 1960s

AFTER HIS SPEECH ON BEHALF OF BARRY GOLDWATER, IT WAS clear that there was a lot of Republican support for Ronald Reagan and his ideas. Yet, questions lingered over the best way for the party to capitalize on this newfound enthusiasm. While some members of the Reagan bandwagon looked to the national stage, it was obvious that their guy would need some elected experience outside of Washington before he could advance to the presidency. Accompanied by a coterie of conservative California businessmen, Reagan explored possibilities at the state level, where his next move soon became apparent.

As the nation's most populated state, with the largest budget and plenty of problems, not to mention millions of people who had welcomed into their homes the friendly host of *GE Theatre*, the California governorship was ripe for the picking. And though Reagan's driving concern was perched much farther away, in Moscow, he was quite interested in domestic politics. California could be a microcosm, a laboratory, for his domestic conservatism, giving him an opportunity to develop policy and score some legislative bona fides for his political resume. It was a plan that seemed seamless, if not for an obstacle that stood in the way: The current governor, Pat Brown, was a popular incumbent. Reagan decided to run regardless, sure of his chances, and his destiny.

Reagan laid out that sense of destiny in a 1965 memoir. While the book voiced many of Reagan's ideas from past speeches, it also gave him the opportunity to express himself in yet another new medium and to appeal directly to voters in California and around the country. Titled, *Where's the Rest of Me?*, the book finished with a revealing flourish. In its final paragraphs, which borrowed from his 1960s speeches, including the "Time for Choosing" speech, lay the logic that compelled him to forcefully reject the coming policy of détente that propelled him to pursue the presidency. Displaying the crisp anti-Communism which would come to define many of his stances toward the USSR, he charged that a "policy of accommodation is appeasement, and appeasement does not give a choice between peace and war, only between fight and surrender." He went on:

> We are told that the problem is too complex for a simple answer. They are wrong. There is no easy answer, but there is a simple answer. We must have the courage to do what we know is morally right, and this policy of accommodation asks us to accept the greatest possible immorality. We are being asked to buy our safety from the threat of the [atomic] bomb by selling into permanent slavery our fellow human beings enslaved behind the Iron Curtain. To tell them to give up their hope of freedom because we are ready to make a deal with their slave masters.[1]

Reagan rejected any "deal" between the United States and USSR that sold into "permanent slavery" those Eastern European captives behind the Iron Curtain. Such deals were not simply wrong and immoral, but the "greatest possible immorality." Insisting that a nation which chose such a course was opting for "disgrace," Reagan said that Alexander Hamilton warned that a nation that prefers disgrace to danger is prepared for a master and deserves one. America, he said, should choose the high road, not the low road. "Should Moses have told the children of Israel to live in slavery rather than dare the wilderness?" asked Reagan. "Should Christ have refused the Cross? Should the patriots at Concord Bridge have refused to fire the shot heard round the world?"[2]

The future president concluded that Americans must choose "courage" over accommodation, telling his compatriots that all of America had a collective rendezvous with destiny. Together they could preserve for their children "this, the last best hope of man on earth," or, rather, they could "sentence them to take the first step into a thousand years of darkness." If they tried but failed, said Reagan dramatically, at least their children and children's children could "say of us that we justified our brief moment here. We did all that could be done." At the very least, this meant summoning the moral courage to reject accommodation with slave masters if and when such deals reared their cowardly heads.[3]

STAYING FOCUSED ON THE CAMPAIGN

While his memoirs provided him with the opportunity to address larger, international issues like Soviet Communism, such grandiose, apocalyptic thinking must have made Reagan's gubernatorial bid at times agonizing. The campaign forced him to set aside the global issues that stoked his passions in favor of the more immediate task of winning the governorship. With the help of a campaign team spearheaded by Phil Battaglia, Bill Clark, Tom Reed, and, among others, Lyn Nofziger, Reagan did his best to stay focused on California questions.

To that end, Nofziger, a California newspaper man who became candidate Reagan's press secretary, was mystified over how Reagan "exuded confidence" during the campaign. He puzzled: Politically, Reagan was a novice who spent his entire life in radio or the movies. He knew little about governing. He had never been through the rough-and-tumble of a campaign. He never dealt with a political press, most of which was registered Democrat. Sure, Reagan ruminated often on the world's daunting problems; those issues, however, were not part of the agenda of the California state assembly. And yet, said Nofziger, "none of this bothered Reagan." "From the beginning he was serenely confident that he could handle himself." During that 1966 campaign, "nobody had more confidence in Ronald Reagan than Ronald Reagan."[4]

Pat Brown certainly had little confidence in him. The Democratic governor proceeded to make the mistake of nearly every left-of-center individual who sized up Reagan over the next twenty-plus years: he dismissed and underestimated him.

In November 1966, Ronald Reagan was elected governor of California by the stunning margin of a million votes, carrying 400,000 Democrat defectors with him, and fifty-five of fifty-eight counties. Numerous Republican legislators were swept in on his coattails. Though that first term saw its ups and downs, his governorship still won encomiums from the state's top (even liberal) editorial boards. The *San Francisco Chronicle* reported that in his first term Reagan had "saved the state from bankruptcy."[5] The *Los Angeles Times* judged Reagan an "accomplished practitioner in the art of government" and a "proven administrator." The national media was also impressed: *Newsweek* dubbed him "one of the most brilliantly gifted politicians anywhere in the U.S. today—a campaigner unmatched for sheer star quality since the departure of Dwight Eisenhower and the arrival of the Kennedys a decade ago."[6]

THE GUBERNATORIAL YEARS—THE BATTLE RAGES ON

While far removed from his concerns for the Soviet bloc, it was clear from these first-term successes that Reagan's strategy to use California as a stepping stone to a national office was working. And though he lacked the ability to shape national policy toward the Soviets, thirty years in radio, film, and television had taught him the power and utility of the spoken word. The realities of holding the highest state office in California made it such that Reagan's "war" rhetoric needed to lessen, but it did not need to disappear. Since he was considered by many to be a potential presidential candidate, Reagan needed to speak out on national and international issues, including the Cold War. This also meant that he routinely was called upon to voice his opinions on the story that was dominating all of the headlines: the Vietnam War.

One episode that embodied such elements, and then vanished into the past, was a fascinating May 15, 1967 debate between Governor Reagan and Senator Robert F. Kennedy (D-NY). The subject: Vietnam. The debate was titled, "The Image of America and the Youth of the World," and was billed by CBS as a "Town Meeting of the World." It was broadcast from 10:00 to 11:00 PM (EDT) by CBS TV Network and CBS Radio Network. It was produced by later *60 Minutes* brainchild Don Hewitt and hosted by CBS News correspondent Charles Collingwood.[7]

The debate was watched by 15 million Americans and served as a wake-up call to those who underestimated Reagan. Revealing the governor to be exceptionally well-informed on the Vietnam issue, there was total agreement—even among media sources such as the *San Francisco Chronicle* and *Newsweek* who revered Bobby Kennedy—that Reagan overwhelmingly won the debate. "To those unfamiliar with Reagan's big-league savvy," reported *Newsweek*, "the ease with which he fielded questions about Vietnam may have come as a revelation." *Newsweek* judged that "political rookie Reagan . . . left old campaigner Kennedy blinking when the session ended," and thoughtfully speculated whether the debate might be a "dry run" for a future set of "Great Debates" between these two promising presidential aspirants.[8]

Historian David Halberstam, a liberal, acknowledged that "the general consensus" was that "Reagan . . . destroyed him."[9] Lou Cannon agreed as well, saying that, "Reagan clearly bested Kennedy,"[10] as did another of Reagan's first biographers, Joseph Lewis. In his 1968 work on Reagan, Lewis recorded that the "tanned and relaxed" Reagan "talked easily and precisely without a hint of uncertainty or hostility," and "deflated" the "anguished" Kennedy, who "gulped in restrained agony" when answering questions. Kennedy, said Lewis, "looked as if he had stumbled into a minefield"—which is a good metaphor, since the hostile questioners treated both him and Reagan like war criminals.[11]

Truthfully, this was not a debate between Ronald Reagan and Bobby Kennedy. Rather, it descended into a venomous America-bashing session by a panel of extremely rude international students, who seemed to bask in their big chance to unleash their torrent of anger on the two available representatives of the country they despised. *Newsweek* rightly described the students as "interrogators." In this atmosphere, Reagan and Kennedy ended up debating the group of students, not one another. And it was there that Reagan was so effective, whereas Kennedy was passive, meek, apologetic, and ineffective. Those looking for a defense of the United States not merely in Vietnam but as anything other than history's greatest purveyor of global misery were frustrated by Kennedy's lame responses but buoyed by Reagan's strong retorts.

In one of the evening's many disturbing moments, the students mockingly laughed out loud when Reagan said (obviously correctly) that the people of Mao's China had never chosen their government. In a more gratifying exchange, a contemptuous British student, who Kennedy permitted to roll all over him, complained that the Diem regime, with the alleged help of U.S. advisers, had incarcerated six million Vietnamese in "forced prison camps." With a smile, Reagan told the angry young man that there was no record whatsoever to confirm the allegation and that there were only sixteen million people in all of South Vietnam. *Newsweek* was impressed by this exchange, writing that Reagan "effortlessly reeled off more facts and quasifacts about the Vietnam conflict than anyone suspected he ever knew."[12]

Especially notable, but forgotten by history, were Reagan's remarks that evening concerning the Berlin Wall. The governor asserted:

> When we signed the Consular Treaty with the Soviet Union, I think there were things that we could've asked in return: I think it would be very admirable if the Berlin Wall, which was built in direct contravention to a treaty,

should disappear. I think this would be a step toward peace and toward self-determination for all people, if it were.

Here was possibly Reagan's first public call for the removal of the Berlin Wall, offered in May 1967, twenty years before his famous plea to Mikhail Gorbachev.

Once the hour had passed, Chris Collingwood jumped in to mercifully stop the spectacle. Despite the unpleasant nature of many of the student interactions, Reagan had performed well—so well that his presidential boosters eventually sought to use clips from the debate during the 1968 Oregon primary, and requested a copy from CBS. Kennedy, however, reportedly did not want the video to be made available; CBS acceded to his request.[13] Kennedy himself conceded defeat to Reagan, telling his aides after the debate to never again put him on the same stage with "that son-of-a-bitch." Kennedy was heard to ask immediately after the debate, "Who the f—— got me into this?"[14]

TEAR DOWN THAT WALL

While the debate was a win for Reagan, he had other (more routine) engagements that addressed the Cold War during this time. In a speech in Albany, Oregon on November 11, 1967, Reagan spoke of the USSR as the "totalitarian force in the world [that] has made plain its goal of world domination." As always, he said this was a goal that had been reiterated by Khrushchev and all present Soviet rulers—"Each one has stated they will not retreat one inch from the Marxian concept of a one-world socialist state." America needed to "fight" that enemy. "If we have the courage to face reality, peace is not so difficult to come by. We can have peace by morning if we do not mind the price."

The price, however, was an impediment; Reagan questioned America's will: "Why are we so reluctant to do this? Because there is a price we will not pay." He pointed to the Soviet tanks that crushed the Hungarian uprising in 1956, and the "echoes" of those who then cried: "People of the civilized world, in the name of liberty and solidarity, we are asking your help. . . . Listen to our cry."[15]

Reagan's words that day were revealing of his Cold War intentions. He continued: We cannot bring peace "by simply refusing to fight." It was "the height of folly" to believe "that we can end the Cold War simply by convincing the enemy of our good intentions." He asked: "How many nations have

backed down the road of good intentions to end up against a wall of no retreat with the only choice to fight or surrender?" He recalled the "sorry" example of Neville Chamberlain. He insisted that America could not "safely rest
the case of freedom with the United Nations."[16] No, that was America's job.

These were strong words. It might have served Reagan well to make clear
to his critics that choosing to "fight" did not mean that he favored armed aggression against the USSR. He did not make that clarification, probably because he felt it was unnecessary to do so; he did not want to nuke or invade
the Soviet Union.

A few months later, in May and June 1968, Reagan made a number of
historically important statements that likewise have somehow slipped through
the cracks of history. In a May 21 speech in Miami, Reagan again, for the second time within a year, and nearly twenty years before his address at the Brandenburg Gate, talked about removing the Berlin Wall: "If Russia needs our
wheat to satisfy the hunger of her people, it might be well to point out that
wheat could be delivered easier if there were no Berlin Wall between us."[17]
Here Reagan essentially suggested that the Lyndon B. Johnson administration, or an incoming administration, link U.S. wheat exports to a Soviet commitment to take down the wall. The very next day, in Cleveland, he again
attacked the wall, denouncing "Khrushchev's contemptuous raising of the
wall around Berlin."[18]

With his travels enabling him to speak to a variety of audiences, these
stops around the country proved valuable to Reagan. Not only did they enhance his campaigning skills and political recognition, but they also provided
him with a forum to air his passions. Though Reagan had been speaking on
these issues for more than twenty years, he had never done so as a politician.
Speaking to disparate groups of individuals gave him a certain confidence
that he would rely upon during his national campaigns in later years.

COMMUNISM KILLS THE KENNEDYS

Most intriguing were unrelated (and likewise forgotten) remarks made by
Reagan two weeks later, on June 5, 1968. On that day, Bobby Kennedy was
again an item for Reagan, though this time in a dreadful way that the governor could not have imagined in their debate a year earlier: Kennedy had just
been assassinated.

Reagan was invited to talk about the tragedy on entertainer Joey Bishop's
television show. A rare transcript of his appearance is today held by Bill Clark,

who, as Governor Reagan's chief of staff, grabbed a copy after the show. When Reagan appointed Clark a judge in San Luis Obispo County in 1970, Clark placed the transcript in a box, where it remains four decades later.[19]

Reagan spoke at length about Kennedy, the loss, and even offered spiritual advice on how to cope with the sadness. Kennedy, said Reagan, had been struck down in a "senseless" and "savage act." "I am sure that all of us are praying not only for him but for his family and for those others who were so senselessly struck down also in the fusillade of bullets. . . . I believe we should go on praying, to the best of our ability, to ask for God's mercy in what has happened to us." The governor said there was a "pall" over his state of California, the state in which the shooting took place.

Particularly interesting was how the Cold Warrior found a way to direct the discussion to America's real enemy: the USSR. Reagan noted that Kennedy's killer, an Arab, committed the crime because of the senator's support of Israel, specifically during the Six Day War that had occurred exactly one year earlier.

Indeed, we learned that the conflict was intentionally precipitated by the Kremlin, which concocted false intelligence reports about alleged Israeli troop movements. Moscow shared the phony information with Egypt and other Arab states for the explicit purpose of creating a military confrontation with Israel, which the Soviets believed would advance their broader foreign-policy interests in the Middle East and the world.[20] The shameless maneuver started a war, and was more evil than Machiavellian. It stands as one of the Soviet Union's most egregious international crimes.

Ronald Reagan knew this. As a result, he linked—not unreasonably—Bobby Kennedy's assassination to the USSR. "The enemy sits in Moscow," he told Joey Bishop. "I call him an enemy because I believe he has proven this, by deed, in the Middle East. The actions of the enemy led to and precipitated the tragedy of last night." Moscow had precipitated the Six Day War, which, in turn, had prompted RFK's assassin. Reagan's next words matched precisely those of his first presidential press conference thirteen years later: That same Soviet power, he said, believed that "the end justifies the means" and that "there is no morality except that which furthers [its] cause."

Although this take on Bobby Kennedy's assassination has not been popular historically, there was justification for it and as such Ronald Reagan did not shy from it. As he would display frequently during his political career, he was unwilling to let the threat of criticism deter him from voicing his quite strong opinions. To Reagan, Robert Kennedy's assassination was a cautionary tale, one that offered a very real sense of the evils that Communism inevitably leads to if allowed to flourish.

A similar extraordinary example, also unnoted in Reagan biographies, took place only eight days later. This time, Reagan connected the earlier assassination of the other Kennedy, John F. Kennedy, to Soviet Communism. Speaking on June 13, 1968 at the Indiana State Fairgrounds in Indianapolis, Reagan promised his fellow midwesterners that "in the days ahead . . . you and I are going to write a page in history." That remark was grand enough; yet, he said much more: a rare transcript shows that Reagan, on the heels of RFK's shooting, was eager to remind Americans of the worldview that had motivated JFK's murderer:

> Five years ago, a president was murdered by one who renounced his American citizenship to embrace the godless philosophy of Communism, and it was Communist violence he brought to our land. The shattering sound of his shots were still ringing in our ears when a policy decision was made to play down his Communist attachment lest we provoke the Soviet Union.[21]

Apparently, in the days since RFK's assassination, Reagan had formulated a new angle in his outrage toward Soviet Communism: Moscow's nefarious ways were leading, directly or indirectly, to the extermination of some of America's leading political figures.

Although on the surface Reagan's stark suspicions might have appeared to some as residual McCarthy-era hysteria, the reality was that both Jack and Bobby Kennedy were equally anti-Communist, a detail not wasted on Reagan. Both Jack and his father had crossed party lines and loyalties to support Republican Richard Nixon in his Senate bid against "Pink Lady" Democrat Helen Gahagan Douglas (for whom no less than an actor and liberal Democrat named Ronald Reagan had been campaigning). In what would later become a great irony in light of the 1960 presidential campaign, a check was mailed from the Kennedy home to Nixon's campaign. Moreover, Bobby had once worked as a staff counsel to Senator Joe McCarthy, and even asked McCarthy to be the godfather of his first child, daughter Kathleen.

With controversial stances such as these on the Kennedy assassinations, Reagan was making himself one of the country's most outspoken and polarizing voices on the Soviet Union. But as the 1960s came to a close, Reagan found himself still unable to impact U.S.–Soviet relations on a national scale. Although there was a Republican administration in the White House, it did not take long for Reagan to see that President Richard Nixon's true stripes toward Communism did not add up to the hard line that Reagan felt was necessary.

While Ronald Reagan had already emerged as a future presidential candidate, the developments of the 1970s would embolden his efforts, giving him a greater imperative and sense of urgency. He had to continue along his political path—which for now was located in California, though it did not dead end in Sacramento. The governorship had been a crucial first step, but now it was time for him to take his statewide actions and achievements and put them on display for the entire nation to judge.

5.

Breaking the Mold: 1970s

IN NOVEMBER 1978, PRESIDENTIAL CANDIDATE RONALD Reagan took a fact-finding trip to partitioned Germany. Accompanying the former governor were Richard V. Allen, Peter Hannaford, and the three men's wives. The six Americans entered East Berlin through Checkpoint Charlie, from where they went to Alexanderplatz and entered a large store, unaware that the East German secret police were watching and photographing their every move.[1]

As they left the store, they stood on the platz, observing the quiet shuffling of passersby, the lack of merriment, the drabness. At that moment, two East German policemen, Volskpolizisten, sauntered past and stopped a citizen carrying shopping bags. In front of the future president of the United States, a Cold Warrior already dedicated to undermining Communism, the police stupidly forced the shopper to drop his bags and show his papers, ignorant of the long-term repercussions of their actions. One of the officers rifled through the bags with the tip of an AK-47 while the other poked his victim with a machine gun. Reagan was enraged, his resolve steeled. Though just a routine shopping trip in the workers' paradise, the incident, remembered Allen, "set Ronald Reagan's blood to boiling. . . . Reagan was livid, and muttered that this was an outrage."[2]

It was also a warm-up. Next the group ventured to the Berlin Wall, a structure that Reagan had remonstrated against since its construction. But now he felt its cruelty firsthand. There was something about being there, something sickeningly real about directly observing a wall in which the East German guards faced East, not West, patrolling their own people, not an outside enemy. Armed East German soldiers faced their own unarmed citizenry, many of whom they shot and killed; the "threat" came from those looking to escape, not enter. *This*, Reagan thought to himself in disgust, was Communism.

Reagan took it all in. The enormity of the inhumanity pressed upon him, heavy like that wall that separated East and West and sat erect like a gray, cold tombstone to human freedom. He stood in stunned silence for several minutes. The presidential hopeful then turned to Allen and company, telling them, "We have got to find a way to knock this thing down."[3]

Reagan's forethought was part prophecy, part proclamation. Years later, Allen said correctly: "I believe the encounter with the wall and witnessing the armed harassment of an ordinary citizen seared into the governor's memory the brutality of the communist system." It "reinforced his dedication to placing it upon the ash heap of history." "It was clear from his reaction," said Allen, Reagan's one-day national security adviser, "that he was determined to one day go about removing such a system."[4]

DURING THE 1970S, RONALD REAGAN BEGAN TO OUTLINE MORE specifically the areas of U.S. Cold War foreign policy with which he disagreed, a process which he undertook forcefully and bluntly.[5] While he continued his ardent criticism of Soviet Communism during his governorship, his rhetoric remained strong even after he left Sacramento in early 1975. In what became a clear buildup to his two presidential campaigns of the decade, Reagan began to highlight the specific flaws of current and past United States' models for dealing with the Soviet Union. Using the voice that he had been cultivating for nearly thirty years, Reagan in the 1970s became the nation's most vocal champion for taking a tough stance on Russia. When many of America's politicians, both Democrat and Republican, seemed ready to extend olive branches of détente to the Evil Empire, Reagan did not sit quiet.

AREAS OF DISAGREEMENT

In the 1970s, a consistent Reagan refrain was that the United States should not accept the existence of the Soviet empire and its subjugation of Eastern Europe. Specifically, there were three concepts related to the Cold War that Ronald Reagan adamantly and vocally refuted during this period: Yalta, containment, and détente.

Yalta took place in the Crimea (Soviet territory) in February 1945 and involved FDR, Churchill, and Stalin. The travesty that resulted from Yalta was that Stalin broke his promise to hold free and fair elections in Eastern Europe, with those nations instead becoming Communist satellites and part of the Soviet bloc. At Yalta, critics charge, FDR and Churchill naïvely sold out Eastern Europe, condemning the historic cities of Eastern Europe to a future of Soviet totalitarianism. These were the enslaved peoples behind the Iron Curtain; the inhabitants of the "captive nations."

Ronald Reagan refused to accept that fate for Eastern Europe. "Reagan rejected Yalta, as if it were irrelevant," said Ed Meese. "He felt it simply wasn't right."[6] He never stopped complaining of Yalta.[7] Indeed, *Pravda* later devoted space to informing Soviet citizens of Reagan's explicit desire to undo Yalta. The Communist organ told readers that Reagan had declared that the dividing line between Western and Eastern Europe could never be legitimate. "We wish to undo this boundary," it accurately quoted him as saying.[8]

For similar reasons, Reagan disliked containment, the doctrine conceptualized in 1947 by George F. Kennan. This strategy sought to contain the Soviet empire within its present reach, but the problem for Reagan was that containment did nothing to free those already captive peoples along the Soviet border. Rather than preserve containment, Reagan wanted to go beyond it, to reverse it, to liberate the Soviet bloc. "For as long as I knew him," said Richard Allen, "Ronald Reagan rejected the doctrine of containment."[9]

Casper Weinberger, who had been with Reagan since 1967 and followed him to Washington as his secretary of defense, reiterated these sentiments, saying "The more he looked, the more he studied, the more he saw, the more he concluded that this [the USSR] was a regime that had to go. And this was certainly his own thinking well before the presidency, not some adviser's." Reagan came to this conclusion "quite early on. He did an awful lot of reading that people didn't realize. He was very well educated on the whole thing."[10] Reagan, then, said Weinberger, determined that, "It wasn't about containment; it was about winning the Cold War."[11]

Reagan, added Weinberger, insisted that Communism was incompatible with freedom and "was ultimately going to have to be destroyed and defeated."[12] He was unambiguous regarding Reagan's intentions:

> He recognized the folly inherent in the policy of "containment" of communism and the Soviet Union. He was not content to rest with the assumption that in eighty or ninety years, the USSR might collapse. He said from the

beginning that communism and democratic capitalism were incompatible, that communism had to be challenged and defeated.[13]

Weinberger stated categorically: "We were going to try to win the Cold War."[14]

Rejecting containment meant seeking victory, and the Soviets knew that this was what Reagan was striving toward. They grasped what Reagan meant when he rejected containment, and as early as 1975, they tried to head him off before he could pick up steam. In April 1975, in what may have been the first shot at Reagan in the Soviet press, commentator Yuri Zhukov complained in the pages of *Pravda*: "The resurrected political dinosaur from California proposes a . . . 'policy of rolling back communism.' It is incredible but true!" Equally amazing, said Zhukov, Reagan "promises the restoration of the old system in the countries of Eastern Europe."[15] Clearly, the Kremlin understood the stakes.

To Reagan, winning meant not only rejecting containment, but also rejecting détente, the policy of so-called "relaxed" U.S.–Soviet relations that the United States began under Nixon and continued through both the Ford and Carter administration. Sure, détente might "lessen tensions" between the superpowers by stimulating treaties and trade. However, in so doing, Reagan believed it condoned the enslavement of Eastern Europeans, not to mention Russians, Ukrainians, Estonians, Latvians, Lithuanians, Georgians, and millions more. Also, by helping the USSR, détente sustained the USSR, and prolonged rather than shortened its totalitarian existence and control.

By this rationale, accepting détente was akin to a schoolyard situation: There is a bully who terrorizes the smaller kids. Only one kid in class can lick the bully, but this would-be hero refuses to act simply so he and the brute can get along. And yet, even that is not a sufficient analogy. To be applicable to Reagan's world, the bully would not just push around the weaker children and take their lunch money, he would beat them and not permit them to say grace (he hates religion, after all) before their meals. Moreover, the reluctant rescuer in this scenario could and should be a beacon, a shining light to others, challenged by the forces of history to dispatch the thug—who looked to expand his thuggery to other schoolyards—to the ash heap of the playground.

Détente allowed the bully to have his way.

To Reagan, rejecting Yalta, containment, and détente was step one on the road to recovery for Eastern Europe—of freedom from the Soviet grip. As

Margaret Thatcher said, his rejection of containment and détente "proclaimed that the truce with Communism was over."[16]

Although these were the three main doctrines which Reagan rejected, there were other areas where he and the U.S. administrations differed, all of which were cut from the same cloth. For one, he rejected Nikita Khrushchev's "peaceful coexistence," because he spurned the idea of Soviet existence. He wanted peace but not coexistence. In Reagan's judgment, coexistence did not engender "peace" for the enslaved inhabitants behind the Iron Curtain, who could be jailed for exercising the most basic civil rights. As long as coexistence remained, it meant that the USSR would continue its position as captor over the peoples beneath its control.

In addition, coexistence fed the notion of so-called "spheres of influence," which maintained that both the United States and USSR had regions of influence or hegemony, in the West and East, and each side ought to respect the other's sphere. To Reagan, "respecting" a Soviet sphere entailed, again, accepting the East's subjugation. On the contrary, the U.S. sphere in Western Europe was free; the United States only governed one country: the United States.

This refusal to draw parallels between the two countries could also be seen in his rejection of moral equivalency, a doctrine dear to the heart of the political left, which claimed, among other things, that both the United States and USSR had legitimate mutual interests and thus an equally justifiable right to pursue their interests. To Reagan, the USSR pursuing its interests meant shackling people from Budapest to Bucharest. As far as Reagan was concerned, the people of Eastern Europe were still under siege, as they had been in World War II, only by a different hostile, totalitarian, occupying power. Western Europe won freedom in 1945, whereas Eastern Europe did not; its nightmare was still ongoing, and Reagan would not passively tolerate such repression for so many.

There was and could be no moral equivalency between America and a country that sought "world domination." "It's a frightening thought," said Reagan in the spring of 1975, "but it should make Americans all the more determined to show Europe that we have no intention of leaving the pages of history with a whimper: that, with or without them, we'll make our stand." When sharing this thought in an interview, he quoted (from memory) Churchill, from when the prime minister readied to take on the Nazis, another moral unequal: "This is only the beginning of the reckoning. This is only the first sip, the first foretaste of the bitter cup which will be proffered to us year by year, unless by a supreme recovery of moral health and martial vigor we rise again and take our stand for freedom as in the olden times." Churchill's words, Reagan instructed, "are now our destiny."[17]

Ed Meese connected these different but intertwined concepts in Reagan's eyes:

> He was not satisfied with the détente idea, and moral equivalency, which was anathema to him, and implicit in détente. He saw détente as a one-way street [that benefited only Soviet interests]. He wanted to reverse that. He believed that throughout the seventies and said it repeatedly when he first ran in 1976.[18]

These beliefs, particularly the real result of détente, came to determine much of Reagan's stated policy during the buildup to his presidential campaigns. They came to represent the backbone of the stance that he would offer when president, much to the dismay of those inside the Kremlin. From Yalta to détente to "spheres of influence," Reagan believed he had documented proof of failed attitudes toward the Soviet Union; at each juncture the United States was in denial about the true nature of Communism. Reagan was not willing to make the same mistake, and these ideas soon became the cornerstone of the principles that he hoped would govern America to greatness.

MORALITY AND FOREIGN POLICY

While the 1970s saw Reagan starting to voice his many critiques of the United States' Soviet policies, they also saw his fervent beliefs take on a new tone, one focused not just on the fall of Communism but on the moral imperative of the West to generate the collapse. This new voice spoke of the necessity of the demise of Communism, drawing a clear and unassailable connection between U.S. foreign policy and morality.[19] To Reagan, accepting détente meant a rejection of a moral foreign policy, an idea which became a Reagan campaign theme for 1976. Richard Allen termed it Reagan's "spring offensive" of 1976—an assault on the "intellectual and moral bankruptcy" of not only the USSR but current U.S. policy.[20]

This morality dimension made Reagan's foreign-policy preferences quite different from those of the previous hard-line anti-Communist Republican president, the *realpolitik* Richard Nixon, who felt that morality-driven foreign policy was nonsense. *Nations are driven by interests,* Nixon protested, *not morality!* "It's ridiculous," he growled, speaking of Reagan: "Foreign affairs aren't about trust. They're about interests and power." He thought Reagan's talk of morality and trust was "naïve."[21]

Despite Nixon's critiques, Reagan had serious reservations with a Machiavellian foreign policy. For starters, he took exception with his friend over how Taiwan was treated once the United States began recognizing Red China. Reagan understood that a rapprochement with China would hurt Moscow and help Washington. In an August 1971 letter, he wrote: "Personally, I think the Red Chinese are a bunch of murdering bums. I think the President probably believes the same; but in the big chess game going on, where Russia is still head man on the other side, we need a little elbow room."[22] Hence, he supported Nixon's shift toward China, keeping in mind, as he told fellow conservative M. Stanton Evans, that "Russia is still enemy number one."[23]

But though Reagan recognized the immense strategic value in a closer U.S. relationship with Communist China, he was highly sensitive not to "sell out" Taiwan in the process. When it appeared America had sold out its loyal ally, Reagan was furious, believing that the United States had not acted properly toward Taiwan. Later, when Taiwan was ousted from the United Nations, Governor Reagan wrote to President and Madame Chiang Kai-Shek, whom the Reagans had earlier visited on a mission for President Nixon: "Mrs. Reagan and I want you to know how deeply shocked and disappointed we were by the completely immoral action of the U.N. General Assembly."[24]

Reagan applied that moral sense to the Cold War. He said in a campaign speech that he wanted to "exert America's moral leadership in the world again."[25] This meant he could not accept any theory or policy that accepted the status quo for Eastern Europe.

He found people in Moscow who agreed: Speaking of the Brezhnev doctrine, which called for Soviet support of Communist guerrilla forces throughout the Third World, Genrikh Trofimenko, the well-known director of the prestigious Institute for U.S.A. and Canada Studies of the Russian Academy of Sciences, said that the Soviet government's opinion in the 1970s was that "there could not be a peaceful coexistence between wicked warmongering imperialists [America] and honorable and peace-loving Communists caring for the well-being of all progressive humanity." This, he said, was the view elucidated by Comrade Brezhnev. To his credit, said Trofimenko, Reagan realized this, recognizing détente and peaceful coexistence for the "shams" they were.[26]

Further vindicating Reagan's allegations were the Soviets themselves, who made it clear that détente did not mean an end to their support of Communist insurgencies. In a January 16, 1976 article in the *Washington Post*, Peter Osnos reported: "Soviet commentators have been saying almost daily that the 'policy of relaxation of tensions [détente] between states with different social systems [US and USSR] cannot be interpreted as a ban on the national liberation struggle of peoples who come out against colonial oppression or as a ban on class struggle.' "[27]

Brezhnev himself said as much. In his address to the twenty-fifth Congress of the Soviet Communist Party in Moscow, he stated candidly: "We make no secret of the fact that we see détente as the way to create more favorable conditions for peaceful socialist and communist construction." Détente enabled the Brezhnev doctrine in supporting Communist ideology and assisting "national liberation movements" around the globe.[28]

Less openly, in a secret speech in 1973, Brezhnev had told fellow Communist leaders in Prague: "We are achieving with détente what our predecessors have been unable to achieve using the mailed fist. We have been able to accomplish more in a short time with détente than was done for years pursuing a confrontation policy. . . . Trust us comrades, for by 1985, as a consequence of what we are now achieving with détente, we will have achieved most of our objectives." Brezhnev's brash declaration was downplayed by most of the major media when it became public four years later. When Reagan learned about it, he devoted a number of his syndicated daily radio commentaries to the speech, alleging that it "should have been front page in every major paper in the land."[29]

In fact, the USSR heartily supported Communist forces that opposed democracy and basic human and property rights. Under détente in the latter 1970s, the Soviets added, according to slightly varying accounts, ten or eleven nations to their list of client states.[30] Such gains alarmed many in the West who did not expect détente to produce such results, and the policy at last began to fall into wide disfavor.

THE 1976 PRESIDENTIAL BID

As the bicentennial of America's independence approached, Reagan had enough of those who robbed independence. He decided to seek the seemingly impossible: to challenge the incumbent president's bid for reelection, an incumbent from his own party. During his vocal anti-Communist campaign in the 1970s, Reagan had attacked not just détente, but also the Republican administration of Gerald R. Ford that favored the policy.[31] While on the campaign trail, Reagan did not hesitate to reiterate his sharp critiques of Ford policy. To cite one, in an October 1975 radio broadcast, he wrote, "We are blind to reality if we refuse to recognize that détente's usefulness to the Soviets is only as a cover for their traditional and basic strategy for aggression. Détente is for the Soviet Union a no-can-lose proposition."[32]

In addition, Reagan opposed Ford's signing of the Helsinki Accords in August 1975, a product of détente which he perceived as a human-rights farce. It was nothing more than a "propaganda plus" for the Kremlin, Reagan wrote in a February 1976 op-ed for *The Wall Street Journal*. By signing the accord, the United States had, in effect, "agreed to legitimize the boundaries of Eastern Europe, legally acquiescing in the loss of freedom of millions of Eastern Europeans." Worse, said Reagan, Helsinki did nothing to constrain the Soviets outside of Eastern Europe. "After Helsinki," wrote Reagan correctly, "the Soviet Union quickly made it clear that the so-called 'wars of national liberation' of which they are so fond, would not be affected by the document."[33] For Reagan, Helsinki was a bitter example of how détente only resulted in further Soviet legitimization and strength. Like détente itself, Helsinki was contemptible.

Throughout the campaign, Reagan hit détente so hard that there was a consensus that President Ford stopped using the term because Reagan had made it a dirty word.[34] *New York Times* columnist Anthony Lewis said Ford not only dropped the word from his vocabulary but fully abandoned it in his foreign policy, and, worse in Lewis' judgment, Ford was beginning to sound like Reagan—"more hysterical, more xenophobic."[35] Similarly, Jon Nordheimer of the *New York Times* reported that Reagan was attempting to place an "indelible stain" on the policy of détente "as exercised" by Ford and his Secretary of State Henry Kissinger.[36] So successful was Reagan that the *Times*, in a May 14, 1976 editorial titled "Mr. Reagan's Veto," claimed that the former governor had "won something approaching veto power over the Ford Administration's foreign policy."[37] In another editorial, titled, "President Under Seige," the *Times* opined: "Governor Reagan has become a credible candidate while President Ford has slipped from almost certain victor to underdog."[38] Reagan was making a dent.

Ford knew he was suddenly vulnerable. After New Hampshire, he had surged to five consecutive decisive victories, at times by big margins. These wins came in mostly eastern states, including the liberal northeast. As Martin Anderson remembered, the unasked question to Reagan by his campaign staff was, "When are you going to quit?" Reagan, however, was adamant. "I'm taking this all the way to the convention in Kansas City," he declared defiantly, "and I'm going even if I lose every damn primary between now and then."[39] Reagan laid out a plan of action, while his doubtful handlers observed in disbelief, dreading the specter of twenty-some painful primaries ahead.

Immediately after that decision, Reagan won North Carolina, claimed a huge triumph in Texas, and followed with victories in Indiana, Georgia, and Alabama. The Ford team began shaking in its boots. In a stunning turnabout, a new question was posed: Could Reagan go to the convention and win enough delegates on the first ballot? Reagan estimated a "very great possibility, if not probability," that he could do just that.[40]

Reagan's affront had been so productive that Ford not only dropped the word "détente" but replaced it with the preferred phrase of Reagan: "peace through strength." This Reagan credo, which became more pronounced in the 1980 campaign and the presidency that followed, proffered that a strong America, equipped with and embarking upon a military and technological buildup, could (among other things) hurt the USSR by prompting it to spend money that its economy did not have. "The Russians," declared Reagan that May, "know they can't match us industrially or technologically."[41]

In a pronouncement that signaled a startling concession, a waffling President Ford declared: "Our policy for American security can best be summarized in three simple words of the English language: peace through strength."[42] Reagan chuckled, noting it was "a slogan with a nice ring to it."[43]

PREPARING FOR THE CONVENTION

As the convention drew nearer, Reagan's rhetorical victories over Ford had become increasingly clear. Not only had he succeeded in changing the Republican perception of the word détente, but his supporters also sought to add a plank titled "morality in foreign policy" to the party platform:

> We recognize and commend that great beacon of human courage and morality, Aleksandr Solzhenitsyn, for his compelling message that we must face the world with no illusions about the nature of tyranny. Ours will be a foreign policy that keeps this ever in mind. . . .
>
> Agreements that are negotiated, such as the one signed in Helsinki, must not take from those who do not have freedom the hope of one day gaining it. . . .

Honestly, openly, and with conviction, we must go forward as a united people to forge a lasting peace in the world based upon our deep belief in the rights of man, the rule of law and guidance by the hand of God.[44]

The plank was widely reported as a repudiation of Ford-Kissinger foreign policy.[45] Fittingly, it quoted Solzhenitsyn, the esteemed Russian dissident who was among the more eloquent and credible critics of détente. In the summer of 1975, Solzhenitsyn had said the following to an AFL-CIO audience in Washington, DC:

I have tried to convey to your countrymen the constrained breathing of the inhabitants of Eastern Europe in these weeks when an amicable agreement of diplomatic shovels will inter in a common grave bodies that are still breathing. I have tried to explain to Americans that 1973, the tender dawn of détente, was precisely the year when the starvation rations in Soviet prisons and concentration camps were reduced even further. And in recent months, when more and more Western speechmakers have pointed to the beneficial consequences of détente, the Soviet Union has adopted a novel and important improvement in its system of punishment: to retain their glorious supremacy in the invention of forced-labor camps, Soviet prison specialists have now established a new form of solitary confinement—forced labor in solitary cells. That means cold, hunger, lack of fresh air, insufficient light, and impossible work norms; the failure to fulfill these norms is punished by confinement under even more brutal conditions.[46]

To Reagan, this complaint was exactly on the mark. Such statements brought him enormous respect for Solzhenitsyn—one that Solzhenitsyn showed was mutual.[47]

Not surprisingly, Solzhenitsyn's jailers disrespectfully disagreed with him and Reagan. And it was during this period that the "California cowboy" drew the full wrath of the Soviet press and became a household name in the Kremlin, as all of Reagan's remarks were agonizingly reported throughout the Soviet press.[48] Commentator-propagandist Valentin Zorin, in a statement issued by the Moscow Domestic Service, screeched that Reagan was seeking to "poison the atmosphere" and "sow doubts about" and even "wreck" détente. He had demonstrated "complete irresponsibility" in suggesting that only the USSR profited from détente, at America's expense.[49]

Separately, a well-sourced *Izvestia* report on the coming 1976 Republican convention, sarcastically titled, "Reagan Applies the 'Corrective,'" tied together clips from the U.S. media, as well as interviews by its own correspondent, in showing how Reagan had influenced Ford. *Izvestia* dubbed the morality plank "the Reagan amendment," and derided its "bombastic, pompous title: 'Morality in Foreign Policy.'" While this harsh language was scathing, it was by no means unique. Indeed, any Reagan talk of morality drew particular ire from the Soviet Communist press, which viewed such statements with a highly inflammatory eye.[50]

AUGUST 19, 1976: THE REPUBLICAN CONVENTION

All of this came to a head on August 19, 1976, when Republicans held their convention at the Kemper Arena in Kansas City, where Reagan, in the end, did not get the nomination. Amid the letdown, Reagan's boosters did not realize that the events of the convention had positioned their man as the Republican front-runner four years later. This was not simply because he had come so close to winning in 1976, but also because of his dramatic performance in Kansas City that evening.

President Ford had just finished speaking, and as a gesture of reconciliation and goodwill, he waved from the podium to the Reagans, seated in a skybox. He beckoned Reagan to come down to speak. The Republican faithful exhorted, "Ron! Ron! Ron!" They chanted, "Speech! Speech! Speech!"

None of this had been rehearsed for the cameras. A blushing Reagan had not planned to speak. He refused, gesturing his hands downward, pushing delegates to sit down and shut up. "It's his night," he muttered to friends, deferring to Ford. "I'm not going down there." Ford graciously pressed on: "Ron, will you come down and bring Nancy?"[51] National television audiences watched in anticipation, as ABC, CBS, and NBC news anchors peered through binoculars with moment-by-moment commentary.

Reagan turned to Mike Deaver: "But what will I say?" He eventually obliged. As he trotted down the corridors on his way to the podium, he said to Nancy, "I haven't the foggiest idea what I'm going to say."[52] This cluelessness was evident in the way he began his remarks. He offered salutations and thanks, to the Fords, to Vice President Nelson Rockefeller, and others. He then offered a couple of rambling, disjointed, customary sentences of appeal to voters, stammering something unmemorable about party platforms.

Then a thought came to him. "If I could just take a moment," he began, taking a pause as he at last decided what he was going to say:[53]

I had an assignment the other day. Someone asked me to write a letter for a time capsule that is going to be opened in Los Angeles a hundred years from now, on our tercentennial. It sounded like an easy assignment. They suggested I write something about the problems and issues of the day and I set out to do so, riding down the coast in an automobile looking at the blue Pacific out on one side and the Santa Ynez mountains on the other, and I couldn't help but wonder if it was going to be that beautiful a hundred years from now as it was on that summer day.

And then, as I tried to write . . . (pause) let your minds turn to that task. You're going to write for people a hundred years from now who know all about us. We know nothing about them. We don't know what kind of a world they'll be living in. . . .

[W]e live in a world in which the great powers have poised and aimed at each other horrible missiles of destruction, that can, in a matter of minutes, arrive in each others' country and destroy virtually the civilized world we live in.

And suddenly it dawned on me. Those who would read this letter a hundred years from now will know whether those missiles were fired. They will know whether we met our challenge. Whether they have the freedoms that we have known up until now will depend on what we do here. Will they look back with appreciation and say, "Thank God for those people in 1976 who headed off that loss of freedom, who kept our world from nuclear destruction?" . . . This is our challenge.[54]

It was an extemporaneous speech—no text, no cards, no teleprompter—and a compelling one. Always believing in the power of a story—the right story, which is not always an easy choice when trying to appeal to the broadest segment—Reagan had found one. Official biographer Edmund Morris later wrote: "The power of the speech was extraordinary. And you could just feel throughout the auditorium the palpable sense among the delegates that [they had] nominated the wrong guy."[55]

Despite last-minute second guessing, Republicans had made their choice. The race for the GOP presidential nomination had come down to the wire, and Ronald Reagan had just ended a remarkable run at the presidency. He had tried something extremely difficult: Usually a loyal party man—

his "eleventh commandment" was "thou shall not speak ill of another Republican"—he had attempted to take the nomination from the Republican incumbent, President Gerald R. Ford. He fell frustratingly short, missing by only 117 votes, grabbing 47.4 percent of delegates in an 1,187 to 1,070 contest. The winner needed 1,130.

Three months later, Gerald Ford lost the presidency to Jimmy Carter.

Michael Reagan recalls that August 1976 evening vividly: "We were upstairs. We just found out that Bob Dole had been picked as the vice presidential nominee. I asked my dad if he was disappointed. He said that what disappointed him the most was that he would not get a chance to look a Soviet leader in the eye and say, 'Nyet.' "[56]

Michael's sister Maureen never forgot a separate response from her father. Maureen described herself as "just devastated" by the defeat, and said she cried for two days—"I just couldn't stop." She remembers how it was traumatic for everyone but her father, who smiled at her and asked, "Are you still crying?" He pulled her aside and shared with her some of Grandma Nelle's theology: "There's a reason for this. . . . Everything happens for a reason. . . . [T]he path is going to open up."[57]

Ronald Reagan might have later found the reason in those words he spoke that evening in Kansas City: Rather than waiting 100 years, future generations needed only a decade or so to learn, mercifully, and with gratitude, that those missiles were not fired. To Reagan, the key to avoiding that exchange would be to end the confrontation altogether. The key would be to end the Cold War, to end the Soviet empire.

DURING THE LAST WEEK OF JANUARY 1977, ONLY DAYS AFTER Jimmy Carter had been inaugurated president, Richard V. Allen boarded a plan on the East Coast destined for the West Coast. Allen was planning to run for governor of New Jersey, and had all the credentials. He flew to California to ask Ronald Reagan to do a fundraiser for his coming campaign. Reagan happily agreed. By Allen's description, the two men ended up "talking and talking and talking" about foreign policy—for four hours. As the conversation in Reagan's Los Angeles home went on, they ate sandwiches.[58]

With the bruising presidential campaign five months behind him, a rejuvenated Reagan shared some candid thoughts with Allen, and none more frank or significant than this: "Dick, my idea of American policy toward the

Soviet Union is simple, and some would say simplistic. It is this: We win and they lose. What do you think of that?"

Allen's eyes flew wide open. He had held a number of foreign-policy jobs, including under Nixon and Kissinger, was a member of the Council of Foreign Relations, worked at the Center for Strategic & International Studies, had written five books on foreign policy, and had never encountered such thinking. "One had never heard such words from the lips of a major political figure," he later remarked. "Until then, we had thought only in terms of 'managing' the relationship with the Soviet Union. Reagan went right to the heart of the matter. . . . [H]e believed we could outdistance the Soviets and cause them to withdraw from the Cold War, or perhaps even collapse."

Allen had heard enough. He was hooked: "I needed no additional information, and resolved to help him, in some way, elevate that thought to the status of national policy. Herein lay the great difference, back in early 1977, between Ronald Reagan and every other politician: He literally believed that we could win, and was prepared to carry this message to the nation as the intellectual foundation of his presidency."[59]

Reagan's resolution was a prediction not merely on Communism's end but on the final endgame between the two combatants, with America vanquishing the Red Menace with his hand at the helm. Allen dubbed it "at once a prophecy and a plan."[60]

Asked twenty-four years later for clarification of this remarkable incident for inclusion in this book, Allen was unequivocal. I asked him: "Are you telling me that on that day in January 1977, Ronald Reagan told you that his goal was to take on and defeat the Soviet empire? That's what you're telling me?" Allen replied: "Yes. That's absolutely right. That's what I'm telling you." When asked to further clarify "that Reagan in fact had a specific intent to take on and defeat the USSR and the Soviet empire before his presidency even began. Was that his intent?" Allen again tersely affirmed: "Yes."[61]

On that January day, Allen was moved to join the cause. He went to Los Angeles to recruit Reagan but instead Reagan recruited him. Allen was so impressed with Reagan and so sold on the Californian's commitment to kill the Soviet empire, that he dropped his gubernatorial campaign to join Reagan's presidential campaign. Now, the world waited—as did the 1980 presidential contest.

THE 1970S WOULD SOON COME TO A CLOSE WITH A SCREECHING halt, ushering in the visible realities of America's weaknesses, which had festered and grown—from a poor economy to unhealthy doses of détente. Americans found themselves in desperate need of hope, a hope they had not felt in years. It was the hope for a strong America and a better world. It was a hope that American presidents had been unable to deliver in years. It was a hope that one man was convinced he could deliver now.

6.

"Let's Make America Great Again": 1980

BY APRIL 24, 1980, FIFTY-THREE AMERICAN HOSTAGES HAD BEEN held for six months at the U.S. embassy in Tehran. President Jimmy Carter had negotiated with Iran but to no avail. Exasperated, he authorized a risky operation to rescue the hostages. "I do not want to undertake this operation," he told his military brass, "but we have no other recourse. . . . We're going to do this operation."[1] Carter judged it was time "to bring our hostages home; their safety and our national honor were at stake."[2]

The adrenaline surged among elite U.S. forces. Carter's plan called for a complex two-night operation involving hundreds of varied personnel, including the elite Delta Force and its 132 commandos and a unit of Army Rangers. Forty-four aircraft would be deployed among six different locations, among them the critical RH-53 helicopters and C-130 Hercules transports rigged with temporary 18,000-gallon fuel tanks. On day one, the helicopters and C-130s would meet in an isolated spot in the Iranian desert where they would refuel the choppers, load the rescue team, and move on to hideouts near Tehran. On day two, undercover U.S. intelligence agents with trucks would escort the elite forces to the embassy, where they would seize the building and evacuate the hostages to a nearby soccer stadium. From there, all would be picked up by the helicopters and taken to the Iranian-controlled Manzariyeh Air Base forty miles away. Army Rangers were to take and hold the air base, shooting anyone who threatened the mission. Once the copters arrived, so would C-141s to fly out the hostages and rescue team.

Unfortunately it never got that far. Failure came in the first critical phase—getting the aircraft to their initial desert destination. Flying low to avoid detection, the helicopters were blinded by dust and sandstorms. Only six of the eight made it to their destination, leaving no room for error. Just then, one of the remaining six succumbed to hydraulic failure. The operation was over; the commander on the scene scrapped the mission.

But the troops were not out of the woods. In their aborted effort, they made quite a bit of noise, shooting at one Iranian truck that ran a roadblock and detaining an Iranian bus with over forty passengers on board. The forty-four troops needed to lift up and out.

First, a single helicopter that was parked behind one of the large C-130s needed to move. As it tried to lift off, it kicked up a dust cloud, forcing it to angle sideways rather than upward. Its blades sliced into the large aircraft, ripping open the huge fuel bladder. Suddenly, these safe havens became giant fireballs. Munitions and explosives ignited. Elite forces tumbled on to the ground, screaming in agony as their flesh smoldered. "The accident was a calamity heaped on despair," said Colonel James Kyle, the on-scene commander who watched in horror. "It was devastating." He called it "the most colossal episode of hope, despair, and tragedy" that he had experienced in nearly three decades of military service. Kyle is still haunted by the nightmares.[3] In the words of Staff Sergeant Taco Sanchez, they had "failed America."[4]

At 7 AM the next day, April 25, President Carter, a man for whom things could not get worse, informed a sleepy nation that eight of its servicemen had died in a secret, failed rescue operation. When the bodies were returned, a memorial service was held at Arlington National Cemetery. Captain J. V. O. Weaver escorted the wife and children of one of the deceased troops. Carter walked over to the family. Said Weaver: "[H]e looked down at those two little boys, and he just got down on his knees and wrapped his arms around them. Tears were streaming down his cheeks. Here's the president of the United States, on his knees, crying, holding these boys."[5]

The ordeal was dubbed Desert One, named for that awful area in the Iranian desert, and like "Bay of Pigs," it became synonymous for a military fiasco. Desert One was not just another humiliation in a long line of U.S. embarrassments; it was a microcosm of the stagnation and low morale that America faced. More than just an aborted rescue mission, this was a cry not just for a new leader but for a new paradigm.

FROM THE FALL OF SAIGON TO THE PROSTRATE ECONOMY, this was a bad time for America. It was the post-Vietnam, post-Watergate era in which the flatlined economy was underscored by a pessimistic measurement called the "misery index." Unemployment and inflation were at double-digit levels. Interest rates approaching 20 percent nearly halted home buying. An energy crisis wrought soaring gas prices and queues. In short, the nation faced its worst economic situation since the Great Depression.

In addition, military morale, still stained from Vietnam, had faced yet another setback with the Desert One fiasco, and U.S. foreign policy was not faring much better. With hostages remaining in Iran, the United States faced hostile regimes in Iran and Nicaragua, and Soviet advances around the globe. Only weeks after Americans were taken hostage in Iran, President Carter met another foreign-policy nightmare when the Red Army stormed into Afghanistan, Moscow's first direct military intervention outside the Warsaw Pact since World War II.

Carter's initial reaction to the Soviet invasion did nothing to slow his slide in opinion polls. In a television interview six days later, he said: "My opinion of the Russians has changed most dramatically in the last week. . . . [T]his action of the Soviets has made a more dramatic change in my own opinion of what the Soviets' ultimate goals are than anything they've done in the previous time I've been in office."[6] His remark was viewed by many as dangerously naïve, as a form of on-the-job education unacceptably late for a president. Ronald Reagan privately said that Carter's assessment "would be laughable, I think, if it were not so tragic."[7] In a separate private comment, Reagan added: "It is frightening to hear a man in the office of the presidency who has just discovered that the Soviets can't be trusted, that they've lied to him."[8]

The public agreed. After the Red invasion, Carter's approval rating fell to 20 percent.[9]

As Uncle Sam reeled, the Soviets hammered Afghanistan, mercilessly pounding the Afghan resistance. Campaigning in Pensacola, Florida on January 9, 1980, Ronald Reagan had a policy recommendation: the United States should supply the guerrillas with shoulder-launched, heat-seeking missiles to shoot down Soviet helicopter gunships.[10]

While that suggestion became a battle that he would wage during his presidency, for now it was clear that one of Reagan's most ardent beliefs had come true: détente was dead. Now, much of America agreed with Reagan that détente was a joke that merely let Moscow pursue expansionary interests with no improvement in behavior; the Soviet invasion was a wake-up call.

Unfortunately, in the eyes of many, America, too, was dead. The nation, we were told, was "in decline"; it had seen its glory days. Edmund Morris captured the mood: "When President Carter wasn't telling us about his hemorrhoids, he was telling us about our national malaise. 'Patriotism' was an embarrassing word. Young people were snickering at men and women in uniform."[11]

Morale was at its lowest point in fifty years, with seemingly little prospect for reversal, and it was this trend that Ronald Reagan sought to reverse. In 1980, more than forty years after he saved seventy-seven people from drowning in Dixon, Illinois, the former Rock River lifeguard appointed himself a new rescue mission: America.

REAGAN'S VISION FOR RESTORATION

In his previous presidential campaign four years earlier, Reagan made it clear what he was against. Now, in 1980, he was especially vocal about what he favored. Ronald Reagan wanted a confident America, a global leader that would protect and promote freedom; the country was to serve as a Shining City Upon a Hill—a light of liberty that would "shine unto the nations," acting not just as a *model* of liberty but a purveyor.[12] To Reagan, spreading freedom entailed rolling back Communist totalitarianism. The Shining City represented not just a lofty ideal; it signaled a future in which America stood proud, a future in which there was not a Soviet superpower.

In U.S. history, the Shining City was invoked in 1630 aboard the Arabella by John Winthrop, the leader of the small group of pilgrim settlers approaching the shoreline. While other Americans cited the metaphor in subsequent centuries, few latched on to it like Reagan. Both Reagan and Winthrop felt that this role for America was divinely ordained: God had chosen their land as an example, to be set upon a hill, aglow, admired by all who covet freedom. An important corollary to Reagan's invocation is that he believed that a leader who recognized this special call needed to encourage fellow Americans to respond to this challenge to lead the free world.

This was a constant theme in a 1980 presidential campaign that arguably began for Reagan with a speech on February 6, 1977—his sixty-sixth birthday—before the annual Conservative Political Action Conference (CPAC), one of his favorite venues. In retrospect, this was one of his most revealing pre-presidential speeches. The address launched a premise he carried through 1980:

It was the Soviet Union, not America, which should hang its head in shame. America was a source of brightness, whereas the USSR was a source of darkness—literally, he told his audience:

> It isn't very often you see a familiar object that shocks and frightens you. But the other day I came across a map of the world created by Freedom House. . . . It is an ordinary map, with one exception: it shows the world's nations in white for free, shaded for partly free and black for not free.
>
> Almost all of the great Eurasian land mass is completely colored black, from the western border of East Germany, through middle and eastern Europe, through the awesome spaces of the Soviet Union, on to the Bering Strait in the north, down past the immensity of China, still further, down to Vietnam and the South China Sea—in all that huge sprawling, inconceivably immense area. . . . If a visitor from another planet were to approach earth, and if this planet showed free nations in light and unfree nations in darkness, the pitifully small beacons of light would make him wonder what was hidden in that terrifying, enormous blackness.
>
> We know what is hidden: Gulag. Torture. Families—and human beings—broken apart. No free press, no freedom of religion. The ancient forms of tyranny revived and made even more hideous and strong through what Winston Churchill once called "a perverted science." Men rotting for years in solitary confinement because they have different political and economic beliefs, solitary confinement that drives these unfortunate ones insane and makes the survivors wish for death.
>
> Only now and then do we in the West hear a voice, from out of that darkness. Then there is silence—the silence of human slavery. There is no more terrifying sound in human experience; with one possible exception. Look at that map again. The very heart of the darkness is the Soviet Union.[13]

America, on the other hand, "must always stand for peace and liberty in the world and the rights of the individual," said Reagan, and "must form sturdy partnerships with our allies for the preservation of freedom." Presaging his Westminster Address five years later, Reagan told CPAC categorically that "the United States has to immediately reexamine its entire view of the world and develop a strategy of freedom." He used the word "freedom" five times in the next six paragraphs of the speech, stating unequivocally that America must develop a strategy to spread freedom, as this is the "moral way" and "we can never go wrong if we do what is morally right." America must speak the "truth" and "cannot be the second best superpower" militarily.[14] He finished the rousing speech with the classic ending for all his CPAC speeches, an ending which would come to define the patriotic ideology of his 1980 presidential campaign: "Then with God's help we shall indeed be as a City Upon a Hill with the eyes of all people upon us."[15]

RESTORING MORALE

The unbridled optimism expressed in the City Upon a Hill image was central to Reagan's goal of restoring American morale. Henry Kissinger, who called it an era of national humiliation, stressed that what Reagan tried to turn around was "very significant."[16] ABC's Jeff Greenfield agreed that the ex-movie actor's project was "Reagan's biggest challenge."[17] In his memoirs, Reagan captured the essence of this fundamental objective:

> During the summer and fall of 1980, there were many problems facing our nation: the tragic neglect of our military establishment, high unemployment and an ailing economy, the continuing expansion of Communism abroad, the taking of the hostages in Iran. But to me none was more serious than the fact America had lost faith in itself. We were told there was a "malaise" in our nation and America was past its prime; we had to get used to less, and the American people were responsible for the problems we faced.
>
> We were told we would have to lower our expectations; America would never again be as prosperous or have a bright a future as it once had.
>
> Well, I disagreed with that. . . . We had to recapture our dreams, our pride in ourselves and our country, and regain that unique sense of destiny and optimism that had always made America different from any other country in the world.
>
> If I could be elected president, I wanted to do what I could to bring about a spiritual revival in America.
>
> I believe—and I intended to make it a theme of my campaign—that America's greatest years were ahead of it, that we had to look at the things that had made it the greatest, richest, and most progressive country on earth in the first place, decided what had gone wrong, and then put it back on course.[18]

Indeed from the very start of his campaign, Reagan expressed this desire, articulating to the public with every appearance the conviction that this was a country on the brink of a rebirth of greatness. When Ronald Reagan formally announced his 1980 candidacy on November 13, 1979, his words were unmistakable: he proclaimed that he could restore the "American spirit"—with "God's help."[19]

Soon, this campaign theme was so clear that the cover of *Time* on March 10, 1980 had an artist's rendering of Reagan's face at a rally that included a hard-hat-wearing man hoisting a sign that read, "Reagan: Let's Make America Great Again,"[20] a slogan which quickly came to embody the campaign message. Not only did the line work its way into campaign paraphernalia (a popular 1980 campaign button featured a picture of a smiling Reagan with those exact words: "Let's Make America Great Again"), but Reagan began ending many of his stump speeches with lines like, "So help us God, we will make America great again."[21]

This issue of morale and many others were front and center in a key March 17, 1980 address by Reagan to the Chicago Council on Foreign Relations. As he had since the GE days, Reagan wrote this speech himself—a rare handwritten draft of which was only recently found.[22] Written on March 13, and the first of two important foreign-policy speeches he gave in Chicago that year,[23] Reagan titled the text, " 'State of the Union' Speech."

In this first Reagan "State of the Union," he argued that regaining U.S. prestige was fundamental to a fruitful foreign policy: "Confronted by so many pressing crises, we would all like to find quick solutions. What can be done, tomorrow, to free our diplomats in Tehran? What can be done now to turn back the Soviet invasion of Afghanistan?" He provided his answer: "We can neither solve these present crises, nor cope with graver, future ones, unless we regain a reputation of reliability toward our allies."[24] To regain that reputation, said Reagan, America had to purge the Vietnam syndrome eroding its prestige and ability to act in foreign policy. Here is an excerpt from his unedited draft:

> We never sought the leadership of the free world but no one else can provide it. And without our leadership there can be no peace in the world. It is time we purged ourselves of the Vietnam syndrome that has colored our thinking for too long a time. Speaking at Notre Dame U. 5 months after he had assumed office Pres. Carter said; "we are free of the inordinate fear of communism which led to the moral poverty of Vietnam."
>
> Possibly Vietnam was the wrong war, in the wrong place at the wrong time. But when 50,000 Young Americans make the ultimate sacrifice to

defend the people of a small defenseless country against the Godless tyranny of communism that is not an act of "moral" poverty. It is in truth a collective act of moral courage.[25]

Working from this draft framework, he began the lengthy speech by blasting the Soviet Union as "an imperialist power whose ambitions extend to the ends of the earth," which "has now surpassed us in virtually every type of weapon. The Soviets arrogantly warn us to stay out of their way." And how did America respond? "We respond," charged Reagan, aiming at the Carter administration, "by finding human rights violations in those countries which have been historically our friends [and] allies. Those friends feel betrayed and abandoned and in several specific cases they have been." Slamming the Soviets, as well as Cuba, and Carter policy, he added: "A Soviet slave state has been established 90 miles off our coast; our embassies are targets for terrorist attacks; our diplomats have been murdered and half a hundred Americans are captives going into the 5th month now in our embassy in Iran."[26]

Reagan contended that America's problem was weakness; it needed to get strong. He telegraphed his "peace-through-strength" credo: "May I suggest an alternate path this nation can take, a change in foreign policy from the vacillation, appeasement, and aimlessness of present policy?" He answered his own question: "That alternate path must offer three broad requirements," he explained. "*First* it must be based on firm convictions, inspired by a clear vision of, and belief in America's future. *Second*, it calls for a strong economy based on the free-market system which gave us an unchallenged leadership in creative technology. *Third*, and very simply we must have the unquestioned military ability to preserve world peace and our national security."[27]

Notably, these would become the three legs to his administration's approach in the 1980s. Underscoring these goals was his intent to restore morale and prestige, a point he made on three different occasions in the speech. Each time he revisited the idea of returning to greatness, Reagan added more detail and emphasis—devoting one sentence to the goal the first time, then one-and-a-half paragraphs the next, and finally a three-paragraph passage at the end of the speech.[28]

Reagan concluded the address by saying: "But while we do all these things and they are essential, we must above all have a grand strategy; a plan for the dangerous decade ahead."[29] Years later, when the former president told adviser Martin Anderson about how he had come to the Oval Office with a "plan," he could have pointed to this prepresidential address.[30]

At all levels, this speech reflected the plan which Reagan would put into motion in less than a year.[31] The strategies for victory laid out in this self-written speech became the foundation from which the future Reagan administration would work. It was a foundation that Reagan alone had laid out.

MARCH–JUNE 1980: THE GAUNTLET OF AN ARMS RACE

In Reagan's eyes, rebuilding the military would not only reinvigorate American confidence but would weaken the Soviets and bring them to the negotiating table. Reagan had started down this line during the campaign against Ford, and picked it up with vigor after the loss, stretching it through November 1980. He recommended that President Carter use the "trump card" of threatening an arms race if the USSR did not agree to acceptable limits on nuclear arms. Carter had "one trump card that has never been used," averred Reagan. "We know and the Soviet Union knows that if there is to be an arms race, they can't even get in the same ballpark as us." If the Soviets "have to compete with us, I'm sure they'll come running to the table and say 'wait a minute,' because they know they can't."[32] He asserted flatly: "The Soviet Union cannot possibly match us in an arms race."[33]

While identical Reagan quotes from the period could be cited over and over,[34] he gave two telling interviews in March and June of 1980 that merit careful attention. The first was with the *National Journal*, in which Reagan estimated: "An arms race is the last thing they want us to do. The Soviet Union, I believe, is up to its maximum ability in developing arms. Their people are denied so many consumer products because it is all going into the military." Thus, "They know that if we turned our full industrial might into an arms race, they cannot keep pace with us. Why haven't we played that card?"[35]

Also intriguing was a two-hour June 1980 interview Reagan gave to a group of *Washington Post* editors and reporters, including Lou Cannon. In the opening of his write-up of the interview, Cannon focused on what he found most remarkable among Reagan's comments: "Ronald Reagan said yesterday that a rapid U.S. arms buildup would be good for the United States because it would strain the defense-burdened Soviet economy and force the Soviets to the arms control bargaining table." Reagan said: "The very fact that we would start [a rapid arms buildup] would serve a notice on the Soviet Union. I think there's every indication and every reason to believe that the Soviet Union cannot increase its production of arms. . . . They've diverted so much to the military that they can't provide for the consumer needs." To Reagan, the United States could not expect to bring the Soviets to the table without a buildup; employing one of his frequent lines, he told the *Post*, "So far as an arms race is concerned, there's one going on right now but there's only one side racing."[36]

Lou Cannon still marvels at this interview, understanding the magnitude of what Reagan said that day. In an oral-history testimony years later at the University of Virginia, Cannon recounted the moment, adding information:

> [W]e asked him if he feared an arms race with the Soviet Union. He replied that he welcomed an arms race because he was convinced the Soviets couldn't compete with us. Reagan believed that showing we were serious about a buildup would lead to negotiations with the Soviet Union because their economy could not support an arms race. Reagan was right. Of course, one never knows what the road not traveled might have held. Still, it seems that the fact that the United States was intent on technological competition with the Soviets was certainly a factor influencing a wise leader like Gorbachev in deciding to negotiate a lower level of armaments. . . .
>
> Many people were surprised by the foreign policy developments and the fact that you finally wound up with Reagan and Gorbachev strolling through Red Square proclaiming a new era. I wasn't surprised with that because I knew Reagan envisioned this sort of rapprochement.[37]

Cannon, a liberal and once a Reagan doubter, also later spoke of the 1980 incident at Hofstra University. He recalled that one of the *Post* reporters had responded to Reagan: "You're going to start this horrible arms race, aren't you? This is going to get worse, [between] the Soviets and us." Reagan calmly answered, "Oh, it won't."[38]

It did, however, get worse before it got better.

JUNE–JULY 1980: THE REPUBLICAN CONVENTION

With the July 1980 GOP convention approaching, the summer of 1980 brought the Reagan campaign to a fever pitch, and this time, Reagan did not need to overcome an incumbent Republican. His path was clear, with only President Carter standing in the way of the Oval Office. Now, it seemed, Reagan was craving the cup of Cold War victory more than ever. In June 1980, he sent a private letter to a New Jersey man named Severin Palydowycz. "[W]e must restore our prestige in the world," wrote Reagan, before commenting:

"We must also keep alive the idea that the conquered nations—the captive nations—of the Soviet Union must regain their freedom."[39]

In his acceptance speech at the Republican convention, Reagan called for a "great national crusade to make America great again." The Crusader urged:

> More than anything else I want my candidacy to unify our county; to renew the American spirit and sense of purpose. . . . They [Democratic politicians] say that the United States has had its day in the sun, that our nation has passed its zenith. . . . My fellow citizens, I utterly reject that view. . . . I will not stand by and watch this great county destroy itself under mediocre leadership that drifts from one crisis to the next, eroding our national will and purpose. We have come together here because the American people deserve better from those to whom they entrust our nation's highest offices and we stand united in our resolve to do something about it. We need a rebirth of the American tradition of leadership.[40]

Yet, what Reagan said privately on the way to the convention was more revolutionary than anything he or his fellow Republicans mouthed for the cameras. Stuart Spencer, a nonideological political guru who was close to Nancy, accompanied Reagan on the long flight from Los Angeles to Detroit. The two men talked about delegate counts, VP picks, convention strategy. After a while, Spencer took the liberty to probe Reagan's deeper motivations. "Why are you doing this, Ron?" he asked. "Why do you want to be president?" Without hesitation, Reagan responded: "To end the Cold War. There has to be a way, and it's time."[41]

HOLLYWOOD CONFIDENCE

The Soviets were aware that Ronald Reagan had drawn a target on their chest. A classified KGB assessment from 1980 made it clear that the Kremlin was already convinced of the unshakable quality of Reagan's core convictions. The KGB noted that unlike more "practical" anti-Communists like Richard Nixon, Reagan was committed to mischief against the Soviet empire and nothing would change him. "Reagan doesn't react to our suggestions," said a frustrated Politburo member. An equally exasperated KGB official tossed in the towel: "We aren't going to change Reagan's behavior."[42]

What the Soviets sensed in Reagan's character was also sensed by another, more benign adversary: Eugene McCarthy. McCarthy's measurement of Reagan became evident in a private meeting one campaign night in Iowa. Reagan and Mike Deaver were astonished when the liberal icon abandoned the Democratic president and informed them he was endorsing the arch-conservative Reagan. Deaver walked McCarthy to the car and prodded him for his reasons. "I'll tell you why," said the Democrat matter-of-factly. "It's because he's the only man since Harry Truman who won't confuse the job with the man."[43]

McCarthy recognized Reagan's self-security, and that the Californian was content in his own skin. He perceived that Ronald Reagan was not looking at the presidency as a means to fulfill some inner ambition. Reagan had long ago received recognition; unless he was unbearably insecure—which he was not—he had easily satisfied any longing. As *GE Theatre* colleague Earl Dunckel put it, "Hell, he doesn't need the money and he really doesn't care about the status of being president." He sought the presidency simply because he wanted to do a job that needed to be done.[44] As more than one Reagan campaign staffer noted of his pursuit of the presidency, he could take it or leave it.[45]

McCarthy sensed that, as did columnist George Will, who conjectured that Reagan was free of the need for applause—as a candidate and later as president—because he had heard enough in Hollywood.[46] Will's point was apt. Whereas many have sought the White House to satisfy an inner appetite, or desire for esteem or applause or recognition, Reagan craved nothing of the sort. His Hollywood confidence was evident after a 1980 debate with President Carter, when a reporter asked Reagan, "Governor, weren't you intimidated by being up there on stage with the President of the United States?" "No," he responded. "I've been on the same stage with John Wayne."[47]

Self-assurance was perhaps one of his greatest assets not just during debates but also when Reagan was blasted by unsympathetic reporters on the campaign trail. For many candidates, that harsh media treatment is a shock that overwhelms them. "The thing about Reagan that was so unusual," said Ben Elliott, who did his part for Reagan's 1980 bid, and later became a speechwriter, "was that he wasn't afraid to offend the liberal elites." Elliott adds: "It took people a long time to realize that Reagan wasn't intimidated by anybody. He had a life full of very rich experiences. He did many things. . . . And no one would intimidate him."[48]

State Department veteran Peter Rodman agreed that Reagan was "totally impervious" to the judgment of the editorial boards of the *New York Times* and *Washington Post*—a truly liberating quality for a conservative Republican.

Rodman notes that another California Republican that recently secured his party's presidential nomination, Richard Nixon, had been constantly tortured by the liberal criticism.[49] That distinction is profoundly important: for Nixon, this resentment fueled an unstoppable hatred that, by his own admission, consumed and destroyed him.[50]

Hollywood had formed that shield, that Teflon veneer. Dinesh D'Souza observed that Reagan was a bona fide celebrity who had starred with beautiful actresses like Ann Sheridan, Ginger Rogers, and Doris Day (and lesser-knowns like Olympe Bradna, Viveca Lindfors, and Rhonda Fleming). His friends were men like Errol Flynn, John Wayne, Jimmy Stewart, and Claude Rains. Fan clubs all over the country idolized him, filled with young girls who begged him for his picture. With all of that, asked D'Douza, what did Reagan care about what some assistant editor at *Time* thought about him?[51]

On November 4, 1980, Ronald Reagan won the U.S. presidency by a comfortable margin, taking forty-four of fifty states from Jimmy Carter, closing a campaign that in many ways began in Hollywood and in Dixon decades earlier. His biographer later found a message sent to Reagan that November 4 by a radio announcer at WSDR, fifteen miles downstream from where Dutch had lifeguarded: "The Rock River flows for you tonight, Mr. President."

DECEMBER 1980: "HE'S GOING TO BRING DOWN THE SOVIET UNION"

What happened that first Tuesday in November 1980 was obviously quite visible to the general public. That was not the case for a dramatic moment that occurred a month later among two individuals at a party in Alabama, an encounter never revealed, until now.

Ollie Delchamps was founder of a supermarket chain. A friend of the late political strategist Lee Atwater, he became closely involved with the 1980 Reagan campaign.[52] A very revealing incident took place between Delchamps and another successful Alabama businessman named Bob Callahan, Sr. in December 1980. It was a moment that Callahan has never forgotten and long sought to share with someone who could record it for the sake of history.[53] Here, in his own words, is what Callahan experienced:

> It was December 1980, the second full weekend, a Saturday. It was a cocktail party in Point Clear, Alabama. There was a man there named Ollie Delchamps. I asked him what Ronald Reagan planned to do once he became

president. He basically gave me this P.R. statement, campaign literature kind of stuff: education, science, taxes, general policy. "No," I said. "I want to know the *real* objectives, not what's in the newspapers." I was a bit forceful and blunt. He turned his back on me. "Oh, no," I thought. "I made him mad!" Then he went over and huddled with some of his colleagues who were also a part of the Reagan side. I don't know what exactly went through his head or what he was saying.

After a while, he turned around, walked back to me, and said, "What is it you want to know?" I told him again, "I want to know what Ronald Reagan *really* intends to do." "Who are you going to tell?" he replied. I said, "No one but me and Ginger [Callahan's wife]." "Let's walk over to the corner," he said. [Delchamps added:] "No one but Ginger, eh?" "Yes," I said. "I promise. No one but Ginger." "Okay," he said. "Ronald Reagan wants to do three things: One, build up the economy. Two, build up defense. Three, he's going to bring down the Soviet Union."[54]

Those words—driven home by the fashion in which Delchamps cornered Callahan and swore him to secrecy—remain seared in Callahan's memory: "Reagan is going to 'bring down the Soviet Union,' he told me. 'Bring it down.' Exact words. I've never forgotten that. And I never told anyone other than my wife. But I'm telling you now. I think it's okay to share the story now." Almost twenty-five years later, Callahan remembered the encounter like it had happened yesterday.[55]

WHILE THESE WORDS HUNG ON THE LIPS OF TWO MEN AT A cocktail party, Reagan himself was getting prepared—not just for his arrival in Washington but for the inevitable struggle that was to come. To Ronald Reagan, this was the moment he had been preparing for since his earliest encounters with Communism; it was the culmination of a life spent standing up for his unflinching belief in the American way of life. Spurred and affirmed by his electoral success, he was ready to take the oath.

PART II

The First Term

7.

First Priorities: January to November 1981

ON JANUARY 20, 1981, RONALD REAGAN WAS INAUGURATED president of the United States. He stood near the podium at the front side of the terrace of the Capitol building. As Chief Justice Warren E. Burger administered the oath of office, Nancy stood aglow next to her husband, beaming, perhaps thinking about how far they had traveled since arriving at the Last Frontier Hotel in Las Vegas three decades earlier. Her husband had told her back then that Vegas was merely a temporary setback on the road to something much better—God had a plan. This January 20 would seem a rather robust vindication of his optimism.

It was a drab, gray day in the nation's capital. The president-elect's hand rested atop his mother's old, wrinkled Bible, opened to Nelle's favorite verse, II Chronicles 7:14, next to which the late Nelle had scribbled, "A wonderful verse for the healing of a nation." And it was that which Nelle's son prescribed for his nation that day—recovery.

The moment he swore the oath, the former lifeguard spelled out his commitment to rescue America. In his Inaugural Address, which, unbeknownst to the press, he wrote entirely himself, Reagan declared: "We're not, as some would have us believe, doomed to an inevitable decline. So, with all the creative energy at our command, let us begin an era of national renewal."[1]

The theme pervaded not only his address but also the ceremony. On the reverse side of the tickets for the inaugural event were photos of Reagan and

Vice President George H. W. Bush alongside the words: "America—A Great New Beginning, 1981." There was no mistaking the message. The next day's headline beamed across the *New York Times*: "Reagan Takes Oath as 40th President; Promises an 'Era of National Renewal.'" Of everything Reagan had said, the *Times* focused on those four words.

It is not an exaggeration to say that the change in mood began at that moment, as the fifty-two hostages were freed in Tehran—prompting the *Times* to continue the Reagan headline into a second line that likewise stretched across the top-of-the-fold: "Minutes Later, 52 U.S. Hostages in Iran Fly to Freedom After 444-Day Ordeal."

NOW, REAGAN SET HIS SIGHTS ON A TARGET THAT FIRST TOOK form in 1950. In those initial days of the presidency, newly christened CIA director Bill Casey told Reagan that the United States had a "historic opportunity" to "do serious damage" to the USSR. This was sweet music to the new chief executive, and Reagan suggested that Casey share his optimism at the January 30 National Security Planning Group (NSPG) meeting.

It was at that meeting that the subject of a covert, strategic offensive against the Soviet Union was first brought up. Reagan weighed arguments presented at the meeting by Casey—as well as counterarguments by Secretary of State Al Haig—before concluding that Casey's course made "the most sense." Reagan agreed with Casey that the administration should launch a concerted effort to play on Soviet vulnerabilities. While the details remained to be hammered out, the initial commitment was made on that January 30, only ten days into the Reagan presidency.[2] It was an auspicious beginning.

What Casey put forth was consistent with what Reagan had been telling his new national security adviser and old comrade, Richard Allen. As the head of Reagan's National Security Council, Allen began instructing his staff that Reagan had a "clear strategy in mind," a "plan" to defeat the USSR, and was now "changing the game plan" in the Cold War confrontation.[3] Relaying what Reagan was saying behind closed doors, Allen told the NSC staff that "the president is determined to do everything possible to destroy the Soviet bloc and end the Cold War with victory."[4]

It was also in early 1981 that a group of Pentagon specialists, convened under the direction of Fred Ikle, Reagan's undersecretary of defense for policy, put together a "defense guidance" that was so secretive that it still remains classified. A copy of the document was obtained by author Peter Schweizer, who has shared some of its contents,[5] which include two remarkable objectives: "Reverse the geographic expansion of Soviet control and military presence throughout the world. . . . Encourage long-term political and military changes within the Soviet empire."[6] Clearly, the Reagan administration was wasting no time in committing itself to huge changes.

More such declarations, made in formal policy documents, would follow. But first, Ronald Reagan faced the ultimate setback—a near fatal one.

MARCH 1981: THE SHOT HEARD 'ROUND THE WORLD

On March 30, 1981, an individual named John Hinckley—a "confused young man," as Reagan described him—in a bid to attract the attention of actress Jodie Foster, tried to take the life of Dixon's former lifeguard, nearly precluding the grand rescue he planned for America and the world. It was a frenetic start to a presidency still in its "honeymoon" period.

We have since seen the image countless times on television: Reagan raises his left arm to ward off a question from an inquiring reporter and then, a second later, the sound of bullets crackle the air; chaos ensues as police scramble for the shooter, and more than one man hits the ground with serious injuries. The president is thrust into the back seat of his limousine by a secret service agent who immediately orders the driver to head to nearby George Washington University Hospital, where emergency surgery finds a dime-shaped, razor-thin bullet only centimeters from Reagan's seventy-year-old heart.

Yet, there was one image that we never saw; it was formulated in the mind of Nancy Reagan, and it haunted her in those initial days after the near assassination. The image is a stunning example, never before reported, of how Nancy was so dedicated to her Ronnie that she was willing to give her life for him. It was shared in February 2006 by Louis H. Evans, the well-known, longtime pastor of the National Presbyterian Church, who kept it to himself for twenty-five years.[7]

The Reagans attended the National Presbyterian Church during their first weeks in Washington. Evans had become their new pastor. The day after the assassination attempt, Nancy was in need of spiritual counseling. She asked Evans to help her track down Donn Moomaw, the Reagans' close friend and pastor at the Bel Air Presbyterian Church in California. As it turned out, Moomaw was at a conference, but he quickly hopped on a plane. Evans picked up him at the airport and brought him to the White House, where they were greeted by Mrs. Reagan in a room that included a small group of friends:

Frank Sinatra and his wife, the Rev. Billy Graham, and a Los Angeles businessman, the name of whom today escapes Evans.[8]

Nancy's words shocked her friends. "I'm really struggling with a feeling of failed responsibility," she confided. "I usually stand at Ronnie's left side. And that's where he took the bullet." Her husband had come perilously close to dying, a fact not known by the public at the time. Nancy was worried, and now she had deep regrets: If only she had been next to her husband as he strolled to that limousine, positioned between him and John Hinckley's pistol, she could have taken the bullet for her beloved Ronnie.

It was always understood that Nancy was Ronald Reagan's protector, the one who played bad cop and watched his back as he trusted everyone, regardless of their loyalty. Yet, this insight adds a heightened level of appreciation for Nancy's commitment. Such fealty must have been a jolt to Reagan—to know that the woman in his life was so utterly, completely devoted to him, even to the extent that she regretted not being there to take a bullet for him that terrible day in March 1981.

Unlike with the Kennedys, Reagan could not point to Communism as a motivation for the shooter. And while he could not blame Communism for the bullet that nearly took his life, his survival meant he could put Communism in the crosshairs, giving Reagan the opportunity to play the role of assassin.

EASTER WEEKEND 1981

After the failed attempt on his life in March 1981, Reagan felt a sense of calling stronger than ever: Believing that God was moving him to something greater, he was certain his life was spared for a special purpose related to the Cold War struggle—to the epic battle against atheistic Communism. The assassination attempt failed to derail his resolve, and a few weeks later, back in the saddle at the White House, on Good Friday, he reflected on the "divine plan" he sensed internally. Mike Deaver summoned New York's Terence Cardinal Cooke, who flew to Washington to counsel the president. "The hand of God was upon you," Cooke told Reagan. "I know," a serious Reagan replied, before confiding: "I have decided that whatever time I have left is for Him."[9]

Two days after this encounter with the prominent Catholic, Reagan had another special meeting with Louis Evans, this time Easter Sunday. Reagan asked Evans, along with his wife, Colleen, to serve communion to him and Nancy in the Yellow Room. Evans agreed, and did not speak of the moment for decades.[10]

After the private service, Pastor Evans told Reagan that while the president had laid in the emergency room, Evans had a kind of "mental image"—Evans himself is not sure of the best word to describe it—that he believes was granted by God, in which he pictured himself standing over Reagan's bed in the operating room, and through which God told Evans that Ronald Reagan would be healed and that God had a plan for Reagan's life. Reagan responded by thanking Evans and saying that he felt that way himself. "God has a plan for my life and I want to find it," he told Evans.

Reagan said more that Easter Sunday: As he gazed pensively out the Yellow Room window toward the Jefferson Memorial, he told Evans that as he struggled for his breath on that table in the emergency room, he felt that if he did not forgive John Hinckley at that very moment, he would not be healed. He forgave him on the spot.

Struggling with conflicting emotions—a sense of grand calling and his mother's humility—Reagan would ultimately conclude that God had chosen his "team" to defeat Soviet Communism.[11] The power of that spiritual component should not be downplayed. His sense of divine purpose now reinforced and amplified his Cold War purpose.

MAY 1981: NOTRE DAME

A similar zeal possessed Reagan in his May 17, 1981 address at Notre Dame University, which, in a profound series of events with enormous implications, came only days after Pope John Paul II was also shot by a would-be assassin. This Notre Dame speech featured Reagan's first public presidential predictions on Communism's demise: "The years ahead are great ones for this country, for the cause of freedom and the spread of civilization. The West won't contain communism, it will transcend communism. . . . It will dismiss it as some bizarre chapter in human history whose last pages are even now being written."[12]

Although no one else said it or expected it, those last pages were at that time being written. This prediction was scoffed at by many liberals; and perhaps even some conservatives figured that Reagan's words were cheerleading—mere rhetoric. In fact, the statement foreshadowed Reagan policy: he would not seek to contain Communism; he would move beyond containment.

In addition to introducing this idea of victory to the public, the speech also demonstrated certain Reagan mainstay themes that would become routine during the 1980s: a spiritual dimension, a duty and obligation, and a great cause. It was the obligation of Americans to seize the cause and take up the fight against expansionary, atheistic Soviet Marxism. In the address, he rallied his audience to the cause, to one "bigger than ourselves," a "common cause" to "attain the unattainable."[13]

If Americans met this challenge, history would look back, Reagan assured, and determine that "the American Nation came of age," that it "affirmed its leadership of free men and women serving selflessly a vision of man with God." He invoked: "It is time for the world to know our intellectual and spiritual values are rooted in the source of all strength, a belief in a Supreme Being, and a law higher than our own."[14]

A month later, a challenge was issued to Reagan: a reporter dared him to stand by his Notre Dame prediction. Reagan went further: He said he believed that recent intrigues in Poland were an added sign of Communism's doom. "I just think it is impossible—and history reveals this—for any form of government to completely deny freedom to people and have that go on interminably," said Reagan. "Communism is an aberration. It's not a moral way of living for human beings, and I think we are seeing the first, beginning cracks, the beginning of the end."[15] Significantly, to Reagan, the Poland situation would be not just an opportunity for another prediction but a chance to fulfill the prediction.

RETURNING TO GREATNESS

In examining the Reagan game plan to win the Cold War, it would be a mistake to start with a specific policy directive or program. The fact is that before Ronald Reagan and his team could create a set of policies to take on the Soviets, the president wanted to get America's domestic house in order, creating a strong national infrastructure to support his campaign against Communism. In Reagan's estimation, this meant that one thing in particular had to be quickly resolved: he had to restore morale to the nation and the presidency. He needed to transmit his own confidence on to his country and countrymen.

This was a process that would not occur overnight; it would require patience, but it would be essential to victory. Without a change in national morale, the country and the plan to defeat Communism would be vulnerable. For now, strength at home was a precursor to strength abroad, and the domestic agenda needed to be top priority before the broader intent to undermine the USSR could commence.

"ECONOMY FIRST"

In Reagan's eyes, there were two key components to boosting America's morale: one was growing the economy and the other was military respectability. Of course, a strong military would be impossible without a healthy economy to support it, and so he chose to focus on the economy first. Indeed there was a consistency to this "economy first" philosophy in his Cold War scheme, one that dated to well before his presidency.

For years, he had been expounding on the value of a strong economy in the fight against Communism, and now he found himself in the position to try to integrate the two. In a July 1968 speech, he asserted: "The peace and security of the world depends on the fiscal and economic stability and the defense potential of the United States."[16] In an October 1972 speech, he said that, "it is America's industrial and economic strength, translated into military potential, that represents the single greatest guarantee of peace for the world."[17] In an April 1975 radio broadcast he wrote, titled simply "Peace," he stated that, "Power is not only sufficient military strength but a sound economy."[18]

Recalling the start of his presidency, he later wrote in his memoirs:

In 1981, no problem the country faced was more serious than the economic crisis . . . because without a recovery, we couldn't afford to do the things necessary to make the country strong again or make a serious effort to lessen the dangers of nuclear war. Nor could America regain confidence in itself and stand tall once again. Nothing was possible unless we made the economy sound again.[19]

The laser-like focus on the economy that first year is evident in the *Presidential Documents* from 1981: Reagan's references to the economy were ubiquitous. Even when asked about foreign policy, Reagan tended to bring up the economy. There were at least a dozen statements in which he maintained that the economy had to get on track before the Soviets could be countered. To cite one, on May 27, Reagan asserted that "the first step in restoring our margin of safety must be the rejuvenation of our economy."[20]

It was in this plan for economic recovery that Reagan put the first stamp on the young administration. Here, his economic philosophy played a key role in shaping the fiscal policy of the nation as he attempted to reign in out-of-control spending, regulations, and taxes, all of which he believed had robbed the American economy of its vitality, and particularly its ability to bounce back after a recession. The economy needed to be freed in order to perform.

The prescription that Reagan recommended rested on four pillars: deregulation, reductions in the rate of growth of government spending, a stable, carefully managed growth of the money supply, and, most significant, tax cuts.[21] Among them, Reagan made great strides in deregulation, a process that had begun under Carter but was expanded and accelerated under Reagan, including deregulation of the trucking industry, the airlines, telecommunications, banking, and oil. He hit the ground running: within ten days of his inauguration, he froze more than 170 pending regulations. By the end of his first term, the total number of pages in the *Federal Register* was cut in half.

The rate of growth in government spending was cut in the domestic area, although Reagan's reductions were sometimes countered by large increases in defense spending. Overall, domestic spending was reduced by roughly a half-trillion dollars in the 1980s. The difficulty for Reagan in controlling government spending was that he faced a Democratic Congress that often disagreed with his policies, and which wanted rises in domestic spending but cuts in defense expenditures—both the opposite of Reagan's preferences. Congress ultimately controlled the purse. By the end of the decade, Reagan, who watched the budget deficit explode on his watch, would be demanding executive privileges like the line-item veto, a budget procedure enjoyed by most state governors.

Reagan made much greater headway in managing the federal money supply as a means to control inflation. Here, he had to rely upon those running the Federal Reserve Board. Nonetheless, he retained and brought in the right men—chairmen like Paul Volcker and Alan Greenspan, who ensured that the first enemy that President Reagan slew was not the Russian bear but corrosive inflation that had killed a presidency or two. This was a huge triumph, dropping the inflation rate down to miniscule levels of which Jimmy Carter could have merely dreamed, and which went a very long way in reversing the malaise and misery index of the 1970s.

Most important, among the various major tax cuts pursued by the new president, the federal income tax reduction was the centerpiece. Reagan secured a 25% across-the-board reduction in federal income tax rates over a three-year period (5%-10%-10%), beginning in October 1981. Eventually, through this and later cuts, the upper-income marginal tax rate was dropped from 70% in 1981, which Reagan believed was punitive and stifling, to 28% before he left office.

While this economic plan was the crucial first step, Reagan knew that re-vitalizing the American economy would not happen over night. Cutting taxes proved less difficult than staying the course during the two years that fol-lowed, as the stimulant effect of the cuts was slow in kicking in. All three of the "troika" that jointly served as Reagan's first "chief of staff"—James Baker, Michael Deaver, and Edwin Meese—as well as future solo Chief of Staff Ken Duberstein, talk about the internal doubt in the first two years of Reagan's economic program. Despite critics outside and inside the White House who pushed Reagan hard to modify course, he remained steadfast. As Baker noted, there was "no way" there would be a change in direction because Reagan was the one individual who would not bend. He was sure his plan would work, and he refused to abandon it, even when pressure bore on him from his closest advisers. He was convinced of the wisdom of his economic strategy and the free-market theories it rested upon.

Reagan's chiefs of staff recall that he incessantly told a specific story dur-ing this period. Each of the four heard it so often that they could recall it ver-batim.[22] It was an anecdote about a father with two boys—a pessimist and an optimist. The father placed the pessimist in a room full of new toys. He placed the optimist in a room with a pile of manure. When he returned, the pessimist was crying and throwing a fit, complaining that he had no toys to play with. He went to the other room and found the optimist digging doggedly through the pile of manure. When the father asked the optimist what he was doing, the boy replied: "I know there's a pony in here somewhere!"

That optimist was Reagan. He went against his staff nonstop in this pe-riod. An exasperated aide told *Time*: "He is absolutely convinced that there will be a big recovery." *Time* noted his certainty that everything would work out, "as it always has in Ronald Reagan's world." Said an aide: "He is an opti-mist. My God is he an optimist!"[23]

BOOSTING MILITARY MORALE

Reagan also felt that shoring up the military was key to regaining respect on the world stage. He complained that paychecks for the armed forces were so small that some married enlisted men were eligible for welfare. He claimed that many military personnel were so ashamed of being in the service that they put on civilian clothes when they left their posts. Hence, he immediately told the Joint Chiefs of Staff that he wanted to do "whatever it took" to make "our men and women proud to wear their uniforms again."[24]

He embarked on a massive military buildup. Only a couple of months after entering office, Reagan asked for an extra 32 billion dollars in defense spending beyond Jimmy Carter's defense request, which itself had actually constituted a notable increase from previous levels. From 1981 to 1985, Reagan's Pentagon sought to raise spending on the procurement of weapons by 25 percent per year, doubling the total by the end of the first term. Reagan pledged to build a 600-ship Navy, restoring old, mothballed destroyers, once proud fighting ships that the president believed still had glorious days ahead of them. He would seek to construct the MX missile, nicknamed the "Peacekeeper," as well as the deployment of 572 Pershing II missiles to be deployed among NATO countries in Western Europe. In addition, the new administration proceeded with a number of impressive new defense technologies, some of which would not become known to the general public until the start of the next decade—weapons systems like the super-accurate Tomahawk cruise missile and the Patriot anti-missile missile, the latter of which would ultimately prove the impossible: that a bullet could hit a bullet.

This is a short list of what Reagan sought militarily, and none of it accounts for the patriotic voice he lent to these efforts, which was an added crucial element to restoring the nation's pride. All of these ventures had the effect of demonstrating a stronger, resurgent America, not only economically but also militarily. Suddenly, the country that had left Vietnam no longer appeared to lack resolve.

OVERALL, THIS GENERAL WILLINGNESS BY REAGAN TO STAY THE course, militarily and with his embattled economic plan, was a sign not only of his faith in his policies but also of strong leadership in the face of overwhelming criticism. Reagan was following the path for victory that he laid out himself, and nowhere was this more evident than during that first presidential year. His prescription for placing America back on solid ground derived from his own steadfast belief in the free-market system and the power of military strength; both would affect overall morale. And the eventual reinvigoration of the economy would become Reagan's proudest accomplishment, not only because it came to turn around the nation but because it made all of his other goals possible.[25]

With 1981 coming to a close, Reagan felt that he had put into place the mechanisms that, by his estimation, would restore America to greatness, that would catalyze America's comeback. He readied for the reward—for a resurgence of American pride. But as fate would have it, his attention would be thrust abroad before year's end, forcing the Crusader to shift his focus to Eastern Europe and express his solidarity with Poland.

8.

Poland Explodes: December 1981

JUST AFTER MIDNIGHT, DECEMBER 13, 1981, A SOFT SNOW FELL lightly on Warsaw, Poland, gently betraying the harshness about to befall the country's long-repressed citizenry. Rudely piercing the peaceful silence were eleven police trucks that raced down Mokotowska Street. In mere seconds, they had blocked off both approaches to the five-story building that housed the headquarters of Poland's fiercely independent trade union.

The structure was rapidly vetted in this literal midnight raid. Outside, riot police wielded clubs at onlookers. Tanks soon followed. When one passerby asked a policeman what was happening, the officer sufficiently explained with just one word: "Solidarity."[1]

The Polish Communist regime, consenting to orders from Moscow, had declared martial law in an effort to stymie the burgeoning labor movement Solidarity, which the government had officially recognized as a legal entity sixteen months earlier. Now unfolding was an attempt to squash the union, an institution that had come to embody the hopes and aspirations of millions of Poles. While government troops were taking over the building, Solidarity leader Lech Walesa and other key labor figures were in Gdansk. Once the phone lines were severed at headquarters, the leaders were cut off from communication and police around the country scrambled to arrest union leaders. All over the country, the police tore down every Solidarity poster that they could find; those not ripped down were doused with paint.[2]

All throughout Poland the chaos raged. The sound of gunfire rang out in the industrial areas of Ursus and the mining lands of southeast Poland, injuring seventy people in the shootings. In Krakow, the hometown of Pope John Paul II, four others were shot.[3] In every corner of the country, the Communists were murdering the workers. The party was smashing the proletariat.

Military traffic was dense, with tanks occupying all Warsaw streets. Army checkpoints were set up and borders were shut down. International flights to and from Warsaw were canceled. Western air traffic over Polish territory was prohibited.

It was "such a shock . . . nothing else was on TV," said Joseph Dudek, a mining engineer from Krakow. "You couldn't move out of your house. Tanks were outside. There were checkpoints everywhere as well as the fear of war approaching."[4]

The final, remaining links between Poland and the outside world were cut on Monday night, December 14, when the elusive Reuters news service was no longer able to report from Warsaw.[5] Six hours later, Communist Party leader General Wojciech Jaruzelski appeared on television to announce a state of emergency: Poland would be governed by something called the Military Council of National Salvation, a committee of fifteen generals and five colonels. General Jaruzelski would be prime minister.[6]

Talking into the camera in typical monotone, General Jaruzelski served up an expressionless twenty-three-minute speech. "I am speaking to all Poles," he said. "Our country is threatened by mortal danger." He said that "anti-state, subversive activities" by "forces hostile to socialism" had pushed Polish society to the "brink of civil war." These "anti-socialist forces," he went on, were "often inspired and financially supported from abroad." The goal of the Council of National Salvation, he explained dramatically, was "to eliminate the danger of the fall of the state."[7]

The formal decree was posted along street corners. Public meetings—the right of assembly—were banned. The new military authorities ordered summary trial for all strike leaders, who were imprisoned. There was also an order that three influential Roman Catholic organizations—the Pax Association, the Christian Social Association, and the Polish Catholic-Social Union—cease activities. All citizens were instructed to carry identification cards, which would be randomly checked in subsequent days. Those who did not have their cards were arrested.[8]

Sources claimed that upwards of 1,000 people were detained in the first sweeps.[9] Likewise arrested were a number of high political figures, including the former prime minister and dozens of former government figures. Governors of four provinces were replaced by military men. This was a comprehensive purge.

On Sunday, Poles gathered in crowds to read the newly issued decree from the new Polish government. It was God's day, and the new regime was smart enough to dare not try to stop church services—the one permissible public meeting place.[10] Yet, before Poles hit the churches, the Soviet leadership issued a statement of support for martial law. "All these steps taken in Poland are, of course, its internal affair," stated the Soviet leadership, as if it had nothing to do with the situation. It was "no secret to anyone," Moscow continued, that the "enemies of Socialism in Poland" were "aiming to overthrow the existing social system." These same forces were striving "by all means to undermine the fraternal friendship between the Polish and Soviet peoples."

In regard to Washington's ultimate intentions, this was actually all true. Making reference to Solidarity and the United States, the Soviet statement concluded: "It is no accident that the enemies of independent Socialist Poland inside the country had the support of certain external circles in the West."[11]

In response to these actions, Solidarity, under siege, issued an appeal to friends everywhere:

> We appeal to you: help us in our struggle by mass protests and moral support. Do not watch passively the attempts to strangle the beginnings of democracy in the heart of Europe. Be with us in these difficult moments.
>
> Solidarity with Solidarity. Poland is not yet lost.[12]

The American president read Solidarity's appeal, and would spend his next seven years in office heeding the precise specifics.

For now, however, there was little Ronald Reagan could do but be angry. Very, very angry. Richard Pipes, the Harvard Sovietologist and the NSC's top expert on Poland, described Reagan's reaction: "The president was absolutely livid. He said, 'Something must be done. We need to hit them hard, and save Solidarity.' The president was gung ho and ready to go."[13]

At that moment Reagan committed to save and sustain Solidarity as the wedge that could splinter the Soviet bloc, as the first crack in the Iron Curtain. He estimated that the labor union held the potential to bring down the whole house of cards. As Richard Allen put it, Reagan right away "thought of Poland as a means to the disintegration and collapse of the main danger, the main adversary, the Soviet Union."[14] Ironically, it turned out that bad news in Poland was good news: In Reagan's mind, the ugliness that was martial law afforded beautiful possibilities.

"HE HAD A PREOCCUPATION WITH POLAND"

Well before the explosive events of December 1981, Ronald Reagan had viewed the Soviet satellite as the linchpin in the Communist empire. In Poland, he saw a potential catalyst, a state whose fall could knock down the Soviet dominoes in Eastern Europe. To Reagan, the Poles were tragic victims of the two totalitarianisms he held in equally low repute—Nazism and Bolshevism. By mere geographic fate, this fine, religious people had been traumatized by villains, and while the Allies liberated Poland in World War II, they sold it down the river to the Soviets at Yalta. Reagan saw no good reason why America should not seek to free the nation from totalitarianism, this time from a despotism colored Russian red instead of German brown. Since 1939, poor, proud Poland had known nothing but foreign-imposed tyranny.

"He had a preoccupation with Poland," said Bill Clark. "He had mentioned Yalta as far back as I go with him as being totally unfair and having to be undone someday."[15] (Reagan's hatred of Yalta was exceeded only by Poles' hatred of Yalta.[16]) Clark added that Reagan "greatly" sympathized with Poland for a number of reasons; among them, no other country lost as high a percentage of its population in World War II.[17] In the 1970s, Reagan wrote several radio broadcasts on Poland's persecution by the USSR.[18]

Reagan was especially affected by the pope's nine-day pilgrimage to his Polish homeland from June 2 to June 11, 1979. The new man in Rome shrewdly chose Poland as his first foreign visit since his election in October 1978. Moscow was scared to death of the pope's visit—another factor, like the emerging Solidarity, which challenged the primacy of the Communist Party, and yet more so because it was about God.

In Poland, Communists picked a bad place to try to further their atheistic empire. Poland is a homogenous nation, with a population 95 percent ethnic Polish and Roman Catholic. The nation is an unshakable bastion of Catholicism. The routine Communist war on religion was a tall task there.[19] The Polish faithful remained deeply devout; the Church there was stronger and more intact than in any other nation in the Soviet bloc.[20]

Thus, it was profound that in 1978 Rome picked its first non-Italian pope in 455 years and its first Slavic pope ever—one from no less than

Poland, the heart of the Soviet bloc. The timing was exquisite. John Paul II, Reagan, and millions of Poles perceived no coincidence in that timing, but instead divine planning. To Reagan, this pope represented the best of both worlds—unwavering faith in God and strident anti-Communism. As a boy, Karol Wojtyla routinely paused after Mass to light candles and offer a series of prayers "for the conversion of Russia."[21] Now, as pope, he was empowered to offer more than prayer.

Reagan paid close attention to the pope's June 1979 trip, where the Holy Father famously and movingly told his brothers and sisters, in words packed with New Testament meaning, "Be not afraid." On June 3, John Paul II openly insisted that all Eastern European governments be allowed freedom of conscience, individual rights, private property, elections, and independence. He asserted: "There can be no just Europe without the independence of Poland marked on its map."[22] Clearly, Reagan had found a kindred spirit, one who shared his own bold willingness to speak out.

The reaction in Moscow was unsurprising, as the godless suggested a solution for the Poles' stubborn faith: more godlessness. "The solution for the Karol Wojtyla problem," Ukrainian Communist Party Chief M. Vladimir Shcherbitsky chimed in from across the Soviet border, "must lie in a renewed and more vigorous propaganda in favor of atheism in the Soviet Union and its 'fraternal socialist societies.' "[23]

In his radio broadcasts on the pope's trip, Reagan blasted this "Communist atheism" that had preyed on Poland following World War II. In one broadcast, an outraged Reagan remarked: "These young people of Poland [who greeted the pope] had been born and raised and spent their entire lives under Communist atheism. Try to make a Polish joke out of that."[24] In another titled, "A Tale of Two Cities"—which he later recycled as a speech in Dallas[25]—Reagan asked:

> Once in the days of Stalin he is said to have dismissed the Vatican by contemptuously asking: "How many divisions does the Pope have?" Well, in recent weeks that question has been answered by Pope John Paul II. It has been a long time since we've seen a leader of such courage and such uncompromising dedication to simple morality—to the belief that right does make might.
>
> On our TV screens we've seen the reaction to this kind of leadership. Wherever he went in his native land the people of Poland came forth in unbelievable numbers. There were crowds of 400,000, 500,000, 1 million,

and then 5 million, gathered from miles around, even though they don't have the easy means of transportation we have, and they gathered knowing there was every possibility they were risking their livelihood and even their freedom.

For 40 years the Polish people have lived under first the Nazis and then the Soviets. For 40 years they have been ringed by tanks and guns. The voices behind those tanks and guns have told them there is no God. Now with the eyes of all the world on them they have looked past those menacing weapons and listened to the voice of one man who has told them there is a God and it is their inalienable right to freely worship that God. Will the Kremlin ever be the same again? Will any of us for that matter? Perhaps that one man—the son of simple farm folk has made us aware that the world is crying out for a spiritual revival and for leadership.[26]

After the pope's visit, Reagan was never the same. He sensed the immensity of what had transpired and recognized that this was a momentous event that threatened Communism's hold on Eastern Europe. Richard Allen sat with Reagan in June 1979 watching these news reports. "He said then and there that the pope was the key figure in determining the fate of Poland," said Allen. "He was overcome by the outpouring of emotion that emanated from the millions who came to see him."[27]

When Bill Clark, another close adviser, was asked if Reagan "realized then and there" that Poland could be the "splinter to break apart the Soviet empire," he went further: "He felt that far before June 1979. He had tremendous interest in Poland and its strategic importance, going back to Yalta and Potsdam, which he felt were terrible days. He knew Poland would be the linchpin in the dissolution of the Soviet empire."[28]

The future president's thinking on the power of Poland and the pope's 1979 visit was not the shared opinion of the West at the time. In a June 5, 1979 editorial, the *New York Times* declared authoritatively: "As much as the visit of Pope John Paul II to Poland must reinvigorate and reinspire the Roman Catholic Church in Poland, it does not threaten the political order of the nation or of Eastern Europe."[29]

While these dissenting opinions were not limited to the *New York Times*, Reagan continued to believe that Poland was the key to the unraveling of the Soviet bloc in Eastern Europe. Inspired both by the potential that the pope represented and by a clear sense of his own role, Reagan found himself in a position to put his desires into play.

AUGUST TO DECEMBER 1980

In the wake of the pope's visit, 1980 proved to be a watershed year for Solidarity. The pope had given new impetus to the workers' movement, and in the summer of 1980, this momentum reached a crescendo when Edward Gierek, the Communist Party secretary and arguably the most powerful man in Poland, felt obliged to resign.

The intensity rose even higher on August 16, 1980, when Lech Walesa, who was then emerging as one of the top antigovernment union activists, called on his fellow workers to leave the Gdansk shipyard, leading to a tense confrontation between strikers and the government. The outside world could not have imagined what happened inside the negotiation room in Gdansk that day, when an unknown member of Solidarity removed the bust of Lenin from the nearby desk and replaced it with a black-and-white photo of Ronald Reagan cut from a newspaper.[30] Considering that Reagan had not even been elected president yet, let alone inaugurated, this was a remarkable gesture. Organizers negotiated an agreement with the Polish government on August 20, 1980, formally creating Solidarity. A few weeks later, in September, Prime Minister Edward Babiuch, in power only seven months, resigned. Things were falling apart everywhere, except within Solidarity's ranks.

Within about three months of the August 1980 agreement, the membership of Solidarity exploded from zero to ten million members, comprising roughly 50 percent of all Polish workers, including farmers. It was a massive grassroots, social movement that became precisely what the USSR most feared—a massive political movement as well.

Lech Walesa was suddenly more confident than ever. Moreover, his confidence had been buoyed by the results of the November election in the United States. On December 7, 1980, a fearless Walesa stood on a snowy, windswept plain on the outskirts of Gdansk and spoke openly about politics and the U.S. election. "It was intuition, perhaps," he said, "but one year ago I envisioned what would happen. Reagan was the only good candidate in your presidential campaign, and I knew he would win." Walesa spoke presciently that December day: "Someday the West will wake up and you may find it too late, as Solzhenitsyn has written. Reagan will do it better. He will settle things in a more efficient way. He will make the U.S. strong and make it stand up."[31]

JUNE 1981

A few months later, the new president returned Walesa's optimism. In a June 1981 press conference, UPI's Dean Reynolds reminded Reagan that at Notre Dame he had claimed that Communism was a "sad, bizarre chapter" in human history whose last pages were in the process of being written. Reynolds followed: "In that context, sir, do the events of the last ten months in Poland constitute the beginning of the end of Soviet domination of Eastern Europe?" Reagan answered:

> Well, what I meant then in my remarks at Notre Dame and what I believe now about what we're seeing tie together. I just think it is impossible—and history reveals this—for any form of government to completely deny freedom to people and have that go on interminably. There eventually comes an end to it. And I think the things we're seeing, not only in Poland but the reports that are coming out of Russia itself about the younger generation and its resistance to long-time government controls, is an indication that communism is an aberration. It's not a normal way of living for human beings, and I think we are seeing the first, beginning cracks, the beginning of the end.[32]

A few minutes later in the press conference, Reagan estimated: "The Poland situation is going to be very tense for quite some time now. The Soviet Union is faced with a problem of this crack in their once Iron Curtain and what happens if they let it go."

SEPTEMBER 1981: FEARS OF SOVIET FORCE IN POLAND

As 1981 moved on, the crack crept to the surface, as Walesa and his union brothers chipped away at the Communist facade. While not a bastion of laissez-faire capitalism, Solidarity was also not Marxist. It was anti-Communist, independent of the USSR, rejected the *nomenklatura*, and represented the majority of Polish workers. More, it was devoutly pro-Catholic; no other institution had so consistently opposed Bolshevism as had the Roman Catholic Church.[33] Furthermore, Solidarity was not only pro-West but favored concepts like free speech, a free press, and free elections—all verboten to Communism.

All of this threatened Soviet Communism's stranglehold on Poland, and in early September 1981, *Pravda* made that clear. Appealing to infallible authority, it noted that the immortal Lenin had insisted on Communist Party control of all trade unions. Indeed, as Arthur R. Rachwald explained: "The idea of an independent labor organization functioning freely is totally incompatible with the Soviet system."[34] Thus, when the Polish Communist government officially recognized Solidarity, it had committed heresy in the Church of Lenin; it had signed away its oxymoronic Communist soul.[35] The supremacy of the Community Party, directed from Moscow, was in jeopardy.

Reagan relished the irony that Solidarity was a disgruntled workers' group "in a so-called workers' state," in an alleged workers' heaven. It was, said Reagan, "a genuine labor movement suppressed by a government of generals who claim to represent the working class."[36] As the former head of SAG, he felt personally connected to Solidarity not just through its anti-Communism but also its unionism. In a speech to the International Brotherhood of Teamsters, he said, "Those of us who know what it is to belong to a union have a special bond with the workers of Poland."[37] Reagan also understood Solidarity's uniqueness at a precarious time for the USSR, noting that nothing like Solidarity had ever existed in the Eastern European bloc. The workers' union "was contrary to anything the Soviets would want or the Communists would want."[38]

By summer 1981, the Kremlin had become extremely worried and wanted Solidarity destroyed. Thus, there was considerable unease in the Reagan administration over the prospect of a Soviet invasion of Poland. These concerns were amplified by a secret source: Washington had a mole in Poland, a brave figure named Colonel Ryszard Kuklinski, a highly respected, high-ranking figure in the Polish Defense Ministry. A liaison between Warsaw and Moscow, Kuklinski was tasked with the grave duty of helping to make preparations for a "hot war" with the West. Kuklinski's real enemy, however, was Soviet Communism and its hold on the colonel's beloved homeland. He secretly supplied a massive volume of material to the CIA, including an explosive disclosure in December 1980: on two sheets of paper inside a smuggled package, Kuklinski carefully outlined Moscow's plans to cross the border into Poland with eighteen Soviet, East German, and Czech divisions by December 8, 1980.[39] The Carter White House had been aware that Soviet troops were massing along the border, but Kuklinski's missive relayed sincerity rather than a bluff.[40] According to authors Jerrold and Leona Schecter, thanks to Kuklinski, the Carter administration knew in advance that the Soviet Union was actually preparing to invade Poland in 1980 in order to crush Solidarity, under the guise of a "peaceful exercise."[41]

The Carter administration had made clear its concerns over Poland, stating that military intervention by Moscow would seriously jeopardize the U.S.–USSR relationship. NATO was placed on a higher level of readiness. Amid continuing fear of a Soviet invasion to "save Poland," Pope John Paul II sent a letter to Chairman Brezhnev in which he boldly made an implicit parallel between a Soviet invasion and the Nazi incursion of September 1939, grimly outlining the moral contours in which he viewed the USSR.[42]

Inheriting the work of Kuklinski, the Reagan administration picked up these fears, and also began receiving new disturbing information. Since the first weeks of the Reagan administration, a colonel from the Pentagon's Defense Intelligence Agency had paid regular visits to the National Security Council bearing satellite photos of Warsaw Pact troop movements along the Polish border. Richard Pipes studied these pictures, and was especially concerned about the Warsaw Pact exercises that took place on Polish territory under the code name "Soiuz-81." These maneuvers could be easily directed into an offensive. By April 1981, after once vacillating over whether an invasion was likely, Pipes thought an invasion was imminent.[43]

Judging from newly released information, Pipes' fears were justified: Over a roughly one-year period up through September 1981, the Soviet leadership carefully considered an invasion, aided by Eastern bloc allies like Czechoslovakia and East Germany. The plan called for a contingent of 14 to 15 divisions to help Polish authorities enforce martial law.[44] However, because General Wojciech Jaruzelski, Poland's prime minister, believed such a scenario would produce mass bloodshed, he persuaded Moscow to allow him to take care of business internally. He saw much greater instability if Soviet soldiers were used rather than the Polish military. Moscow, desiring deniability for any role, went with his advice.[45]

The Kremlin agreed with Jaruzelski and in mid-December 1981, the USSR opted not to employ military force and chose the next closest thing: martial law.

U.S. MILITARY FORCE IN POLAND

With concern over the Soviets possibly on the verge of committing military force to quell the Poland situation, Ronald Reagan entertained the notion of a U.S. military response. Suddenly, the Reagan commitment to Poland had reached a higher level than ever before—one that which, remarkably, over two-and-a-half decades later, the world still remains in the dark.[46]

Unbeknownst to the world, Ronald Reagan considered plans to *actually invade* Polish territory if the Soviets invaded Poland. As an indication of how few people knew this fact, as well as the volatility of such a notion, when I asked Richard Pipes if U.S. force was considered in Poland to counter the Soviets, he tersely replied: "No." He paused before raising his voice to express shock at the mere thought: "No, no. That would have unleashed World War III! And that's not something anybody wanted. No." He continued: "I don't think we would have [used military force] in any case," said Pipes. "No. We thought of diplomatic measures, of economic measures, but never thought of military intervention. I can say that quite categorically."[47] Likewise, Richard Allen, who was national security adviser at the time, stated: "I do not believe such an option [of U.S. military force] was given any serious consideration. Generally, Ronald Reagan was against the use of force."[48] And, indeed, Reagan was ultimately against American force in Poland.

But that does not mean he did not at least privately contemplate it; in fact, he had contemplated the option a year earlier. In December 1980, the incoming secretary of defense, Cap Weinberger, discussed with incoming president Reagan the possible use of U.S. force in Poland. Reagan himself brought up the idea after one of the preinauguration security meetings, which were held at Blair House and at the State Department in the immediate weeks after the November 1980 election.

Looking back over twenty years later, Weinberger recalled the intensity of the situation in Poland then, a full year before martial law. "There was very considerable worry," said Weinberger, "that the Soviets, with two divisions inside Poland, that had been there since the end of the war [World War II], and the constant military exercises and threatening moves around the borders of Poland, that they might very well decide to wander in there without any fear of adverse results or reprisals." Reagan wanted no signals to the Kremlin that such action would be acceptable. "The president was very firm about that," said Weinberger.[49]

The two men talked about a U.S. invasion of Poland. Weinberger remembered:

> I talked to him after one meeting in a session before he was inaugurated. I said, "You know, Mr. President, this talk about [U.S.] military force against the Soviet Union if it went into Poland is all well and good . . . but you must know that we do not have the military strength capable of doing that

now. We don't have the ability to project our power that far and we could not, without very substantial help, successfully come to the aid of the Poles if they were invaded."

And he said to me, "Stop." He turned to me and said, "Yes, I know that Cap. But we must never again be in this position. We must never again *not* take action that we think is essential because we're not strong enough to do it."[50]

And that simply reemphasized to me [the need] to get additional military spending that would get us out of that position. . . .

He certainly considered all ways. But he didn't argue for a moment with my assessment that we weren't able to do that [employ U.S. military force in Poland] at the time. We didn't have all that many divisions and the divisions that we had were under strength. We didn't have adequate spare parts or supplies and couldn't do adequate training. The military that we took over needed an enormous amount.[51]

Weinberger added that Reagan was aware of the risk that the Soviets could build on their two divisions in Poland and could use that troop presence "to intimidate Poland from carrying out any of the things that Walesa and others were talking about at that time." This bothered Reagan greatly, and he very much wanted to reverse that situation. Again, however, *want* and *can* were quite different.[52]

Though a year had passed without any military action in Poland, by either side, Reagan's thinking remained open to the consideration of force, particularly as tensions rose in Poland during December 1981. According to Weinberger:

Force may have been considered. There were some people in the White House and on the NSC staff . . . who were always proposing all kinds of military actions without, I think, having any awareness of the capability of the military to carry them out. . . . People didn't realize that our military strength had been very substantially eroded [in the 1970s]. . . . He [Reagan] very reluctantly accepted that idea and he certainly wasn't going to launch two or three divisions into a battle that would be almost a certain loss.[53]

Taken together, Weinberger's remarks at least *suggest* that Reagan might have thought more aggressively if he had the defense power behind him. However, the strongest statements on this issue come from Bill Clark, the future national security adviser, who knew Reagan's thoughts on Poland better than anyone. In a November 1999 interview, Clark stated:

> We [the Reagan administration and Vatican] also worked together to gener-
> ate strong diplomatic pressure upon the Soviet Union, to convince the lead-
> ers in the Kremlin that they must refrain from invading Poland—from
> doing what the Soviet Union had done to crush the earlier freedom move-
> ments in Hungary in 1956 and Czechoslovakia in 1968. *We in the Reagan*
> *administration were prepared to recommend the use of force if necessary to stop*
> *such an invasion* (emphasis added) following the imposition of martial law.[54]

Five months later in a March 2000 speech in San Francisco, Clark added: "The Soviets and their proxies in Poland declared martial law and started in the summer moving troops up to the border, which looked like another situa-tion as had occurred in Hungary and Czechoslovakia. The President said this just simply cannot happen, even if it means meeting force with force."[55]

In a 2001 interview, Clark, a man who has always been cautious with his words, was careful in discussing the force matter. Still, his words were powerful:

> Measures were discussed. Some might call them extreme, but they might
> have been necessary. The instructions were that we cannot allow a [repeat of
> the] Hungary or Czechoslovakia invasion. . . . And it was touch and go. The
> Soviets did move the troops up to the line. We decided that, effectively,
> force would be met with force.[56]

But the question remained, was the administration willing to use force, if necessary? "Well, that certainly was inferred," said Bill Clark. Importantly, however, Clark stressed that, "Anyone familiar with decision-making processes understands that you consider a full gamut of options. That was one that was considered. But I don't want it to seem or sound more dramatic than it was."

In July 2003, I posed the question to Clark yet again; he was consistent with previous statements: "I'm confident that if the Soviets had crossed the line—if it [the situation] had come to reasonable necessity to use force—he [Reagan] was ready to do it; he was prepared. He wouldn't *want* to [use mili-tary force] but he was prepared." By Soviets crossing the line, Clark said he meant a Red Army invasion of Poland: "He [Reagan] would not tolerate a re-peat of the Hungarian or Czechoslovakian incidents."[57]

Judging from some of these accounts, certainly Clark's, Reagan may have been willing, if need be, under the worst circumstances, to go toe-to-toe with the Soviets on Polish territory, meaning the use of real military force—and assuming he had sufficient firepower.[58] On the other hand, others have no memory of such a Reagan consideration, and reject it entirely. Ed Meese, a close adviser, stated: "I have no recollection of that at all. I think that even the word 'considered' [the use of force] is too strong based on my recollection of our policy. I don't remember any time when that was proposed as even an option." Yet, Meese continued, "That doesn't mean it didn't happen at any time. Maybe it was recommended by lower-level staff in the NSC. I just don't recall it." Meese adds this caution: "It [the idea to use force] doesn't make much sense to me. Of all the things Ronald Reagan did, he didn't want a major war in Europe, or a nuclear war, or a third world war. He could be tough but he didn't want to be overly provocative. And *that* would've been very provocative."[59]

Provocative, indeed. In the end, Reagan obviously never pursued the force option, precisely for all the right reasons—as listed by Pipes and Meese in particular. It would have been too dangerous. Also, of course, the Red Army never invaded Poland, sparing him the decision. Still, this was an extremely volatile, high-stakes notion that Reagan once thought through and discussed with his closest advisers. Moreover, it drives home the gravity of the Poland situation in the 1980s.

DECEMBER 14, 1981—REAGAN'S RESPONSE TO MARTIAL LAW

While military force was out, Reagan did have a number of other options to counter the declaration of martial law. One of his first responses was to call Pope John Paul II on December 14 to discuss the situation. In addition to expressing his outrage, he told the pope: "Our country was inspired when you visited Poland, and to see their commitment to religion and belief in God. It was an inspiration. . . . All of us were very thrilled." Reagan said he looked forward to a time when the two men could meet in person, and the two began seeking ways to cooperate in these immediate days following martial law.[60]

Reagan's anger toward the Soviets flowed over into two separate letters sent to the pope on December 17 and 29, 1981, neither of which was declassified until July 2000.[61] In the December 17 dispatch, Reagan asked the pope to urge General Jaruzelski to hold a meeting with Walesa and Archbishop Glemp. In the second letter, he explained the countermeasures his administration was taking against the USSR, while asking the pope to use his influence with the Polish Church to lift martial law, gain the release of detainees, and resume a dialogue with Solidarity. In addition, he requested that John Paul II press other Western countries to join the United States. "If we are to keep alive the hope for freedom in Poland," Reagan told the pope, "it lies in this direction."[62]

Although he was clearly disturbed by the events unfolding, Reagan grasped the uniqueness of this historical moment. In his diary entry for December 15, 1981, two days after martial law was declared, he noted that in that day's NSC meeting, "I took a stand that this may be the last chance in our lifetime to see a change in the Soviet empire's colonial policy re Eastern Europe."[63] Nonetheless, he also noted that his options were limited and fraught with danger. Among these, he did not desire a reoccurrence of what had happened under President Eisenhower, whom Reagan greatly respected, during the Hungarian uprising of 1956. He later remembered: "Although we wanted to let the Polish people who were struggling for liberty know that we were behind them, we couldn't send out a false signal (as some say the United States did before the doomed 1956 uprisings in Hungary), leading them to expect us to intervene militarily on their side during a revolution." Hence, "as much as we might want to help, there were limits on the actions our people would support in Poland, especially if, as was likely, there was a charade in which the Polish government appeared to request intervention by Russian troops."[64]

Within the Reagan administration, the situation was tense and filled with peril. The State Department, said Richard Pipes, assured the NSC that the Soviet Union was not involved—a foolish assumption that Reagan must have instantly deemed ludicrous. Pipes said that the NSC meetings of December 19, 21, 22, and 23 were emotionally charged—"inspired largely by Reagan's mounting fury at the communists." He said that Reagan's mind raced back to the 1930s when the democracies failed to halt German and Japanese aggression, a history which he had long ago vowed never to repeat.[65]

At the December 21 meeting, Pipes said that Reagan, "spoke eloquently and in great anger," claiming that the events in Poland were the first time in sixty years that something of this magnitude had happened. Referencing a 1937 FDR speech that advocated a "quarantine" of aggressor states, Reagan suggested that diplomatic and economic relations with the USSR be reduced to a minimum. Moreover, he recommended that if U.S. allies did not join in, America should review its alliance. Reagan even maintained that the United

States should be prepared to boycott nations that continued to trade with the Soviets.[66]

As talk escalated, Reagan began to get impatient with developments, insisting at the December 22 meeting that the White House faced "the last chance of a lifetime to go against this damned force." In response to Reagan's lead, Pipes said that the rest of the cabinet, "fell in step, although with varying degrees of enthusiasm." Secretary of State Haig worried about the reaction of Western European allies to any form of U.S. sanctions, while cabinet members whose departments involved the economy—the departments of Commerce, Treasury, and Agriculture—were also concerned. Nonetheless, said Pipes, "on Reagan's insistence, quite severe punitive measures were adopted"—measures soon to be announced.[67]

DECEMBER 23, 1981: A CANDLE AND A PRAYER

With feelings escalating on both sides, tempers reached critical mass by December 23. The day began with a Genrikh Borovik piece for *Literaturnaya Gazeta* titled "Plot Against Poland," with the subtitle, "Lies, Lies, and More Lies—In Both American Newspapers and Washington's Official Statements."[68] If Borovik was angry when he filed his article, by the end of the day he would be screaming. On the other side, the Reagan team started the day with another dramatic NSC meeting on the crisis, in which Secretary of State Haig warned that any U.S. sanctions against the USSR would incense Western Europe, especially West Germany, which was led by the leftist Helmut Schmidt. "*So be it,*" was Reagan's reply. If this was the case, then America would "go it alone."[69]

After the meeting, Reagan reached for his favorite, surest weapon—the rhetorical missile. He told reporters that "if ever there was an example of the moral bankruptcy of communism," martial law was it.[70] Yet, he had much more to say, and later that evening, on national television, he gave a major presidential speech on the Poland situation. In the speech, he noted that ten million of Poland's thirty-six million citizens were members of Solidarity. Taken together with their families, he rightly said that these Solidarity members accounted for the "overwhelming majority" of the Polish population. "By persecuting Solidarity," said Reagan, the Polish government, which he saw as acting as an extension of the Soviet government and broader Communist movement, "wages war against its own people," a point he made repeatedly in days ahead.[71]

With Christmas only two days away, Reagan connected the spirit of the season with Poland: "For a thousand years," he told his fellow Americans, "Christmas has been celebrated in Poland, a land of deep religious faith, but this Christmas brings little joy to the courageous Polish people. They have been betrayed by their own government." Using words like "terror tactics," "tyranny," and "crime," he described the actions of the Communists in stark detail, weaving his frustration and anger into his carefully chosen words.[72]

Working his way through the text, Reagan then made an extraordinary gesture, one that defined the speech and produced quite a ripple effect. The idea itself was simple, but its ramifications were profound: The president asked Americans that Christmas season to light a candle in support of freedom in Poland. Stemming from a private meeting that Reagan had with the Polish ambassador, Romuald Spasowski, and his wife, Reagan felt that the symbolic action would unite America and show Poland that the people of the country were behind them. At the meeting with the president the previous day, both the ambassador and his wife had resolved to defect to the United States. Michael Deaver witnessed the moving meeting:

The ambassador and his wife were ushered into the Oval Office, and the two men sat next to one another in plush leather wingback chairs. Vice President Bush, and the ambassador's wife, sat facing them on a couch.

The ambassador had in his hand a pocket-sized note pad with wire rings and lined paper, and he was obviously referring to notes he wanted to give to the president of the United States. Meanwhile, his wife, a tiny, delicate-looking woman, kept her head in her hands the entire time, while George Bush put an arm around her shoulders to comfort her.

The ambassador said, "It is unbelievable to me that I am sitting in the office of the president of the United States. I wish it were under better circumstances."

He begged the president never to discontinue Radio Free Europe. "You have no idea," he said, "what it meant to us to hear the chimes of Big Ben during World War Two. Please, sir, do not ever underestimate how many millions of people still listen to that channel behind the Iron Curtain."

Then, almost sheepishly, he said, "May I ask you a favor, Mr. President? Would you light a candle and put in the window tonight for the people of Poland?"

And right then, Ronald Reagan got up and went to the second floor, lighted a candle, and put it in the window of the dining room.

Later, in what I still recall as the most human picture of the Reagan presidency, he escorted his guests through the walkway and out to the circular drive on the South Lawn of the White House. In a persistent rain, he escorted them to their car, past the C-9 Secret Service post, holding an umbrella over the head of the wife of the Polish ambassador, as she wept on his shoulder.[73]

That candle might have brought to mind those lit after Mass by a young Karol Wojtyla. Then and now, they burned bright for Russia's conversion.

Of course, the Communist press was not quite so sentimental. Enraged by the religious symbolism, even the slightest American invocation of God sent the Soviets seething. "What honey-tongued speeches are now being made by figures in the American administration concerning God and His servants on earth!" fumed a correspondent from Moscow's *Novoye Vremya*. "What verbal inventiveness they display in flattering the Catholic Church in Poland. Does true piety lie behind this?"[74]

Unable to grasp the sincerity of this display of faith, the Soviet press doubted that piety was the motivation for Reagan. The next day, on Christmas Eve, commentator Valentin Zorin dashed before the TV cameras to question the "rather doubtful Christmas gift" Reagan had just given to Americans. The president had delivered a speech that "flagrantly distorted" events in Poland and the Soviet role.[75] Zorin spoke for the collective state's collective judgment: Reagan had yet again egregiously lied about the Soviet role in Poland, and had lied at Christmas time no less! Zorin apologized that he could not be cheerful on such a holiday occasion.

This speech by Zorin was restrained compared to the large Polish Army daily, *Zolnierz Wolnosci*. The newspaper of the Communist military said that the Christmas address by Reagan could not be regarded "as anything else than a blatant interference in the internal affairs of independent and sovereign Poland"—the standard line in press accounts from Moscow to Prague, usually accompanied by the ever-present Communist claim that there was no Cold War, but instead only nonstop Reagan belligerent attempts to restart one.[76]

Titled, "Blatant Interference in Our Affairs," the *Zolnierz Wolnosci* piece found hypocrisy in the Christmas candle, and was especially vicious toward Ambassador Spasowski. Moscow's TASS news agency heartily enjoyed this angle; it eagerly re-circulated the *Zolnierz Wolnosci* article, but only after padding it with additional inflammatory agitprop. Here is an excerpt from the TASS write-up:

Speaking of the dirty role of traitor Spasowski, who played a role in determining the United States' present stand with regard to Poland, the newspaper [Zolnierz Wolnosci] says further, it is worthwhile quoting his impudent
slanderous verbiage in front of U.S. television cameras, the verbiage which
testifies to that man's moral degradation. The traitor asked the President
that a candle be lit in a White House window on Christmas Eve. . . .

Undoubtedly, there should be a candle burning in a White House window, Zolnierz Wolnosci points out, and not only on the Christmas but every
other evening as well—in memory of those hundreds of thousands of victims of military intervention and U.S. bombing raids in Korea, Vietnam,
Kampuchea and Laos . . . in honor of tens of thousands of victims of different forms of U.S. interference in Nicaragua and El Salvador, and in
honor of the Palestinian women and children killed by U.S. weapons being
used by Israel.[77]

While scathing critiques of Spasowski such as these were broadcast
throughout the Communist press, inside Poland the Polish people took Reagan's candle gesture to heart: "When he did this, it was quite a special moment for Poland," said one Polish woman of Reagan's action, once she was
free to speak two decades later. This was a wonderful symbol that Reagan was
going to help us until we could be free like the United States. He was going to
end this.[78] Similarly, Jan Pompowski, a Solidarity member from the city of
Wroclaw, reciprocated the gesture: "Most of us Catholics saw the future with
hope. We prayed for Reagan, that God would give him wisdom." To Poles,
said Pompowski, Reagan was "a man of truth, who acted according to his beliefs, which were the same as ours. We knew he would not betray us."[79]

At the time, Radek Sikorski, who later held high office in the Polish government in the 1990s, was a political refugee living in England, where he attended graduate school at Oxford. Today, he speaks of that White House
candle as a major symbolic action. He remembers well the vitriol from the
Polish Communist press, from Zolnierz Wolnosci to the principal party organ
Trybuna Ludu. In Polish, the latter meant Tribune of the People; the former
meant Soldier of Freedom, a laughable title that, as one Solidarity member told
me, reflected "typical Communist mendacity."

And yet, said Sikorski, the harsh media attack on Reagan backfired on the
Communists. "The Communist press despised such gestures and the Reagan
administration itself," recalled Sikorski over twenty years later. "So, this told
the Polish people that these people—Reagan, his administration, and the

ambassador—must be okay if the Communists hate them so much." Refer-ring to Ambassador Spasowski, he added: "Poles would read that the man was a traitor, an enemy of socialism, a friend of the Reagan administration, et cetera, and thus they would like him."[80]

What the Communists and their press had not figured was the contempt the Polish people had for them and everything they said. Poles had long ago recognized what Czech dissident-turned-president Vaclav Havel referred to as "the communist culture of the lie"; they had become conditioned to the rot-ten smell of deceit.

In a genuine display of hypocrisy—from the Communist side—the Marxist-atheist press fought back at Reagan's invocation of the Christmas season in Poland by attempting to manufacture its own little Christmas snow globe with greetings from the Communist bloc: The Czech media pushed the boundaries of even Communist propaganda by running fantas-tic accounts of how the martial law crackdown had ushered in a new era of bliss for Poland, and fittingly just in time for the Christmas season. A De-cember 28 piece in the Czech party organ *Rude Pravo* idyllically captured the scene in a Polish village: a "brightly lit Christmas tree of peace" on the square in Knurow; joyous children with caps down to their brows play as parents glance at their frolicking offspring; skiers traverse the tricky slopes; and general "good humor and happiness" abound. Such was the merriment that the "impudent" Reagan was trying to destroy by his blatant interfer-ence.[81]

Though the Communists placed great faith in their ability to invent their own reality, Reagan's words and actions had carried with them the weight of truth. As for the truly faithful, Reagan's speech directed a special line at those Poles who lit candles after Mass. Candles had burned earlier in the century, reminded Reagan, when "an evil influence threatened that the lights were go-ing out all over the world." "Let those candles remind us," he shared, "that these blessings [of freedom and abundance] bring with them a solid obliga-tion, an obligation to the God who guides us, an obligation to the heritage of liberty and dignity handed down to us by our forefathers and an obligation to the children of the world, whose future will be shaped by the way we live our lives today."[82]

This last line was not a platitude, since Reagan personally felt a divine ob-ligation in all of his actions. "We can't let this revolution against Communism fail without our offering a hand," he wrote in his diary. "We may never have an opportunity like this in our lifetime."[83]

REAGAN'S LETTER TO BREZHNEV

On that same busy, Poland-packed December 23, Reagan also decided to deliver what he rightly called a "strong message" to the Soviets. Since the NSC meeting that morning, he had wrestled with a draft of a letter to Brezhnev. The draft was completed by evening. He sent the message to Chairman Brezhnev over the Washington-Moscow cable "hot line" known in the White House as the "Molink." The cable was not declassified until October 1999.[84]

The letter condemned the Soviet role in the crackdown. Reagan began it by listing the violation of "elementary rights" of Polish citizens underway: the incarceration of trade union leaders and intellectuals, their detention in overcrowded jails and "freezing detention camps," the "massive arrests without any legal procedures," and the generally "brutal assaults" by security forces. He then appealed to Brezhnev directly and sternly:

> The recent events in Poland clearly are not an "internal matter" and in writing to you, as the head of the Soviet government, I am not misaddressing my communication. Your country has repeatedly intervened in Polish affairs during the months preceding the recent tragic events. No clearer proof of such intervention is needed than the letter of June 5, 1981, from the Central Committee of the CPSU to the Polish leadership which warned the Poles that the Soviet Union could not tolerate developments there. There were numerous other communications of this nature which placed pressure on the Polish government and depicted the reform movement as a threat to the "vital interests" of all socialist countries. These communications, accompanied by a steady barrage of media assaults as well as military exercises along Poland's borders, were coupled with warnings of intervention unless the Polish government sharply restricted the liberties and rights which it was granting its citizens. . . .
>
> Our two countries have had moments of accord and moments of disagreement. But since Afghanistan nothing has so outraged our public opinion as the pressures and threats which your government has exerted on Poland to stifle the stirrings of freedom. Attempts to suppress the Polish people—either by the Polish army or police acting under Soviet pressure, or through even more direct use of Soviet military force—certainly will not bring about long term stability in Poland and could unleash a process which neither you nor we could fully control. . . .

The consequences of each of these courses for our relationship should be clear.[85]

By insisting that he was not misaddressing his communication, Reagan conveyed his understanding that the Soviet government was responsible for events in Poland. He clearly blamed the Soviet leadership for the crackdown, and his language about unleashing "a process" that neither side could control was especially dire, and a deliberate attempt to convey the severity of the situation.

In the cable, he went on to lecture Brezhnev that the intervention was in direct violation of international agreements the USSR had signed. He cited the Helsinki Final Act, "which you, Mr. President, personally initialed on behalf of your country in 1975."[86] This was feigned surprise, since Reagan had always thought Helsinki a sham. In the next to last paragraph of the three-page letter, he reiterated the blame, underscoring "the heavy responsibility of the Soviet Union for the present situation in Poland."[87]

Here, he also reminded Brezhnev that in their previous correspondence both nations had expressed a desire for better Soviet–American relations. That, however, was being jeopardized by "the present atmosphere of political terror, mass arrests and bloodshed in Poland . . . the Soviet Union must decide whether we can move ahead with this agenda [of improving relations] or whether we will travel a different path." He linked improved U.S.–USSR relations, including negotiations on "critical regional and arms control issues," to Soviet behavior in Poland, specifying that, "representatives of the spiritual, political, and social forces in Poland need to be promptly released from detention."[88]

Needless to say, this was a tough letter. The Polish democracy movement would have been elated to read such a forceful declaration from so high up in Washington. Reagan hoped that this would have the desired impact and demonstrate to the Soviets that he was serious about democratic reform in Poland. They were hard words for a hard situation, but nevertheless they were words that Reagan had been waiting to send for over forty years.

Reagan concluded the cable: "I hope to hear from you in the next few days."[89]

DECEMBER 25

The next few days coincided with Christmas, a holiday that demanded no time off for the Soviet leadership.[90] On Christmas morning, after he opened gifts around the family tree in the White House, Reagan was handed a reply to his missive, a terse Brezhnev note sent via Molink claiming it was the United States, not the USSR, which was interfering in Polish affairs. "What a good Christmas present, I thought," said Reagan sarcastically. Yet, the message conveyed to him that, "I'd made my point."[91]

Reagan's next move remains a bit cloudy. Reflecting the high priority of the incident, Reagan said that he wrote back to Brezhnev that same day. He said he told Brezhnev the United States would not intervene in Poland if the Soviets did not, and proposed that the Polish people "only be given the right of self-determination that had been promised to them by Joseph Stalin himself at the Yalta Conference." Reagan was revisiting that old outrage of his: "At Yalta, I reminded him, Stalin had promised Poland and all the countries of Eastern Europe the right of self-determination but the Soviets had never granted it to any of them."[92]

While Reagan wrote of this letter in his memoirs, the Reagan Library has no record of it. It does, however, hold a Reagan response he wrote and circulated internally, in which he dissects Brezhnev's letter. This internal letter seems to match the December 25 letter that he described and said he sent to Brezhnev. In that internal letter, Reagan wrote:

> On p. 3, he [Brezhnev] says we are dictating to the Poles and that no one should interfere with what the Poles and the Polish authorities are doing in their own home. It seems to me that we are supporting the *right* of the Polish people to vote on the govt. they'd like to have. Mr. B [Brezhnev] is supporting the right of the govt. to deny the Polish people a voice in their govt.
>
> Incidentally, didn't the Yalta Pact call for the people having the right to vote on what govt. they would have? The Soviets violated that pact.[93]

Significantly, there may have also been in this Reagan letter another Reagan reference to the consideration of using U.S. military force. Somewhat cryptically, he wrote: "Mr. B [Brezhnev] says we are intervening—we know the Soviets are—maybe we should tell him we won't if he won't."[94] Though on the surface this might read like a possible reference to potential military conflict, in reality, this was certainly not an explicit threat of U.S. armed force. In this context, Reagan could have also meant intervention in a general political sense and not necessarily in a military way.

DECEMBER 29

On December 29, Reagan backed up his written and spoken words with action. He announced sanctions on Poland and the Soviet Union in an expression of "our displeasure over the crushing of human rights in Poland." Regarding the USSR, he suspended negotiations on a new long-term grain-sale agreement; banned flights to the United States by the Soviet airline, Aeroflot; canceled several agreements on energy and science and technology; suspended a number of key high-tech export licenses; and imposed an embargo on shipment to the USSR of critical American-made products, including the crucial pipe-laying equipment to be used in the construction of the Siberian gas pipeline. In all, seven categories of sanctions were issued.[95]

These moves suited Reagan's purposes, since he wanted to strangle the Soviets economically and martial law offered a perfect excuse to heighten the stakes of economic warfare. Many of these were steps he hoped to do regardless of martial law; martial law merely provided a pretext more palatable to much of the international community.

The Soviet and Eastern European Communist press correctly surmised that martial law simply gave Reagan and his fellow "slanderers" a "pretext" to undermine socialism in the Communist bloc.[96] A *Novoye Vremya* piece circulated by TASS said flatly, "Washington has decided to use the events in Poland to implement its old idea—to roll back socialism."[97] This was the Communist consensus: As Vitaly Kobysh of the International Information Department of the CPSU Central Committee said during a typical interminable discussion on Moscow's "Studio 9" program, the Reagan administration's "main object" in Poland was "to weaken socialism" and generally "to push back communism."[98]

On December 29, Reagan announced the sanctions to the American public, proclaiming openly the words that he had uttered privately in his cable to Brezhnev six days earlier. He stated that the USSR bore "heavy and direct responsibility" for the repression in Poland, specifying the pressure and intimidation by the Soviets and listing specific examples, just as he had in the cable. Moreover, he accused, the Soviets "now openly endorse the suppression which has ensued."[99]

Reagan held forth that December 29 from the executive suite at the Century Plaza Hotel in Los Angeles. Reporters, who were startled by such a confrontational move against the USSR, asked Reagan what he hoped to achieve with the sanctions. One purpose, said Reagan, was to "speak for" those who had been silenced and rendered helpless. When asked what he might do if Communist authorities failed to respond to the sanctions and continued to repress Solidarity, Reagan said cryptically: "Well, there are further actions that could be taken that we have withheld."[100]

The most telling Reagan remarks of December 29 came in a conversation that afternoon with reporter Laurence Barrett. Barrett found Reagan "strangely bullish" on Poland. The president told Barrett he saw Poland as representing something much bigger:

> There is reason for optimism because I think there must be an awful lot of people in the Iron Curtain countries that feel the same way [as the Poles]. In other words, the failure of Communism to provide that workers' utopia that they have talked about for so long has been made evident in Poland. Our job now is to do everything we can to see that [the reform movement] doesn't die aborning. We may never get another chance like this in our lifetime.[101]

While Reagan adamantly believed that American pressure could change Soviet behavior toward Poland, he wanted to ensure that his sanctions helped rather than hurt the Polish people. This was evident in his reaction to a December 31 Western Union telegram sent to the White House by five signers: Lane Kirkland of the AFL-CIO, George Stone of the National Farmers Union, Archbishop John R. Roach of the National Conference of Catholic Bishops, David Preuss of the American Lutheran Church, and Rabbi Alexander Schindler of the Union of American Hebrew Congregations. These men commended Reagan for his "firm and unmistakable" position, but expressed concern about the sanctions weighing "more heavily upon the Polish people then upon the ruling government."[102] Kirkland and crew laid out a list of emergency food supplies that could be shipped to Poles. "America," the telegram concluded, "must not let them down." Reagan's immediate response is seen in his handwritten one-line reaction at the top of his copy of the telegram: "I believe we should give this urgent consideration. RR."[103]

It is also important to understand that the sanctions harmed economically pressed Americans. The Reagan NSC estimated that sanctions would exact a $500 million loss to the U.S. economy in 1982 alone—an economy in a recession.[104] Nonetheless, Reagan felt that the domestic sacrifice in the short-term was well worth the long-term gain.

Reagan later said that he did not think the decision to impose martial law and crush Solidarity "could stand" because of "the history of Poland and the religious aspect and all."[105] But regardless of the immediate outcome, there were larger issues at stake. The most important element to December 1981, as Cap Weinberger put it, was that the administration "decided on the need to make a stand in Poland—not only to prevent an invasion, but to seek ways to undermine [Communist] power in Poland."[106]

REAGAN'S UNMISTAKABLE INTENTIONS

Sensing Reagan's broader objectives, the USSR and its Communist bloc regimes reacted with fear and loathing, as they pumped their propaganda machines for a response. One spin by the Communist press was to attempt to argue that the aggressor was somehow to be found in Washington, not in Moscow or Warsaw, from where martial law was imposed. TASS argued: "The U.S. administration has undertaken a provocative act with the aim of still further poisoning the international climate, increasing tension, making confrontation harsher and pursuing a harder militarist line in foreign policy."[107]

A key component of this spin was to frame Reagan as the aggressor. Driven to a "blind rage" that removed "every vestige of common sense," TASS said that Ronald Reagan had personally lashed out at the USSR when announcing, "a whole range of unilateral discriminatory measures against the Soviet Union." TASS dubbed these U.S. sanctions "blackmail and pressure," saying that by justifying this "unprecedented crude diktat," Reagan "resorted to direct forgery and lies." "The White House boss clearly needed these unsubstantiated accusations in order to distort generally known facts," said TASS, such as the "fact" that the "trials" which had befallen the Polish people "are primarily the result of direct interference in Poland's affairs by American imperialism." "It is precisely the United States," said TASS, "which reared the Polish counterrevolution."[108]

Much to Reagan's chagrin, in their clamor over U.S. interference, Iron Curtain Communists could count on occasional support among America's Western allies, which they in turn used as a central facet of their PR war against the United States. Such support was most salient among the West German left, from the rank-and-file protester to top government officials. The Communist press frequently quoted West German Chancellor Helmut Schmidt and Foreign Minister Hans Dietrich Genscher, the latter of whom, according to a UPI quote borrowed extensively by the Soviet media, said he opposed "external intervention" in Poland's internal affairs—meaning strictly *American* external intervention, not Soviet.[109] The Communist press also cited the leftist Pierre Trudeau government of Canada, especially relishing a quote from the prime minister's spokesman, who said: "We do not stand at attention whenever Reagan speaks."[110]

What Moscow's anti-Reagan campaign proved was that Reagan certainly had the attention and great concern of Communist headquarters. Moreover, all of the bluster and hyperbole in these statements revealed something significant: as early as December 1981, the Kremlin accurately understood what much of America had yet to learn: Reagan wanted to subvert Soviet Communism in Poland and, more so, to spark a chain reaction throughout the Communist bloc.

"Washington is behaving as the patron of Polish counterrevolution [with] plans to overthrow the socialist order," asserted a January 1982 *Pravda* piece by Aleksey Petrov. "The White House figures would like to see in Poland a kind of rallying point in order to attempt to introduce a split in the socialist world and, at the same time, to shake the whole existing system of international relations." The White House, said Petrov, saw Poland as a "fulcrum" to split the Communist bloc.[111] TASS correspondent Yuri Kornilov agreed, saying it was "quite obvious" that the United States was seeking a situation "in which a weak Poland, torn by internal contradictions, is weakening the whole socialist camp."[112]

From inside Poland, Adam Rostowski wrote in *Zolnierz Wolnosci* that the Reagan team held "hopes that the events in Poland would trigger an avalanche of evolutionary or, under suitable circumstances, revolutionary (read counterrevolutionary), changes in other socialist countries in East Europe, and particularly Czechoslovakia, the GDR, Romania, and maybe—for the second time—in Hungary."[113]

It was as if Rostowski, Kornilov, and Petrov had been reading Ronald Reagan's diary—or perhaps his Christmas list that December 1981.

BY THE END OF DECEMBER, REAGAN'S SANCTIONS WERE IN place. He had faced the first clear test of his anti-Communist resolve. Neither international nor domestic pressure was capable of dissuading him from the course that he deemed proper, and the sanctions would remain in place until he deemed it appropriate to lift them. In Poland he had witnessed the true nature of Communism reveal itself, a nature he had known for forty years.

While the dramatic events of that December loomed large over the new year, they were merely a precursor for the exciting year that lay ahead. It was a year which would set the stage not only for Reagan's bold initiatives in 1982, but for the years of his presidency that lay ahead. It was a year in which the glove which fit so comfortably during 1981 would come off.

9.

Commencing the Crusade:
January to June 1982

ON JANUARY 1, 1982, THE FRONT PAGE OF THE *WASHINGTON Post* featured three lead stories of high interest. In the top right corner was a dispatch from Solidarity's Zbigniew Bujak, smuggled from his undisclosed hideout. Bujak wanted to assure the world that Solidarity had not been vanquished and that "final victory" would come. In the middle of the top-of-the-fold was an interview with another Solidarity activist that the *Post* identified only by pseudonym. He expressed betrayal by his own military. "[I]t's our own Army," he despaired. "We're confused and we don't know how to react."

Most significant of these three stories was the one in the top left column, which reported that Richard Allen's job as national security adviser was expected to go to Bill Clark, a fervently anti-Communist Catholic with a special sympathy for Poland, who was about to become Ronald Reagan's indispensable point man in laying the groundwork for Cold War victory.[1]

Reagan awoke as usual at 7:00 AM and read these stories among his standard stack of newspapers—the *Washington Post, Los Angeles Times, New York Times, USA Today,* and the *Washington Times.* He then attended his regular morning briefings. On this New Year's Day, he had planned two public appearances, both via video link. The first concerned football, which he hoped to catch on TV for a few minutes that afternoon. Speaking with actor Michael Landon, who was covering the Tournament of Roses Parade in Pasadena for

NBC, Reagan helped ring in the new year for the millions of Americans watching the parade.

In his other appearance, Reagan gave a short talk broadcast worldwide by the U.S. International Communication Agency. "We look forward to the coming year as a time of opportunity," said the president. "We hope and pray . . . that mankind will be a little better for things which we as individuals and as a nation will do in the year ahead." He noted that in the previous year Anwar Sadat, "a great man of peace," was murdered, and that attempts were made on the lives of Pope John Paul II and himself. This gave him "occasion to realize that we must use what time we have to further those values which will last after we as individuals are gone."[2]

Those wishful words applied to what he envisioned for Poland. In the last year, said Reagan, Poland's workers battled bloodshed—to "edge their country closer to freedom." The president denounced fascism and Communism and said that both deny the existence of God and those God-given liberties "that are the inalienable right of each person on this planet."[3] He closed with a wish of goodwill and offer of prayer.

All of this was a bellwether of the year to come; it was fitting that Reagan could not avoid Poland that first day. The year 1982 would be decisive, playing a crucial part in his strategies for Poland and the fall of the Soviet Union. While the year 1981 had given the president the opportunity to get the "domestic house in order," he was ready to look beyond national policies to a broader international agenda. At the start of 1982, judging that the bricks and mortar were in place, Reagan called his foreign-policy staff into the Oval Office and, in Bill Clark's words, told them, "Gentlemen, our concentration has been on domestic matters this year, and I want to roll the sleeves up now and get to foreign policy, defense, and intelligence."[4]

BILL CLARK AND THE NSC

Bill Clark had been Governor Reagan's chief of staff before Reagan began appointing him up through the various levels of the California court, eventually all the way to the California Supreme Court, where he remained in January 1981 when his old friend was sworn in as president. At the inaugural, Clark was one of Reagan's special guests, and Reagan leaned on him to join the new administration. A rancher at heart, content with his wife, kids, and land near Paso Robles, Clark did not want to go to Washington. Reagan pushed, and eventually Clark relented.

Sensing that he would need an "America desk" at the State Department, Reagan asked Clark to become the number two at Foggy Bottom. There was no member of the administration more loyal and close to Reagan. Edmund Morris would dub Clark the "most important and influential person in the first administration," as did *Time*, the *New York Times Magazine*, and any remotely knowledgeable source.[5] Now, after a year as deputy secretary of state, Reagan needed Clark to run his national security policy, to head up his National Security Council, to help him win the Cold War.

Already devoted to the pope and Poland, the devout Catholic Clark seemed literally linked to both from the outset. On January 4, the first day on the job, Clark accepted a letter from John Paul II, a response to two Reagan letters written to the pope over the previous two weeks. The pope wrote to say that he supported the administration's countermeasures against martial law in Poland, considering them complementary to the moral pressure he was willing to apply. Excited by the response, the new national security adviser wrote two memos (both not declassified until July 2000) to Reagan summarizing the pope's reaction and happily reporting that John Paul II's letter "makes it clear" that he backed Reagan's policies and goals, seeing "his actions as complementary to ours."

But there were limitations to what the pope could and could not do publicly, and Clark's memos reflected this point. He cautioned that the pope said he could not be "as publicly forthcoming in expressing this support as we would wish." This lack of public support created a problem with Western Europeans, notably the West German government, which in its push for a more accommodating policy toward the Poland situation had declared its stance in line with that of the Vatican. To the contrary, said Clark, the pope's letter made it clear that the Vatican was closer to Washington than Bonn.[6]

In turn, Reagan replied to John Paul II's January 4 letter. His response remained classified for nearly twenty years. He told the pope that the pontiff's January 4 missive, in which John Paul II (in Reagan's words) recounted the "tragic history" of the Polish people and their "unquenchable thirst for liberty," had moved him deeply. Reagan lamented the "terrible crimes" and "unspeakable afflictions" borne by Poles over the years. Then, in a remarkable religious judgment not typical of Reagan—who was more apt to speak of God's forgiveness than wrath—the president asserted:

> Tragically, the leaders of the Western democracies were too often ready, even in the present century, to condone by their silence many terrible consequences

in the political life of nations. I tremble to think of God's verdict on those who acquiesced in these deeds, as well as on those who perpetrated them. . . .

The ultimate responsibility of the Soviet Union for this tragedy is indisputable, however artfully the Russians may disguise their involvement. The Soviet action in Poland is not an aberration of policy. It is an act of brutality.[7]

Bill Clark had long known Reagan's intense interest in Poland, and now he was in a position to feed that interest. Clark had a protocol to ensure Poland stayed on the front burner. To the PDB, the super-sensitive President's Daily (Intelligence) Brief, a section was devoted strictly to Poland.[8] Clark recalls how Reagan craved that regular morning update on Poland:

Bill Casey would, by courier, send the President's Daily Brief [PDB] each morning at about 5:00 a.m. to our war room downstairs in our [National] Security Council. It was a very limited edition, five colors showing the activity across the globe for the preceding 24 hours. It would be delivered to the President in his residence before he came over [by 7:00 a.m.]. His first question for a long period of time was usually, "What is happening in Poland this morning?"

He'd write questions all over the margins about things that weren't clear in the briefing. And, of course, the agency [CIA] would come down with further explanations. He called Poland the hub of the Soviet empire—of all of the seven adjoining countries, second in priority only to Russia itself. He watched closely John Paul's activity. . . . He followed Lech Walesa closely and that wonderful movement, realizing that once that hub began to unravel, the whole thing would come apart. He thought that when Poland would go, so would the empire. Poland was an opportunity for what he called "the unraveling of the empire."[9]

Clark had no doubts as to Reagan's intentions. "It certainly was his policy to undermine the USSR," said Clark. "Without a doubt, as early as the gubernatorial days, that was his thinking: whether by undermining the Soviet Union or by helping the Soviet Union undermine itself. That's a given. That was his policy. No question."[10] Reagan assuredly told Clark and his staff "several times both as governor and many times later as president," that, "The wall around atheistic communism is destined to come down within the Divine Plan because it lives a lie."[11] So frequently did Reagan speak of this

Divine Plan that Clark today casually abbreviates the two words as simply "the D.P."

Thus, knowing what Reagan wanted, Clark said, "It was up to us as his cabinet and staff to convert the President's early vision into underlying policy, policy into strategy, strategy into tactics, and on to implementation and ultimate success."[12] According to Clark, Reagan's "strategy to accelerate the demise of the Soviet Union" consisted of five pillars: "economic, political, military, ideological, and moral."[13]

While these were the general contours of Reagan's strategy to undermine, it was up to Clark, as the new national security adviser, to corral the manpower and craft a series of policy directives to form this strategy to undermine. To execute this plan, Clark assembled a collection of NSC staffers, people like Tom Reed, Roger Robinson, Paula Dobriansky, Ken deGraffeinreid, John Lenczowski, Richard Morris, Sven Kraemer, David Laux, Norm Bailey, Robert "Bud" McFarlane, John Poindexter, and others.[14]

Also significant were the NSC members that Clark chose to retain. Chief among them was Richard Pipes, who since 1950 had been a highly respected Harvard professor of Russian history. In 1981 Richard Allen brought Pipes into the Reagan NSC to start a two-year leave from Harvard that would end in December 1982. Widely regarded as the NSC's senior Kremlinologist, Pipes served as the NSC's political affairs directorate and European and Soviet affairs directorate. His opinion was respected throughout the White House, and whenever Soviet subjects came up in the Oval Office, the president frequently asked, "What does Dick Pipes think?"[15]

Naturally, drawing respect from Reagan meant drawing ire from the Soviet propaganda machine. Pipes was described in *Pravda* as "one of the ideological mentors of the U.S. administration,"[16] while The Moscow Domestic Service was more strident, excoriating Pipes as an "odious figure" that had "long plowed the furrow of ardent anti-sovietism and anti-communism." The Ivy League professor was a "dyed-in-the-wool reactionary, hysterically fighting for nuclear confrontation with the Soviet Union."[17]

But despite the Communist remonstrance, Pipes found himself at home in the Reagan administration. To Pipes, Reagan possessed "a keen grasp of the vulnerabilities of the Soviet regime," displayed "great discernment and the instinctive judgment of a true statesman," and had been inspired by "a strong moral sense and a sound understanding of what it is to live under tyranny." Reagan, said Pipes, "acted with the conviction that the Soviet Union was not strong but weak, that its power rested on police terror at home and nuclear blackmail abroad, and that, being in the profoundest sense unnatural, it did not have long to live."[18] Pipes added that Reagan "instinctively understood, as all great statesmen do, what matters and what does not, what is right and what is wrong for his country." This quality, said Pipes, cannot be taught: "like perfect pitch, one is born with it."[19]

JANUARY TO MARCH 1982: RHETORIC AND ECONOMICS

In early 1982, rhetoric and economics became central to the effort to undermine, and both were put on display by Reagan in those first months of the year. Of the former, it seemed like every couple of days Reagan made a comment about Poland in a speech or interview or press conference. Every few weeks he made a lengthy statement of support, assailing the Communist government or the Kremlin, while continuously issuing memorials and proclamations of support for Solidarity and the Polish people.

The first concrete step came on January 20, 1982, when the president officially designated January 30 as Solidarity Day in the United States, a time for Americans to show "special affinity" with Solidarity. "I urge the people of the United States, and free peoples everywhere," exhorted Reagan, "to observe this day in meetings, demonstrations, rallies, worship services, and all other appropriate expressions of support. We will show our solidarity with the courageous people of Poland."[20] Though on the surface this seemed to be merely a symbolic gesture, it served the important role of demonstrating to Moscow and the rest of the world that Poland had not been forgotten.

There were many notable Reagan remarks from this period.[21] Among them, Reagan returned to the Conservative Political Action Conference on February 26, 1982, where, before the faithful, he quoted the writing of Whittaker Chambers, a man he and they admired. Chambers had said thirty years earlier that within the next decades history would decide whether all mankind was to become free or Communist, and that it was their fate to live upon that turning point in history.[22] Chambers was pessimistic; he felt that America was on the right side of history in the Cold War struggle, but believed it was on the "losing side." Despite his utmost respect for Chambers, Reagan told his fellow conservatives that the turning point had arrived and, unlike Chambers, he was an optimist. He told the gathered that he and they had "already come a long way together," beginning with that speech for Goldwater. Now, he had more to ask: "Join me in a new effort," he urged, "a new crusade."[23]

These were grand words. Rhetoric, however, could only go so far. If Reagan was truly going to expedite and exploit the fissures in Communism, he knew that he would have to employ economics as part of his campaign against the USSR. It was this thinking that led NSC member Tom Reed to prepare a study on the USSR's economic crisis.

Reed's report was discussed at an intriguing March 1982 White House meeting, where Reagan wasted little time: "Why can't we just lean on the Soviets until they go broke?" the president asked.[24] Some of the moderate Cabinet members objected, insisting, among other things, that the USSR was stable. Henry Rowen, the brilliant chair of the National Intelligence Council, voiced disagreement with the pragmatists, saying that, quite the contrary, he believed the Soviet Union to be on the verge of collapse. Reagan thanked Rowen for his input and then, with a simple nod of his head, said to Reed: "That's the direction we're going to go."

THE ECONOMIC WAR

Reagan had long viewed economics as a means to facilitate political change in the Soviet Union. In a speech made in the early 1960s, he averred: "If we truly believe that our way of life is best aren't the Russians more likely to recognize that fact and modify their stand if we let their economy come unhinged so that the contrast is apparent?"[25] To Reagan, Soviet disadvantages were America's advantage. Lenin's inferior system ought to be permitted to unhinge by its own inferiority, and it was up to the United States to hurt rather than help the Soviet system.

As president, Reagan's long-held views were now reinforced by the classified national-security briefings he received each morning. These detailed briefings shed much light on the Soviet situation and buoyed Reagan's suspicion that the USSR was in trouble. In addition, Bill Clark and CIA director Bill Casey worked intently to provide Reagan with hand-delivered raw intelligence on a regular basis, an unprecedented step for a president. NSC staff member John Poindexter (later a national security adviser) said that Reagan "loved seeing the raw intelligence on the Soviet economy," a sentiment which Chief of Staff Donald T. Regan echoed: "He just loved reading that stuff. He would take that big stack and read them one by one over the weekend."[26]

In these briefings Reagan saw what he called "great opportunities" for Soviet destruction.[27] During a March 26, 1982 classified briefing, the president was told that his adversary was hobbling badly. He shared the dire news with his diary: "Briefing on the Soviet economy. They are in very bad shape, and if we can cut off their credit they'll have to yell 'Uncle' or starve."[28] He found the bad news inspiring, as it supported his long-held beliefs about the Communist system and the ways in which he could exploit the ruptures in the system for his own end:

> Now, the economic statistics and intelligence reports I was getting during my daily National Security Council briefing were revealing tangible evidence that Communism as we knew it was approaching the brink of collapse, not only in the Soviet Union but throughout the Eastern bloc. The Soviet economy was being held together with baling wire; it was a basket case, partly because of massive spending on armaments. In Poland and other Eastern-bloc countries, the economies were also a mess, and there were rumblings of nationalistic fervor within the captive Soviet empire.
>
> You had to wonder how long the Soviets could keep their empire intact. If they didn't make some changes, it seemed clear to me that in time Communism would collapse of its own weight, and I wondered how we as a nation could use these cracks in the Soviet system to accelerate the process of collapse.[29]

Reagan was learning that the Soviet economy "was in even worse shape than I'd realized," further assuring him that "Communism was doomed." Not only did the system lack the free market incentives that spawned economic growth, said Reagan, but, also, history was full of examples showing that any totalitarian state that deprived its people of liberty would ultimately fail. The Soviet system, he determined, "could not survive."[30] "The situation was so bad," concluded Reagan, "that if Western countries got together and cut off credits to it [the USSR], we could bring it to its knees."[31]

While the USSR might be hampered by its own cracks and economic inconsistencies, Reagan felt that his administration needed to identify the fissures and drive a wedge between them. His team would seek to exploit these internal contradictions unlike any administration since the start of the Cold War.

The next steps were to furnish the sword and summon the will to plunge it deep. For that, he could count on not only Bill Clark and Clark's team at the NSC but also on the vital assistance of Bill Casey, the highly unorthodox intelligence director. Casey was more the top mole, the operations officer-in-chief, than a bureaucrat running an agency.[32] Under Casey, the CIA played a central role in economic warfare, with the analysis division launching a systematic campaign to identify Soviet economic weaknesses.[33] "What we realized is that the CIA [prior to the Reagan administration] was monitoring Soviet strengths," said Herb Meyer, special assistant to Casey. "It was not looking at Soviet weaknesses." Now, under direct orders from the president, Casey began conducting Soviet vulnerability assessments.[34]

Because of his vehement anti-Communism and strategic insight, Reagan found a soul mate in Casey, a man who like Reagan saw the Soviet Union as vulnerable to pressure. With Casey at Langley, Reagan's intent to undercut the Soviet Union now found a formal method, one that would use intelligence collection as an offensive tool. In this capacity, Reagan unleashed Casey, picking his brain for ideas, placing tremendous confidence in him, and then signing off on dangerous initiatives that exploited Soviet vulnerabilities. He did this quite discreetly, often in closed-door meetings that involved only him, Casey, and the national security adviser (Clark in the most crucial years): As Bill Clark acknowledged: "Few of these initiatives were discussed at cabinet meetings. The president made his decisions with two or three advisers in the room."[35]

Moscow immediately recognized Casey's prominence inside the Oval Office, and was quite concerned, as evident by the way he was profiled in the Soviet press. He was referred to as the "Queens Gangster," as "Casey the Untouchable," or as simply "the Crook."[36] The Soviets were enraged when Casey, in a speech in San Antonio, colorfully exclaimed that Marxism-Leninism had unleashed the Four Horsemen of the Apocalypse.[37] In response, Vitaliy Korionov mocked the CIA director in *Pravda*: "Casey stated, voice atremble: 'The Marxist-Leninists have unleashed the four horses [sic] of the apocalypse—famine, plague, war, and death.' Are you not scared, reader?"[38] And as for Casey's and Reagan's confidence that they could reverse Communist gains from Afghanistan to Nicaragua, Korionov warned: "But just you try, gentlemen!" Well, Casey and Reagan were ready to try.

Casey shared Reagan's optimism that Communism was not the future; aside from Reagan, he was the rare administration member who publicly articulated that vision. In a speech at Ashland University, Casey estimated: "Now for the good news. The pendulum of history slowly but surely is swinging away from Soviet Marxism as a model of Third World countries, and toward the concepts of democracy and free market economies."[39]

And from the start, Bill Casey was ready to do his part to help Reagan rock the pendulum.

TWO-TIERED APPROACH

Though they were clearly putting together the tools to wage an economic war, Reagan understandably shied from publicly using those very provocative words and regularly reminded zealous staff to refrain as well.[40] "Certainly it was economic warfare," said Reagan defense official Richard Perle, "although we had to deny it at the time."[41] Similarly, Secretary of Defense Caspar Weinberger explained that it would have to be "a silent campaign."[42] Clearly, there was no doubt within the staff over what they were doing, and together they sensed that while they would not be able to publicly celebrate the individual victories of each battle, the ultimate victory of the war would be the sweetest of them all—if and when it came.

This war would come to manifest itself in what Clark and others called the president's "roll back strategy," which, as Clark put it, aimed at "changing the Soviet system from within by further destabilizing their economy through strict export controls [on] high technology transfers, by accelerated arms competition, and by exposing the Soviet system for what it was."[43] This broad declaration could be divided into two categories: First, there was an effort to restrict Soviet cash flow and revenues by depriving the USSR of vital trade and technology. Second, there was an attempt to spend the USSR into a corner—if not into oblivion—by challenging it to a costly arms race and technological competition that Reagan felt it could not match and would die trying.

RESTRICTING CASH FLOW

Depriving Moscow of money quickly became a team effort, and one of the unsung heroes was Roger W. Robinson, who joined the Reagan team that March 1982 at age thirty-one. When he was thirty-two, Robinson became senior director of U.S. international economic affairs at the NSC, where he ran the bulk of international economic policy—that which involved foreign policy and security dimensions. Like Bill Casey, he was a rare find in that he had both Wall Street and foreign-policy backgrounds. Robinson had previously worked at Chase Manhattan, where for five years he was a vice president in charge of Chase's loan portfolio for the Soviet Union and Eastern and Central Europe.

It was Robinson and his colleague Norm Bailey who, under the leadership of Clark and Reagan, focused the administration's attention on going after the Soviet hard-currency flow—and Reagan personally "worked closely" in configuring this security-minded economic strategy.[44] The consistent approval of Clark and Reagan, emphasizes Robinson, was alone a "major breakthrough" in his work—in that another president and national security adviser might have vetoed his suggestions. It was through Robinson's work that the administration came to see that Soviet hard-currency income in 1982 was only about $32 billion, much of which came from gas and oil exports.[45] This was a stunning degree of dependence on merely two resources.

THE PIPELINE

In 1982, natural gas was one of the Soviets' top exports and as such it was the lifeblood of the Soviet economy. The Soviets had already begun a huge $35 billion project that would transfer natural gas from Siberia to Western Europe. The project called for two 3,600-mile pipelines, each of which would be fifty-six inches in diameter, extending from the Yamal Peninsula into Western European cities via the Soviet gas grid. While the construction of these pipelines would give Western Europeans a gas supply that would last them into the twenty-first century, it would also provide the Soviets with billions of dollars in revenue.

Before Ronald Reagan took office, Western Europe had committed to cooperating with the Soviet Union in the construction of the pipeline. Neither the previous U.S. administration nor Western governments had resisted the enormous project; quite the contrary, Western Europe, grappling with very high unemployment, favored construction because it would bring jobs. In 1981, the French president, Francois Mitterand, entered office with a number of large contracts already signed and approved by his predecessor. Each of these French companies needed the money and Mitterand said he felt legally bound to them.[46] Even Margaret Thatcher supported the pipeline, as several large British firms were assured big contracts.

Despite the Western nations' support for the project, Reagan was convinced that it needed to be stopped. If America was going to be successful in this economic war, the administration needed to stifle each and every opportunity for the Soviets to generate cash flow and this meant stifling the pipeline. Secretary of Defense Weinberger explained Reagan's take on this issue: "He knew that what the Soviet Union needed very much was hard currency. And he knew that the construction of the pipeline would give them that. He felt very strongly that you didn't want to assist them in any way in getting hard currency."[47]

It was a position that he knew would be unpopular with America's allies, but he also knew that it would be unpopular with his enemies as well. As construction neared, Reagan began considering his options to prevent assembly of the pipeline. It was merely the beginning of a fight that would last for the next year and a half.

THE FAREWELL DOSSIER

While many at the NSC were preoccupied with exploiting the pipeline, other members of the Reagan administration were busy finding economic alternatives that could also adversely impact the Soviets. One such alternative was an extraordinary effort that insiders dubbed the "Farewell Dossier," a supersecretive initiative, known to only a handful of men, and entrusted to one person: a brilliant, enigmatic NSC staffer named Gus Weiss. Weiss was a behind-the-scenes foot soldier in the Reagan economic war, who had served in both Nixon's and Ford's NSCs, as well as on Carter's U.S. Intelligence Board. He held a master's degree from Harvard and a doctorate in economics from New York University. Apolitical and nonideological, Weiss, who tinkered with game theory as a hobby, was married to public service and dedicated his life to it. Over the course of his career, he won a number of merit medals for his exploits, including accolades from the CIA, NASA, and the National Security Agency. Even NASA awarded him the Exceptional Service Medal.

Interviewed shortly before he died, Weiss spoke of the thinking inside the Reagan NSC:

> We actually got word that he [Reagan] was going to try to "prevail" [over the USSR]. . . . So, we spent all of our time looking. Where can we find the Achilles' heels? That's what we looked for. We went on full economic warfare. One thing we did was the pipeline. But there was far more than that. There are things that no one knows about.[48]

The Farewell Dossier was one such "Achilles heel" that Weiss uncovered, and it proved to be a highly successful piece of the economic war's espionage.[49] Soviet intelligence worked around the clock to steal critical technical and scientific knowledge from the West. For years, the U.S. government suspected that the USSR had such a coherent, organized effort underway, but it was not until 1981 that the Soviet program was discovered. It was then that French intelligence obtained the services of a 53-year-old high-level Soviet defector named Colonel Vladimir I. Vetrov. Vetrov became known as "Farewell."

During the course of his service, Farewell photographed 4,000 KGB documents, fully revealing the Soviet program. On July 19, 1981, French President Mitterand—in a generous but unreported example of French cooperation—told a grateful Reagan about Vetrov and offered to give the intelligence to the United States.[50] Reagan eagerly accepted the gift.

Once the Farewell Dossier arrived in Washington, Reagan asked CIA director Bill Casey to consider how to best use the material. In the fall of 1981, Weiss was cleared to read the material and given the go ahead to brainstorm ideas. "It literally was dropped on my lap," he said later. Weiss got to work, trying to make sense of the massive volume of material, soon learning that the KGB had created a new unit called Directorate T, which was tasked to plumb the R&D programs of Western nations. The operating arm of Directorate T was given the name Line X. Through this apparatus, said Weiss, "a master plan" was put in place to acquire American high-tech products and know-how.[51]

Upon reading the material, Weiss said his worst nightmares came true: Line X had been so successful "that the Soviet military and civil sectors were in large measure running their research on that of the West, particularly the United States. Our science was supporting their national defense." Radar, computers, machine tools, semiconductors—everything was an open book for them.[52] As Tom Reed, who was privy to the Farewell Dossier, put it, it was as if "the Pentagon had been in an arms race with itself."[53] In the dossier, Colonel Vetrov had spilled the beans on Directorate T, divulging the names of over 200 Line X officers stationed throughout the West and laid bare more than 100 leads on Line X activities.

In response to this wealth of information, Weiss planned an ingenious response, which he presented to Casey in December 1981: Because of Farewell, noted Weiss, Reagan's NSC was suddenly in possession of a Line X shopping list of still-needed technology by the Soviets. Weiss offered a suggestion: U.S. counter-intelligence could intervene and supply some of these technologies, and even add enticing new technologies to the shelf, but with a fatal catch: these technologies would appear genuine but would later prove defective and destructive. Using the Soviet need for new technology as a weakness, the United States could sabotage the Soviet program.[54]

Excited by the potential, Weiss shared these conclusions with Casey, and in January 1982, an impressed Casey took Weiss' plan to Reagan. From the onset, Reagan was enthusiastic, immediately giving the go-ahead.[55] There were no written memoranda on this meeting or on the entire project. Reagan wanted the matter kept tightly under wraps to ensure that White House moderates did not find out about it and leak it to the press in order to kill the effort, which they often did when they disagreed with a policy.[56]

The plan required close cooperation among the CIA, FBI, DOD, and suppliers who would modify certain products and make them available to Line X. While it would take time for these efforts to come to fruition, an important covert measure was now underway, adding an intriguing new front to the economic campaign. With these weapons in hand, it was now time for the administration to turn attention to other fronts of the economic war, and to make undermining a stated part of official policy.

THE FORMAL DIRECTIVES TO UNDERMINE

Though undermining had been the goal from day one of the Reagan administration, it was not until early 1982 that the administration began the series of National Security Decision Directives (NSDDs) which would eventually help precipitate the fall of Communism. Springing from the NSDDs were at least a dozen actions—formal and informal—that the Reagan team pursued for the purpose of crippling and even trying to kill the Soviet Communist empire. Obviously, some of these actions could occur only as world events developed; the key was to recognize and grasp opportune moments as they presented themselves—whether in Poland or the pipeline in Siberia.

The NSDDs began with the arrival of Bill Clark in January. The directives would ultimately conceptualize the plan of attack in the economic war.[57]

With the sleeves up, Clark said that "at the president's direction" the administration began "the study of the overall Soviet situation and what our existing relations were and what they should be." These study directives developed into "decision directives."[58] This initial effort, said Clark, led to the development of NSDDs 2 through 120, which "created the national security policy" of the Reagan administration, a policy of which Clark spoke of Reagan's close personal role:

> In all the 'DDs,' he was totally involved. This wasn't a workshop situation where elves sat around pounding out shoes for the king. . . . In my daily briefings, I would give him progress reports on the NSDD system and keep him abreast by thumbnailing where we were in all these important DDs.

We would have Situation Room settings where all the national security principals would come together on these. He was very much a part of it. Not
just signing, but also progression, briefing, discussion, and guidance.[59]

Richard Pipes also testifies to Reagan's involvement: "As someone involved in the formulation of Soviet policy . . . I can attest that the direction
of this policy was set by the president and not by his staff, and that it was vigorously implemented over the objections of several more dovish secretaries."[60]
Among the dovish secretaries, Secretary of State George Shultz, hired in mid-
1982, likewise confirms that Reagan "was very much involved. He set the
policies. He set policy for how we should think about our approach to the
Soviet Union."[61]

The first major step in this policy was NSDD-24, which was signed
February 9, and bore the lengthy title, "Mission to Certain European Countries Concerning Oil and Gas Equipment Exports to the Soviet Union and
Restricting Credits to the Soviet Bloc Countries." Despite this rather dull title, the document contained some rather striking initiatives designed to halt
the Soviet pipeline construction. Specifically, NSDD-24 directed a delega-
tion made up of high-level officials from State, Treasury, Defense, and
Commerce to negotiate with the governments of Italy, France, West Germany, and the UK "with reference to the export to the Soviet Union of oil
and gas equipment manufactured in their territories by subsidiaries
and/or licensees of U.S. companies, as well as the question of restricting
and/or raising the costs of credits to the countries of the Soviet bloc."[62] An
attempt to deter the Western European governments from involvement with
the pipeline, NSDD-24 put into motion the covert diplomatic efforts
that sought to limit Western Europe's role in construction.[63] By utilizing
the departmental powers of the executive branch, Reagan would reach out
to the participating governments, trying to convince them to abandon the
pipeline.

Ultimately, the significance of this NSDD lay not only in its concentrated
diplomatic efforts to oppose the pipeline, but also in its clear articulation of
government policy. This NSDD became a standard by which others would be
judged, as it successfully demonstrated both clear objectives and decisive action,
a combination that would be the hallmark of the most important NSDDs that
lay in the weeks and months ahead. Distilling months of planning and internal
policy debate into a concentrated document, the NSDDs which began with

NSDD-24 eliminated the need for guesswork among the administration's security officials and provided them with a distinct roadmap.

APRIL TO MAY 1982

While Clark and his team at the NSC kept the economic strategy on track, Reagan pressured the Soviets on another front. On April 5, 1982, he gave a speech to the AFL-CIO in which prayer and Poland were very much on his mind: "Poland's government says it will crush democratic freedoms," said Reagan, before shaking his finger: "You can imprison your people. You can close their schools. You can take away their books, harass their priests, and smash their unions. You can never destroy the love of God and freedom that burns in their hearts. They will triumph over you."[64] It was an opinion that Reagan echoed throughout that spring, and in May he shared a similar view in Hambach, West Germany on May 6, 1982: "You know some may not like to hear it, but history is not on the side of those who manipulate the meaning of words like revolution, freedom, and peace. History is on the side of those struggling for a true revolution of peace with freedom all across the world."[65]

Jetting back home—truly home—to his alma mater, Eureka College, on May 9, 1982, the Crusader made some emphatic statements. Handing little Eureka some big publicity, Reagan insisted that the "course" Soviet leaders had chosen would "undermine the foundations of the Soviet system." He identified the ingredients he saw as setting the stage for that unraveling:

> The Soviet empire is faltering because rigid centralized control has destroyed incentives for innovation, efficiency and individual achievement. But in the midst of social and economic problems, the Soviet dictatorship has forged the largest armed force in the world. It has done so by preempting the human needs of its people and, in the end, this course will undermine the foundations of the Soviet system.

He went on to talk about the proper relationship between the United States and Soviet economies, speaking more candidly than ever about how the West should deal with the USSR:

> We recognize that some of our allies' economic requirements are distinct from our own. But the Soviets must not have access to Western technology

with military applications, and we must not subsidize the Soviet economy. The Soviet Union must make the difficult choices brought on by its military budgets and economic shortcomings.[66]

Here he was sending a message—not just to the Soviets but to the Western nations as well. To Reagan, the influx of Western money into the USSR was one of the greatest buoys to the Soviet system. If America was going to undermine successfully, he would have to convince the American public and the world to avoid assistance or dependence of any kind on the Soviets.

Within a week and a half of that speech, Reagan's and Clark's NSC produced yet another highly significant NSDD, one that made NSDD-24 seem like the calm before the storm. Labeled "U.S. National Security Strategy," a title as flat as the blade of a sword, NSDD-32 was signed on May 20, 1982. It was the first of the Reagan administration's formal intents to actually reverse Soviet expansion and encourage democratic change within both the Soviet bloc and the USSR itself.[67] With its unmistakable language, it stands as a remarkable demonstration of a very early desire to produce historical change. NSDD-32 set the precedent for additional consequential NSDDs, laying the crucial groundwork for NSDD-66, NSDD-75, and others.

NSDD-32 began by stating that the Reagan national security plan "requires development and integration of a set of strategies, including diplomatic, informational, economic/political, and military components." It then listed ten bulleted points which laid out the "global objectives" of U.S. security policy and established the Reagan intent to undermine. Incorporating a variety of components, the sum total of the objectives constituted a plan for rollback laid out in multiple parts. The document's language was rife with strong wording, such as this objective on page one, which was reiterated later in the document, as if for good measure: "To contain and reverse the expansion of Soviet control and military presence throughout the world, and to increase the costs of Soviet support and use of proxy, terrorist, and subversive forces."

This phrasing was enormously significant, as it expressed Reagan's goal of not merely containing the USSR, but going beyond containment and actually rolling back positions and territory already controlled by the USSR. Furthermore, these words also sought to reverse positions over which the Soviet military threatened control, meaning most of Eastern Europe and numerous other spots around the world. To "increase the costs" meant U.S. support of counter-Soviet forces like the Mujahedin rebels in Afghanistan, among others.

Additionally significant was the fact that NSDD-32 authorized clandestine support for Solidarity, allowing for secret financial, intelligence, and logistical support to ensure the survival of the trade union as an explosive force in the Soviet empire. The straightforward wording in NSDD-32 belies the large internal debate over this issue. While NSC members such as Richard Pipes thought the support of Solidarity was essential, they were strongly opposed by Secretary of State Al Haig, who deemed the plan "crazy," and Vice President George Bush, who worried about inflaming Moscow and counseled against clandestine operations. Pipes said that Commerce Secretary Malcolm Baldrige and Chief of Staff James Baker also disagreed, thinking that the policy "wasn't realistic."[68]

Nevertheless, all were vetoed by Reagan, who liked the idea and included it in NSDD-32.[69] Pipes recalled: "The president talked about . . . how we had to do everything possible to help these people in Solidarity who were struggling for freedom. . . . Reagan really understood what was at stake."[70] The president instructed Bill Casey to draw up a covert plan; the DCI enthusiastically complied. In fact, Casey had moved to prepare plans for aiding Solidarity as early as April 1981, a half year before martial law, and as such he became the principal architect of the operation.[71]

Throughout the document there were continual references to places where Reagan wanted to discourage the Soviets from active expansion. Page two alone featured three calls to curb the USSR and its activities. The first desired to "neutralize the efforts of the USSR to increase its influence." The second called for economic pressure by the Reagan administration to "foster" restraint in Soviet military spending, to discourage Soviet "adventurism," and to "weaken the Soviet alliance system by forcing the USSR to bear the brunt of its economic shortcomings, and to encourage long-term liberalizing and nationalist tendencies within the Soviet Union and allied countries." Within those allied countries, meaning the Soviet bloc, the directive called for the "independent evolution" of "popular movements and institutions in Poland and other East European countries."[72] The third objective sought to limit technology and resources sent to the USSR, and to stymie "Soviet military capabilities by strengthening the U.S. military."

In short, NSDD-32 constituted a formal assault on Soviets interests, intended to rupture the empire, a declaration of which the USSR was quite aware: Writing in *Pravda*, Yuri Zhukov told citizens that the aim of the directive was to place "massive pressure" on the Soviet Union, with an intent to "bring about internal reforms" in the country.[73] A later article in *Pravda* derided NSDD-32 as "adventuristic," "insane," and as the "administration's bible." *Pravda* directly attributed the directive to Reagan, asserting that he "personally put forward his observations" on each section.[74] It is not clear how *Pravda* knew such a Reagan imprint; nonetheless, it grasped NSDD-32's importance as a reflection of Reagan personally and as a barometer of the USSR's rocky road ahead.

LATE MAY 1982

While NSDD-32 had laid the groundwork, much of the administration's thinking remained elusive to the public until the days immediately following NSDD-32's authorization. In a speech at Georgetown's Center for Strategic and International Studies on May 21, Bill Clark offered a rare public admission by a Reagan official. In that speech, Clark explained the administration's aims as candidly as then possible: "We must force our principal adversary, the Soviet Union, to bear the brunt of its economic shortcomings." This was a cornerstone of a "new strategy" by the United States, reported Richard Halloran in the *New York Times*. Halloran noted that the strategy was based on the classified NSDD-32, authorized the previous day. (Not providing the exact number of the directive, the *Times* referred only to "an eight-page National Security Decision Memorandum.") That strategy laid out by Clark on that day, said the *Times*, "made official a theme that several administration officials have hinted at, that of exploiting Soviet economic weaknesses." The *Times* reported that nine drafts of the strategy document had been examined in previous months. It quoted Clark as saying that "the President played an extraordinarily active role" in commenting on all nine drafts. "When it was done," said Clark, "the study and the decision were the President's."[75]

One of the chief contributors to Clark's Georgetown speech was Tom Reed, who delivered a second speech on NSDD-32 to the Armed Forces Communications and Electronics Association a few days later. In that unnoticed address, Reed shared the ambition of the administration that NSDD-32 would start the process to "one day convince the leadership of the USSR to turn their attention inward, to seek the legitimacy that comes only from the consent of the governed."[76]

Despite these public declarations, it was not until the release of the five-year defense plan that the public began to feel the larger significance of NSDD-32. On May 29, a top-secret document known as the "five-year defense plan" was released to the public. Pentagon officials termed it the "first complete defense guidance of this Administration," drafted for Secretary Weinberger's signature.

It was a document designed to "form the basis," reported the *New York Times*, for DOD's budget requests for the next five years. It was also a "basic source" for NSDD-32, which, added the *Times*, was now "the foundation of the Administration's overall strategic position."[77]

Although the release of this "overall strategic position" was captivating, the public was more attuned to the plan because of its examination of strategy for fighting a nuclear war with the USSR. Importantly, examining options for fighting a nuclear war did not mean an endorsement of the policy. On the contrary, it was merely standard procedure for military planners to devise options for potential scenarios. This was true in May 1982 as well, though one might not have imagined judging by the hysterical reaction of those who feared that the plan meant the Reagan administration favored nuclear war. The 125-page unpublished document instructed the armed forces to draw up scenarios for defeating the USSR at any level of conflict.[78]

More significantly, the DOD plan, formally titled, "Fiscal Year 1984–1988 Defense Guidance," echoed NSDD-32's economic-warfare aspects. In a front-page article on May 30, the *New York Times* reported that the plan stated that Western trade policies "would put as much pressure as possible on a Soviet economy already burdened with military spending."[79] The *Times* was candid with readers: "As a peacetime complement to military strategy, the guidance document asserts that the United States and its allies should, in effect, declare economic and technical war on the Soviet Union."

The DOD plan spoke of challenging the USSR with military systems the Soviets could not afford to match, advocating that the United States develop weapons that "are difficult for the Soviets to counter, impose disproportionate costs, open up new areas of major military competition and obsolesce previous Soviet investment." This emphasis on "putting economic pressure on the Soviet Union," asserted the *Times*, in the article's biggest nonrevelation, was a marked departure from the Carter administration.[80]

Moreover, reported the *Times*, the plan made clear that, "Particular attention would be given to eroding support within the Soviet sphere of Eastern Europe." Again, here was more formal intent to stir dissension within the Soviet empire, and the Communist bloc in particular. The plan said that the Reagan administration should "exploit political, economic, and military weaknesses within the Warsaw Pact." American "special operations forces" would conduct operations in Eastern Europe with the aim of diminishing Soviet support. The *Times* explained that special operations were a euphemism for guerrillas, saboteurs, commandos, and "similar unconventional forces."[81]

While the administration had put a portion of its cards on the table through the release of the DOD plan, Reagan officials still kept a few aces up their sleeves. Despite their strong language and clear purpose, NSDD-24 and NSDD-32 were merely the beginning of the directives handed down by Reagan that would change the face of the Cold War. Carefully crafted to maximize their potential impact, these initiatives would search new fronts for confrontation with the Soviet Union, always seeking ideas and means to erode Soviet infrastructure.

Though confident in these endeavors, Reagan hoped to enlist the aid of a crucial figure. Economics had constituted one front; religion was another. It was thus that he decided to keep true to his words of the previous Christmas and engage Pope John Paul II in helping him elicit change in the Soviet empire.

10.

The Vatican and Westminster: June 7–8, 1982

ON MONDAY, JUNE 7, 1982 RONALD REAGAN WAS IN ROME. HE was there as part of a brief trip to Europe. It was a straightforward trip lacking many stops, but in its simplicity, it contained unparalleled steps in the rhetorical and symbolic war against the Soviet Union. For two seminal days in June 1982, Reagan made some of his strongest gestures to date, signaling to America and to the world his belief that Communism's days were running out.

With a media crush outside, as reporters jockeyed for position and at times literally tripped over one another, Reagan and the pope met at the Vatican, a little over a year after assassination attempts that almost took their lives.[1] The day he was shot the pope had received a cable from Reagan, in which the president expressed his shock and prayers.[2] Since then, the staffs of the two men had worked diligently to arrange a meeting between them. "It was always assumed the president would meet with the Holy Father as soon as feasible," said Bill Clark, among those most excited about the prospects, "especially after they both took shots ... only a few weeks apart. I don't know if any one person said 'we have to see the pope.' It was just assumed because of their mutual interests that at some point the two men would come together and form some sort of collaboration."[3]

Reagan had long coveted such an idea, and the events in Poland the previous

December merely reinforced the importance of such a meeting. Not only had he long viewed the pope as the key to Poland's fate, but among his earliest goals as president was to officially recognize the Vatican as a state "and make them an ally."[4]

Now, for the first time, the men spoke face to face inside the venerable Vatican Library. The subject of the shootings was broached. Pio Cardinal Laghi said that Reagan told the pope: "Look how the evil forces were put in our way and how Providence intervened." Bill Clark said that both men referred to the "miraculous" fact that they had survived; indeed, only later did we learn that both men had come perilously close to dying.[5]

The Protestant and Catholic, said Clark, shared a "unity" in spiritual views and in their "vision on the Soviet empire," namely, "that right or correctness would ultimately prevail in the divine plan." That day, each shared their view that they had been given "a spiritual mission—a special role in the divine plan of life." Both expressed concern for "the terrible oppression of atheistic communism," as Clark put it, and agreed that "atheistic communism lived a lie that, when fully understood, must ultimately fail."[6]

Together they expressed a common vision to end the Cold War. As Reagan said, "We both felt that a great mistake had been made at Yalta and something should be done. Solidarity was the very weapon for bringing this about."[7] It was an important unity, and in his dramatic 1992 story for *Time* magazine, Carl Bernstein reported that it was at this meeting where Reagan and the pope secretly joined forces not only to strengthen Solidarity and pressure Warsaw "but to free all of Eastern Europe." In that first meeting, wrote Bernstein, they consented to undertake a clandestine campaign "to hasten the dissolution of the communist empire." The two men "were convinced that Poland could be broken out of the Soviet orbit if the Vatican and the United States committed the resources to destabilizing the Polish government and keeping the outlawed Solidarity movement alive after the declaration of martial law in 1981."[8] Reagan told the pope: "Hope remains in Poland. We, working together, can keep it alive."[9]

Both leaders were convinced that a free, non-Communist Poland would be, in Bernstein's words, "a dagger to the heart of the Soviet empire." They were certain that if Poland became democratic, other Eastern European states would follow.[10] A cardinal who was one of John Paul II's closest aides put it this way: "Nobody believed the collapse of communism would happen this fast or on this timetable. But in their first meeting, the Holy Father and the

President committed themselves and the institutions of the church and America to such a goal. And from that day, the focus was to bring it about in Poland."

THE PUBLIC FACE

While the two figures clearly shared a bond over their recent experiences and distaste for Communism, there was very little that they were able to share openly. In his subsequent remarks to the press, Reagan said that he left the encounter with a feeling of hope and dedication, knowing that a world which produced such "courage and vision" from a man like Karol Wojtyla, a survivor of adversity, was capable of "building a better future." Telling the gaggle of frenetic media that he felt a dedication "to do all in one's power to live up" to the faith and values of the "free West," he then pointed to general "certain common experiences" of him and the pope which "gave our meeting a special meaning to me."[11]

In his press remarks, Reagan also stated that God had blessed America with a freedom and abundance that had been denied to less fortunate brethren. Since the end of World War II, he said, America did its best to provide those less fortunate with billions of dollars in food, medicine, and materials. "And we'll continue to do so in the years ahead." Shifting to contemporary Poland specifically, he pledged, "While denying financial assistance to the repressive Polish regime, America will continue to provide the Polish people with as much food and commodity support as possible through church and private organizations." Like the Church in its spiritual role, America would "seek to pursue the same goals of peace, freedom, and humanity along political and economic lines." Applying his Christian faith to the fight against Communism, Reagan cited a Scriptural rationale for U.S. aid: "Americans have always believed that in the words of Scripture, 'Unto whomsoever much is given, of him shall much be required.' "[12]

He concluded by asking the pope for his prayers "that God will guide us in our efforts for peace on this journey and in the years ahead." For his part, the pope closed: "With faith in God and belief in universal human solidarity may America step forward in this crucial moment in history to consolidate her rightful place at the service of world peace."[13] It was a bargain that Reagan was happy to try to meet.

"HOLY ALLIANCE"

Although much of the correspondence within this historic partnership remains classified, with numerous documents fully redacted or not released at all, leaving a multitude of unanswered questions, this much is certain: The meeting launched a deliberate and coordinated effort on behalf of both the White House and Vatican to end Communism. The major players included Clark, Casey, Pipes, Ambassador Vernon Walters, Pio Cardinal Laghi, and Agostino Cardinal Casaroli. Clark's deputy at the NSC, Robert McFarlane, says that almost everything that had to do with Poland was handled outside of normal State Department channels and went through Casey and Clark. He adds, "I knew that they [Casey and Clark] were meeting with Pio Laghi [the apostolic delegate to Washington], and that Pio Laghi had been to see the president, but Clark would never tell me what the substance of the discussions was." Clark and Laghi met regularly to discuss developments in the Polish situation. Crucial decisions on funneling aid to Solidarity and responding to the Polish and Soviet regimes were made by Reagan, Casey, and Clark, in consultation with Vatican officials.[14]

Working in close proximity to each other in Washington, a close relationship developed between Casey, Clark, and Laghi. "Casey and I dropped into his [Laghi's] residence early mornings during critical times to gather his comments and counsel," says Clark. "We'd have breakfast and coffee and discuss what was being done in Poland. I'd speak to him frequently on the phone, and he would be in touch with the pope." On at least six different occasions, Laghi came to the White House and met with Clark and Reagan. Each time, he discreetly entered the White House through the southwest gate to avoid the notice of the press.[15]

Former Carter National Security Adviser Zbigniew Brzezinski, a strong anti-Communist and native of Poland, talks of Casey's role in the effort: "Casey ran it . . . and led it; he was very flexible and very imaginative and not very bureaucratic. If something needed to be done it was done."[16] Casey shared Reagan's sense that they faced a grand juncture, and that the pope could help enormously. In his biography of John Paul II, which he coauthored with Marco Politi, Bernstein wrote:

> Intuitively, they [Casey and Reagan] understood what the pontiff might accomplish and how his acts might push forward their own global policies. . . .
> The highest priority of American foreign policy was now Poland, he [Casey] informed the Pope. In Washington, Reagan and Casey had discussed the possibility of "breaking Poland out of the Soviet orbit," with help from the Holy Father.[17]

Intelligence, namely information-sharing, played a central role in the Reagan-Pope collaboration. The Reagan administration fueled an intelligence shuttle between Washington and the Vatican, through which Casey and Walters clandestinely briefed the pope on a regular basis.[18] Between them, they paid fifteen secret visits to John Paul II over a six-year period. Walters visited at roughly six-month intervals.[19]

Both Reagan and the pope eagerly anticipated the information gained from these briefings. The pope benefited from the mighty, long arm of U.S. technical intelligence, receiving some of the nation's most guarded secrets and sophisticated analysis. He was able to pour over satellite imagery that was detailed beyond his conception.[20]

Vatican representatives and the pope were consulted on U.S. thinking in world affairs.[21] This consultation, and subsequent influence, swung both ways: To cite just one later example, in February 1984, Vice President Bush held a fifty-five-minute meeting with John Paul II at the Vatican. He briefed the Holy Father on Lebanon, on his meeting with Soviet General Secretary Konstantin Chernenko the previous day, and more. Following those items, reported Bush in a cable sent to the White House Situation Room, declassified in July 2000, "I then asked him if he had any advice for us on Poland."[22] Bush said that John Paul II reacted by discussing the Poland situation "for some time." "The Holy Father said people are getting hurt," said Bush, referring to the punitive effect of U.S. sanctions. The pope told him, "This must be changed." The pope said that both he and the Holy See agreed there should be a change in sanctions policy.[23]

The pope also told Bush of how Poles had suffered under totalitarianism. Even if Polish Communist leader Jaruzelski wanted to improve the situation, he was limited in what he could do. "He is limited by the neighborhood—namely GDR [East Germany], Czechoslovakia, etc.," Bush translated. That meant the Reagan administration and Vatican needed to assist. Relaying the pope's words, Bush continued: "We must do something to help the [Polish] people. So many times in the past they have defended themselves against oppression."[24]

As evidence of how Bush's report influenced Reagan, the president wrote a February 22 follow-up letter to John Paul II, also declassified in July 2000, which began:

Vice President Bush has informed me of his recent meeting with you. I want you to know that I deeply appreciate your counsel and that, following upon it, I have decided to take the following steps concerning Poland. Our Charge in Warsaw will inform General Jaruzelski, through a most confidential envoy, that we are prepared to lift the ban on regularly scheduled LOT flights to the United States, permit the resumption of travel under the Marie Sklodowska-Curie Travel Fund, and begin an official—but highly confidential—dialogue. This dialogue would include all aspects of our relations and set forth what we would expect from the Polish side in return for positive actions on our part.[25]

Clearly, the briefings, advice, and influence between Washington and the Holy See swung both ways. In the letter, Reagan linked his sanctions modifications to a positive Polish response;[26] he also listed his own specific human-rights concerns:

These steps would, however, depend upon Warsaw's willingness to release the eleven KOR and Solidarity activists without any onerous conditions or harassment. I am also conveying to the General my strong interest in the situation of Lech Walesa and his family and my concern that they not be subjected to officially-inspired harassment. . . . I sincerely hope General Jaruzelski will respond positively to our approach and agree to take steps which will lead to the reconciliation in Poland for which we all hope and pray.[27]

Finally, there were several meetings, not to mention an obvious special bond, between not just the principals but the two men at the top. Ronald Reagan and Pope John Paul II met together probably at least seven times.[28] Some of the most telling moments from those meetings can be encapsulated into a single enduring image that Nancy Reagan describes as one of her favorites, captured by a photographer: "The pope is sitting with his head bent, listening, and Ronnie is half way out of his chair and talking to the pope, and his hand is out and his finger's out. Obviously he's telling him something. And you wonder, what in the world is Ronnie saying? The pope is listening very carefully to him."

It was a common image derived from two men who respected one another and engaged a world they yearned to transform. John Paul II told Nancy that there was a psychological and emotional tie between the two that he never had with another president.[29] And it all started with that meeting on June 7, 1982.

In the years following the end of the Reagan administration, much of the discussion regarding the pope centered on the nature of this relationship between the two prominent men. The defining characteristics of the collaboration became a subject of dispute after Carl Bernstein's article, "The Holy Alliance," ran in 1992. Some on both sides of the Atlantic resented the implication that there was collusion unifying the two sides. One high-level Reagan foreign-policy official was so contemptuous of Bernstein's characterization that he unfairly maligned the former Watergate reporter as a "slimeball" and "scumbag."[30] This official apparently feared that such a term did a disservice by trivializing what happened, perhaps making it less believable to serious observers. He insisted there was no Holy Alliance, nor "conspiracy," but mainly "shared interests"—"two groups going down the same path." Likewise, one historian of this episode, a sympathetic one to both the pope and Reagan, personally told me that Bernstein's claim of a conspiracy of two is "horseshit."[31]

Many others also rejected the notion that there was a conspiracy between the two sides. Richard V. Allen was particularly hostile to the term, saying that when he once referred to the partnership as an alliance, he meant it as a metaphor not as a statement of fact.[32] Similarly, John Paul II biographer George Weigel calls Bernstein a "journalistic fantasist," and states categorically that there was "neither alliance nor conspiracy," though there was "a common purpose born of a set of shared convictions."[33] The gentlemanly Bill Clark also rejects the notion of a Holy Alliance or conspiracy: "[T]he idea that this was some sort of 'Holy Conspiracy' is overreaching a bit. There was no plot or plan between the two sides. . . . We knew we were both going in the same direction and so we decided to collaborate, particularly on intelligence issues regarding the Eastern Bloc." He adds: "There was a natural convergence of interests, which led officials at the White House to work together with their counterparts at the Vatican." "Primarily," Clark continued, "that cooperation involved the sharing of intelligence information. But no, there was not a formal alliance as such."[34]

In the end, Clark dubbed the mutual effort a "successful collaboration" which took place "under Ronald Reagan's direction." This mutual effort was encouraged to flourish during Clark's tenure at the NSC, bringing the White House and Vatican closer together than at any other point.

Though some clearly do not like the term, it seems that this mutual effort was a "conspiracy" of sorts, especially when clandestine priests were on the ground and mutual aid was secretly provided by the two sides. It certainly seems fair to characterize it as an alliance, and even a "holy" one, being that it was one that the spirit-filled Reagan and spiritual father John Paul II pushed for the purpose of undermining Communism.

Regardless of the specific language used to describe it, the end result was that their meeting on June 7, 1982 forged an indelible bond, a sacred pledge to share information that would be mutually beneficial to both sides. And despite the clear obstacles, both men persevered in the hope that they might one day live to see the walls of Communism come tumbling down. It was a unique connection between two unique individuals, the like of which have rarely been seen in modern politics.

JUNE 8, 1982: WESTMINSTER

The day after his historic meeting with the pope, Reagan left the Vatican reinvigorated with a spiritual zeal to undermine Communism. Filled with a sense of grander purpose, he flew to London, where on June 8 at Westminster he gave the most prescient speech of his presidency.

While Tony Dolan was the speechwriter for the Westminster address, Reagan's hand—literally his pen—was apparent in every line as Reagan played a key role in the speech's language, ensuring that it embodied all that he preached. An early draft of the speech, probably the first, is dated May 19, 1982, and is on file at the Reagan Library. That draft, written by Dolan, was overhauled by Reagan, with the president removing numerous words, lines, and paragraphs and adding so much text that he could have received a coauthor credit.[35] The May 19 draft was twenty-four pages long, with twenty-seven entire paragraphs that were removed by Reagan in addition to dozens of sentences and hundreds of words the president slashed.[36]

Throughout the draft process, Reagan played this editing role, adding significant, well-known lines to the text. Few sections of the Westminster address were more memorable than the one Reagan penciled in to page fourteen of the May 24 draft: "What I am describing now is a policy and a hope for the long term—the march of freedom and democracy which will leave Marxism-Leninism on the ash heap of history as it has left other totalitarian ideologies which stifle the freedom and muzzle the self-expression of its citizens."[37] Very tellingly, Reagan opted for the word "policy" in addition to "hope."[38]

The speech offered more: The president called upon Western allies to encourage democratic developments in Eastern Europe by assisting inchoate democratic institutions behind the Iron Curtain. He said he wanted Eastern Europeans "to choose their own way to develop their own culture, to reconcile their own differences through peaceful means." This effort, said Reagan, would constitute a "crusade for freedom."

In the 1950s, he had signed onto the Crusade for Freedom; now, he was resurrecting and spearheading it. "[L]et us move toward a world in which all people are at last free to determine their own destiny," he urged. Forecasting a simultaneous splurge in democracy, he assured people that "around the world today, the democratic revolution is gathering new strength." He went on to say that "in the Caribbean and Central America, sixteen of twenty-four countries have freely elected governments. And in the United Nations, eight of ten developing nations which have joined the body in the past five years are democracies." He offered: "This is precisely our mission today: to preserve freedom as well as peace. It may not be easy to see, but I believe we live now at the turning point."

As he had at CPAC six months earlier, Reagan spoke of a crossroads—a "turning point." He then turned his attention to what he saw as the source of darkness:

> In an ironic sense, Karl Marx was right. We are witnessing today a great revolutionary crisis. . . . But the crisis is happening not in the free, non-Marxist West, but in the home of Marxism-Leninism, the Soviet Union.
>
> It is the Soviet Union that runs against the tide of history by denying freedom and human dignity to its citizens. It also is in deep economic difficulty. . . . The dimensions of this failure are astounding: a country which employs one-fifth of its population in agriculture is unable to feed its own people.

Calling on more evidence of the economic crisis, Reagan borrowed material he had used in his radio addresses from the 1970s:

> Were it not for the tiny private sector tolerated in Soviet agriculture, the country might be on the brink of famine. These private plots occupy a bare three percent of the arable land but account for nearly one-quarter of Soviet farm output and nearly one-third of meat products and vegetables.
>
> Overcentralized, with little or no incentives, year after year, the Soviet system pours its best resource into the making of instruments of destruction.

The constant shrinkage of economic growth combined with the growth of military production is putting a heavy strain on the Soviet people.

What we see here is a political structure that no longer corresponds to its economic base, a society where productive forces are hampered by political ones.

The decay of the Soviet experiment should come as no surprise to us. Wherever the comparisons have been made between free and closed societies—West Germany and East Germany, Austria and Czechoslovakia, Malaysia and Vietnam—it is the democratic countries that are prosperous and responsive to the needs of their people.

Here again was Reagan's foreknowledge of the calamity coming to the USSR, served up in 1982 when others were claiming the USSR was fine. He perceived economic failures that were in fact abounding and getting worse, and he was hoping to exacerbate the problems through his administration's burgeoning economic-warfare efforts.

Next, he shared his mutual goals of spreading democracy and reversing Communism:

> We must not hesitate to declare our ultimate objectives and to take concrete actions to move toward them. We must be staunch in our conviction that freedom is not the sole prerogative of a lucky few but the inalienable and universal right of all human beings. The objective I propose is quite simple to state: To foster the infrastructure of democracy—the system of a free press, unions, political parties, universities—which allows a people to choose their own way, to develop their own culture, to reconcile their own differences through peaceful means. . . .
>
> It is time that we committed ourselves as a nation—in both the public and private sectors—to assisting democratic development.
>
> I do not wish to sound overly optimistic, yet the Soviet Union is not immune from the reality of what is going on in the world. It has happened in the past: a small ruling elite either mistakenly attempts to ease domestic unrest through greater repression and foreign adventure, or it chooses a wiser course—it begins to allow its people a voice in their own destiny.

Then came the talk of that "plan" that Reagan had penciled in, that "hope"; it was followed by a Reagan prediction and a challenge:

What I am describing now is a plan and a hope for the long term—the march of freedom and democracy which will leave Marxism-Leninism on the ash heap of history as it has left other tyrannies which stifle the freedom and muzzle the self-expression of the people. . . .

For the ultimate determinant in the struggle now going on for the world will not be bombs and rockets, but a test of wills and ideas—a trial of spiritual resolve: the values we hold, the beliefs we cherish, the ideals to which we are dedicated.

I have often wondered about the shyness of some of us in the West about standing for these ideals that have done so much to ease the plight of man and the hardships of our imperfect world. . . . Let us be shy no longer—let us go to our strength. Let us offer hope. Let us tell the world that a new age is not only possible but probable. . . . What kind of people do we think we are? And let us answer: Free people, worthy of freedom and determined not only to remain so, but to help others gain their freedom as well. . . .

Let us now begin a major effort to secure the best—a crusade for freedom that will engage the faith and fortitude of the next generation. . . . [L]et us move toward a world in which all people are at last free to determine their own destiny.

Lou Cannon, who spent his career in newsrooms, rightly notes that the Western press derided the Westminster Address as "wishful thinking, bordering on delusional."[39] In London, Andrew Alexander, a *Daily Mail* columnist, protested: "To be invited to defend ourselves against Communism is one thing. To be asked to join a crusade for the overthrow of Communism is quite another."[40]

There was no doubt about how the Soviets interpreted Reagan's words. Official spokesmen described the address as a declaration to destroy the USSR.[41] Writing in *Sovetskaya Rossiya*, S. Volovets said the speech showed that Reagan had "declared an ideological 'crusade' against atheistic communism."[42] Volovets parroted the line of the Moscow Domestic Service, which said the "notorious speech" threatened to impose upon the rest of the world the American way of life—that is, it sought to "impose" freedom.[43]

The pages of *Pravda* were likewise certain about what Reagan had in mind. Vitaliy Korionov warned that Reagan and his merry band of "latter-day crusaders" had malicious intentions toward the Soviet bloc: "The new 'crusade' is essentially a whole system of methods geared to undermining the socialist community."[44] He was right, and had merely mimicked the just-released CPSU Central Committee report to the twenty-sixth Party Congress, which explained the U.S. effort this way: "It employs an entire system of means geared to undermining the socialist world and causing it to disintegrate."[45]

Pravda did not like this American bravado, and responded with some swagger of its own. Bolshakov issued a warning to the Reagan team: "[I]f the U.S. Administration supposes that by means of force and ideological sabotage it will succeed in 'changing history,' let it remember that its numerous predecessors in the sphere of organizing 'crusades' against communism finished up in the very place where Washington resolved to dispatch Marxism-Leninism—the garbage heap of history."

Indeed even before Reagan made the speech there were rumblings about his new strategy in the Communist press. The day of the Westminster Address, Poland's Domestic Service released an appraisal of Reagan's overall strategy, in which it judged that Reagan had made an unequivocal choice "to go on the offensive," as part of a "global campaign for democracy," which included the development of "appropriate institutions for the realization of this task."[46] Warsaw's Communists saw this as a bad thing. Little did they know that in their zeal to frighten Poles with Reagan's words, they had uplifted them.

Each word in this analysis by Poland's Domestic Service was right on the money, as was a July 1982 editorial in *Pravda* that said that Reagan specifically had issued "an open call to 'undermine' socialism in the USSR and the other socialist countries."[47]

In short, Reagan subversion was becoming the theme in the Communist press for 1982. It was that year, summed up *Pravda*, that Reagan publicly launched his crusade against the USSR.[48]

Despite the criticisms that would be lobbed at him in both the Western and Communist presses, Reagan remained resolute. There had been nothing accidental in the Westminster Address, but instead from start to finish it was a deliberate attempt to send an unmistakable message to the Soviets and to the world. It was a message that he had been advancing for four decades, and it was a speech that would be the embodiment of that image for years to come. That moment at Westminster saw Reagan uncover the full extent of his rhetorical arsenal, as he went headfirst in announcing his intentions to attempt to change history.

Though the Soviet press found itself up in arms over Reagan's confrontational language, words were merely a precursor of events to come. Emboldened by the speech and his meeting with the pope, Reagan returned home deter-mined to take the next step in his crusade. The first half of 1982 was just the beginning of his assault on the USSR's Communist infrastructure during that year. Over the course of the next six months, the economic campaign against the USSR would shift to high gear, as the Crusader's war against Communism would move from speeches to action.

11.

Plans and the Pipeline: Mid-June to December 1982

THROUGHOUT THE REMAINDER OF 1982, THE REAGAN ADMIN-istration continued to focus its economic attack, sensing ripe opportunities. Bill Clark noticed that the Soviet leadership also began sensing something, namely that the USSR's slippery slope had begun, and, most ominously, at a time when an economic affront from Washington was going full throttle. This was made clear to Clark in a very personal encounter that summer. At a diplomatic function, Soviet Ambassador Anatoly Dobrynin, speaking for KGB head Yuri Andropov and General Secretary Brezhnev,[1] whispered to Clark, "You have declared war on us, economic war." It was not until almost twenty years later that Clark could respond candidly, "Yes, we had."[2]

And Moscow had seen nothing yet. The first half of 1982 was merely the start of Reagan's assault on the Communist infrastructure of the USSR. The second half of 1982 would see the economic attack on the Soviets rise to unprecedented levels.

During the summer of 1982, Reagan redoubled his efforts to block the construction of the Siberian pipeline, an endeavor in which he had the support of his NSC, DOD, and the CIA. The Departments of State and, pre-dictably, Commerce, did not want to obstruct it; likewise, Congress faced considerable pressure from American business, which wanted a piece of the action, as a number of companies saw a chance to make big profits by helping to build the project.

Secretary of State Al Haig feared that U.S. allies could not be brought along as participants in the obstruction, and that any attempt to lean on them would only result in alienation. Moreover, he told Reagan that he had spoken with "experts" on the Soviet Union who maintained that it was "crazy" to think the United States could "bust" the USSR through this kind of economic warfare. Sure, said Haig, the USSR was in trouble, but its system would not be changed by economic means.[3] Instead, the secretary of state suggested a compromise: the administration could acquiesce to the pipeline in exchange for an agreement to consult with U.S. allies on restricting future technology transfers to the USSR. In other words, America and its allies could (at some point) look for future cooperative means to deprive the Soviets of key technologies.

Haig's offer did not sit well with Reagan and those united with him in his effort to undermine. Lou Cannon makes clear that it was Reagan himself who was the primary obstacle. Reagan, reported Cannon, was "intransigent," dedicated to the "quixotic goal" of halting the pipeline, and was "naively convinced that he could talk the allies out of an economic decision they had already made by warning them of the danger of depending on the Soviet Union."[4]

In addition to wanting jobs from the pipeline construction, Western Europe also desired the lower natural gas prices that promised to flow from the USSR. As a result, the White House tried to conceive other natural gas options for Western Europe to entice it away from the pipeline. Even Reagan himself went so far as to lay out publicly other natural gas possibilities, saying in an interview that the White House would work with Western Europe to investigate sources of energy closer to home. He pointed to the North Sea, Norway, and the Netherlands. "We would be happy to help them with the development of those," offered Reagan.[5] Later, he added another alternative, saying that his administration was "taking steps domestically to improve our competitiveness in coal exports to Europe."[6]

While Reagan used many rationales to combat the pipeline, his main argument was that Western Europeans would assume two principal risks as primary customers of Soviet natural gas: they could become dependent upon the Soviet Union for a precious resource and, by extension, could position themselves to be blackmailed by the USSR at some future point.[7] This was no imaginary product of Cold Warrior paranoia. In fact, Western Europe was already highly dependent on Soviet natural-gas supplies, and the successful completion of this project could result in Europe becoming 50 to 70 percent dependent on Soviet supplies within ten years.[8] Reagan repeatedly warned Western Europeans that this dependency would leave them vulnerable to Soviet blackmail. Roger Robinson, the NSC staffer who was one of the chief architects of the economic assault, stated: "We feared the USSR would not hesitate to use the spigot to compromise NATO and Europe politically through this subterranean flow of energy."[9]

Worse, Reagan felt it was tragically selfish that Western Europeans might proceed with a project that helped themselves but prolonged the captivity of their Eastern European neighbors, which would remain satellites to Moscow as long as the Soviet system existed. The pipeline would nourish the Communist economy and allow it to continue; those Westerners would be selling their Eastern brethren down the river.

And yet, they were undeterred. Western Europe was willing to help the Soviet Union—and thereby boost the Soviet economy—so long as Western European firms could reap profits and citizens could spend less on natural gas. "The consensus in Western Europe was that economic appeasement of the USSR should continue," said Roger Robinson, who worked round the clock to help Reagan and Bill Clark block the pipeline. "However, we had a president and a national security adviser who weren't Atlanticists and weren't playing that détente-oriented game anymore."[10]

Reagan applied a favorite parable to this situation. He was fond of the quote in which Lenin had allegedly said that when the time came to hang the capitalists the capitalists would vie with each other to sell the Communists the rope.[11] Reagan vowed to be one capitalist who would not sell the rope. One of his oft-told lessons was a story about a large factory known as the Kama River truck plant, which the Russians reportedly could not have built without U.S. financial and technological assistance. Motors made in the Kama plant ended up in Soviet armored personnel carriers and assault vehicles. This was a Reagan fear, made more loathsome when the U.S. bureaucracy was complicit in the malfeasance, as he believed had been the case with the Carter State and Commerce departments in the Kama deal.[12] Reagan refused to see his administration, State and Commerce included, complicit in a pipeline deal that helped the Soviets at U.S. expense.

Reagan later recounted how he spent 1982 trying to tighten the noose around the pipeline and the Kremlin:

Throughout most of 1982, I tried to persuade our European allies to restrict credit to the Soviets and join us in imposing other sanctions aimed at halting construction of the trans-Siberian natural-gas pipeline. I eventually had a little success. I was unable, however, to persuade them to apply as much economic pressure on the Soviet Union as I thought we should to accelerate the demise of Communism; many of our European allies cared more about their economic relationships . . . than tightening a knot around the Soviets.[13]

At the time, Reagan was getting confirmation during NSC briefings of what he long believed: Soviet Communism was on the precipice, and simply needed the right pokes and prods.[14] He said as much openly in an interview with the editorial board of the *New York Post*. "I think the economic signs are there," he said of the Soviets. "They are in deep trouble, and so are their satellites." Thus, he explained, it was important that America and its Western allies "curtail" credit to the Soviets "because I think they've run out of hard cash and they economically are very vulnerable right at the moment."[15]

This was not the prevailing view, and Reagan was belittled for saying such things. Harvard's Arthur M. Schlesinger, Jr. declared after a 1982 visit to Moscow: "Those in the U.S. who think the Soviet Union is on the verge of economic and social collapse, ready with one small push to go over the brink, are . . . only kidding themselves."[16]

TIME TO "TAKE A STAND"

What really riled Reagan was the willingness of Western Europe to help the Soviets while a crackdown on Solidarity was underway. Lech Walesa, after all, was in prison—a metaphor for the whole of the Communist bloc, a living martyr. As the Poland situation festered, Reagan reviewed his options. He was bucked up by a persuasive, well-timed two-page memo written by Richard Pipes (urged by Norm Bailey) and handed to Reagan by Bill Clark on the morning of May 24, shortly before a definitive NSC meeting.[17] That memo seemed to further convince him to hang tough. In his diary later that evening, Reagan wrote of the NSC meeting that day: "We had a session on sanctions [over Poland], limiting Soviet credit and the Versailles [June economic summit] meeting. There was a lot of talk about . . . our allies. I firmly said to hell with it. It's time to tell them this is our chance to bring the Soviets into the real world and for them to take a stand with us, shut off credit, etc."[18]

The French, in particular, would hear none of this. On June 15, the *Washington Post* carried an exclusive front-page interview with the French president. The article, pointedly titled "France Refuses to Wage Economic War on Soviets," began: "France will reject efforts by the Reagan administration to enlist Western Europe in a campaign of economic warfare against the Soviet Union, President Francois Mitterand has declared." The *Post* reported that Mitterand was prepared to cooperate with the United States in "defensive measures" against the Soviets, such as, for instance, "to contain their ambitions." However, he "firmly" rejected an offensive strategy based on trade and financial restrictions intended to "undermine" Soviet strength. "We are not going to wage any kind of war on the Russians," declared Mitterand. "You have to be serious about such a course. It could lead to a real war. If economic embargo is a first act of war, it risks being caught up by a second. No, it is not the right move."[19]

The front-page article was a coup for the Kremlin, for Western Europe, and for the U.S. State Department. Perhaps now, thought State, President Reagan would drop his stubborn crusade against the Soviet pipeline. The article was met with anxiety at Bill Clark's NSC; the pressure on Reagan to reverse his course was mounting.

All of this came to a head on June 18, 1982. At a meeting that day, the bulk of Reagan's Cabinet pressed him to abandon his position on the pipeline, and quickly it became clear that this meeting would prove to be the decisive moment in this debate. Most of the Cabinet, including the secretaries of Commerce, Treasury, Defense, and State were represented, as was the CIA, USTR, OMB, plus members of what Ed Meese called "The Establishment," that is, the soft-liners who held out hope that the president would think reasonably and agree with them.

At the end of a heated discussion, everyone listened in hushed anticipation as Reagan spoke up: "Well, they can have their damned pipeline." He paused and glanced at the enormous sighs of relief exhaling from pragmatists around the table. The Establishment was relieved—but not for long. Just then, the former actor hit the table with his fists and finished: "But not with American equipment and not with American technology!" He stood and left the room.[20]

THE SOVIETS RESPOND

The State Department was quite displeased. Al Haig, for this and many other reasons, offered to resign, and Reagan accepted his resignation. By June 22, Haig was out.

Likewise, the Soviets were not happy with the president's decision. "In general, we have noticed that the current American administration is very successfully blowing up bridges, one after another, that were built over decades," fumed Foreign Minister Andrei Gromyko at a New York press conference. "As soon as Washington notices that some bridge or other is still intact, it at once plants a mine and blows up that bridge."[21]

Complaints from Moscow were omnipresent, and sometimes vicious. The publication, *Za Rubezhom*, took special umbrage; it seemed to carry the banner as the flagship Soviet source for expressing outrage over the pipeline obstruction. Reagan's statements were alternately denounced in *Za Rubezhom* as "absurd, stupid fabrications," just plain "stupid fabrications," or "fantastic lies," or "crude" lies, or "disgraceful" lies, or, as the title of one resourceful article relayed, "Lies as Policy, or the Policy of Lies."[22]

These accounts from *Za Rubezhom* reveal how protests by Western European governments played into Soviet hands. A page-one editorial in *Za Rubezhom* accused Reagan of a sort of U.S. imperialism, treating West European allies like American states:

> With rare unanimity, almost all of the West European countries have risen up against official Washington's crude arbitrariness. Indeed, having decided to prevent at all costs the construction of the Siberia-West Europe gas pipeline in order to damage our country, U.S. President Reagan deemed it possible at the same time to "punish" the non-American firms that are co-operating with the Soviet Union. He announced that the "sanctions" provided for by U.S. domestic legislation would be employed against them, as though the FRG (West Germany), Britain, France, Italy, Japan and other sovereign states were now part of the United States or had to submit to U.S. jurisdiction. . . .
>
> The intention of the most aggressive U.S. circles to impede socialist building in our country is nothing new. Attempts to finish off socialism by means of an economic blockade have been made repeatedly in the past. . . .
>
> Washington's intention to delay the commissioning of the gas pipeline, if only for a short time, will not be realized. The American President's arrogant statements have had the opposite effect in the Soviet Union to what he was expecting. Instead of the pace of construction work slowing, it has quickened.
>
> Finding himself in a mess with his venture against the gas pipeline and having reduced relations between the United States and its allies to their

lowest point in the entire postwar period, Reagan has begun feverishly patching up the holes he himself has created in the capitalist community.[23]

Also unhappy with Reagan's decision were America's Western European allies, who quickly sought to circumvent U.S. sanctions on Moscow. On June 30, 1982 a reporter told Reagan that even the government of his good friend Margaret Thatcher, to whom he had delivered his Westminster Address only twelve days earlier, had just taken steps to enable British companies to get around the U.S. embargo on pipeline equipment to the Soviet Union. Reagan's reply at a news conference reflected his wider thinking:

> This is simply a matter of principle. We proposed that embargo back at the time when the trouble began in Poland—and we believe firmly that the Soviet Union is the supporter of the trouble in Poland. . . . [T]hese sanctions were imposed until [the United States sees a relaxation of] the oppression . . . of the people of Poland. . . . Now, if that is done, we'll lift those sanctions. But I don't see any way that, in principle, we could back away from that.
>
>   I understand that it's a hardship [for Britain and Western Europe]. We tried to persuade our allies not to go forward with the pipeline for two reasons. One, we think there is a risk that if they become industrially dependent on the Soviet Union for energy—and all the valves are on the Soviet side of the border—that the Soviet Union can engage in a kind of blackmail when that happens. The second thing is, the Soviet Union is very hard pressed financially and economically today. They have put their people literally on a starvation diet with regard to consumer items while they poured all their resources into the most massive military buildup the world has ever seen. And that buildup is obviously aimed at the nations in the alliance. And they, the Soviet Union, now hard-pressed for cash because of its own actions, can receive anywhere from ten to twelve billion dollars a year in hard cash payments in return for the energy when the pipeline is completed— which I assume, if they continue the present policies, would be used to arm further against the rest of us and against our allies and thus force more cost for armaments for the rest of the world.[24]

A month later, Reagan reiterated this thinking with a bit of an edge in a July 22 interview with KMOX-TV in St. Louis, saying of the Soviets: "They're up against the wall. They don't have cash . . . the way they did." The pipeline, he repeated, would give Moscow 10 to 12 billion dollars per year in "cold, hard cash," aided and abetted by "cash customers" in Western Europe.[25]

Reagan sounded this message wherever he could, campaigning for his cause like a wartime president.[26] Even though his decision had been made, he still faced an uphill PR battle. Through the remaining summer and into the fall of 1982, he traveled to American towns that could have gained badly needed jobs from the pipeline construction. Despite encountering crowds that were displeased with his policies, Reagan stood his ground, explaining his rationale and ensuring that those Americans most directly affected by his policy understood the stakes of his decision.[27]

In addition to these towns that could have benefited, large companies like Caterpillar and GE were slated to receive lucrative contracts from the construction of the pipeline, inciting a massive domestic lobbying effort to overturn Reagan's decision and leading to congressional legislation which was poised to do precisely that. This was an inherent difficulty of Reagan's hard stance and one that he understood from the outset. At the meeting where he made his decision to block the pipeline, he had expressed regret: "You know, this is a tough one. I know what's right in the foreign policy aspect of it, but I hate like hell to hurt American business." But, he insisted, "A matter of great principle is at stake."[28]

Later, Dwayne Andreas, chairman of Archer-Daniels-Midland and director of the U.S.–USSR Trade and Economic Council, was in Sokolniki Park in Moscow when he was nabbed by an *Izvestia* reporter, to whom he complained that sanctions and other policies against the USSR were generating major losses for U.S. firms. He told *Izvestia*: "They can hardly be calculated. I can say with great assurance that the loss of deals having to do with the construction of the gas pipeline totaled $2 billion." The losses, he worried, "will continue in the future as well, since the U.S. is losing its monopoly position in certain areas of specialized machinery." American companies were being supplanted by competitors in Western Europe and Japan, but Reagan thought it was worth the loss.[29] Although he was adamantly probusiness, his hatred of the Soviets took precedence in this matter. It was a wartime mentality that the home front had to make sacrifices in order to ensure victory, and in the end, Reagan was convinced that the long-term benefits would outweigh the short-term costs.

THE FRENCH FACTOR

While the British were not pleased with the pipeline embargo, the French continued to provide the most difficult opposition. French resistance became so vocal and so fierce that on October 27 Bill Clark made a secret trip to Paris to talk with President Mitterand.[30] The French had been publicly disrespectful of Reagan, calling him the "cowboy from Hollywood" and other names. Mitterand and his ministers openly sniped at Reagan in the press, and Reagan, who disliked nobody, was developing a healthy disdain for the French leadership.[31] For the sake of Franco-American relations, he thought Mitterand should learn to bite his tongue. Clark remembers: "The French wouldn't stop. He [Reagan] leaned over to me and told me, 'Bill, talk to them.'" Clark was dispatched "back channel" to Paris to discuss the French attitude toward Reagan, and to the USSR.[32]

Mitterand lectured Clark: By pushing allies to block the pipeline, America was guilty of "hegemony" and violating sovereignty. (He was unwittingly iterating the Moscow line.[33]) Believing Reagan's prognostications of Communist demise to be greatly exaggerated, Mitterand assured Clark that nothing would come of the democratic movement in Poland; Polish society would not become more "liberal."[34] All of the allies, insisted the French president, must tone down their rhetoric; but only Reagan must yield. Clark told Mitterand that neither Reagan's personality nor principles would permit that.[35]

Over the next few weeks, testy rebuffs transpired, including a November 12 phone call from an amiable Reagan that an angry Mitterand refused. That day, reported *Time*'s Laurence Barrett, might have been "the most uncivil day" in Franco-American relations since FDR and Charles de Gaulle sparred in World War II.[36]

The next day, in a November 13 radio address, Reagan might have had Mitterand in mind when he proffered:

The balance between the United States and the Soviet Union cannot be measured in weapons and bombers alone. To a large degree, the strength of each nation is also based on economic strength. Unfortunately, the West's economic relations with the U.S.S.R. have not always served the national security goals of the alliance. The Soviet Union faces serious economic problems. But we—and I mean all of the nations of the free world—have helped the Soviets avoid some hard economic choices by providing preferential terms of trade, by allowing them to acquire militarily relevant technology, and by providing them a market for their energy resources, even though this creates an excessive dependence on them. By giving such preferential treatment, we've added to our own problems—creating a situation where we have to spend more money on our defense to keep up with Soviet capabilities which we helped create.

Since taking office, I have emphasized to our allies the importance of our economic, as well as our political, relationship with the Soviet Union.[37]

Disgusted by Western Europe's leftist leaders, it had become clear that Reagan had few friends in this endeavor.[38] It was a sign of his dedication and his unwavering belief in his instincts that he stayed the course.

FAREWELL PAYS OFF

In the end, Reagan did manage to achieve a major blow against the pipeline—literally—although it did not come about through diplomacy. Instead, deliverance came in the form of Gus Weiss's good works with the Farewell Dossier.

By mid-1982, shipments of defective products—manufactured through Line X, born of the dossier—were arriving in the USSR. The products included contrived computer chips that found their way into Soviet military hardware, flawed turbines, defective plans for chemical plants, and much more. Weiss claimed that the Soviet Space Shuttle was actually built from a design rejected by NASA and fed through Line X.[39] In one dramatic example, rigged software bought by the Soviets triggered a huge explosion in the new Siberian gas pipeline. The software was specially designed to pass Soviet quality-acceptance tests, to work for a while, and then to malfunction. Specifically, the software ran the pumps, turbines, and valves in the pipeline but was programmed to eventually produce pressures beyond the capacity of the pipeline's joints and welds.

All this came to a head during the summer of 1982, when a giant explosion occurred in the pipeline. According to Tom Reed, it was "the most monumental non-nuclear explosion and fire ever seen from space," and U.S. satellites picked up the explosion, which was so enormous that NORAD feared that a small nuclear device had been detonated. As Reed recalled, initially there was a crisis-like reaction at the NSC, until Gus Weiss came strolling down the hall to tell his colleagues not to worry. "It took him [Weiss] another twenty years to tell me why," said Reed later.[40]

Although there were no human casualties from the explosion, the economic fallout was considerable.[41] Ronald Reagan, with Gus Weiss's unique help, had found his own device for obstructing the pipeline, regardless of Western Europe's veto—and, perhaps most ironically, with French help, since it was the French who had handed over the dossier in the first place. In all, the Farewell Dossier project was, as Weiss estimated, a "great success." The Soviets' economic espionage backfired when they needed it most.

As for Farewell himself—that is, Colonel Vetrov—his activities were eventually discovered by the KGB and he was executed. It was Weiss's hope that when historians sort out the reasons for the end of the Cold War, Colonel Vetrov would receive a well-deserved footnote.[42]

THE ECONOMIC WAR CONTINUES

Amid all the controversy surrounding the pipeline embargo in 1982, Reagan continued to make prognostications on the demise of Communism,[43] as well as a number of moves to try to advance freedom. He launched "Project Democracy," a program designed to place freedom on the offensive, particularly in Eastern Europe, and convened a cabinet-level meeting to discuss how to carry out the project overtly and covertly. Among other things, this led to the creation of the National Endowment for Democracy.[44]

Meanwhile Reagan's team—once again, led by Clark's NSC—continued to put together a string of NSDDs committed to undermining the Soviet empire. A series of critical directives surfaced beginning in late June, expanding the economic war to different fronts. Eventually these would prove to be instrumental in the implementation of his antipipeline agenda.

Approved by Reagan on June 22, 1982, NSDD-41 expanded the sanctions on oil and gas exports to the USSR that had been imposed the previous December after martial law was declared in Poland. With the expansion in NSDD-41, the sanctions would now include "equipment produced by subsidiaries of U.S. companies abroad as well as equipment produced abroad under license issued by U.S. companies." This language was a direct reference to the pipeline, stating a desire to advance the Reagan administration's objective of "reconciliation in Poland." Predictably, NSDD-41 would face major obstacles from Western Europe, creating, in the words of the new secretary of state, George Shultz, a "monstrous problem" in cross-Atlantic relations, which would hamstring the directive's efficacy from the outset.[45]

In the same spirit but with greater success, on July 23, Reagan authorized NSDD-48. Symbolic of the economic warfare campaign, it was written in vague language and carried out silently. Inconspicuously titled, "International Economic Policy," NSDD-48 did one simple thing: it created the Senior Interdepartmental Group–International Economic Policy (SIG-IEP). The purpose of the group, said the single-page NSDD-48, was, "To advise and assist the NSC in exercising its authority and discharging its responsibilities for international economic policy as it relates to our foreign policy."[46]

This sounded innocent enough, like it could have been referring to exporting fluffy pillows to New Zealand, but in reality, it was far from benign. As scholar Christopher Simpson stated, the directive related "to the administration's de facto economic warfare against the USSR." Established by Bill Clark under Reagan's direction, SIG-IEP created a formal NSC mechanism for linking economic and foreign policy.[47] The chief author of NSDD-48 was Roger Robinson, who rightly describes the directive as an "essential but unrecognized" tool in the Reagan arsenal.[48]

According to Cold War historian Derek Leebaert, SIG-IEP was monumental in that it provided the only occasion during the Cold War in which the heads of the CIA, NSC, and DOD belonged to a "top policy-formulating organization" under a mandate of integrating economic and financial affairs with national security. Though the cabinet-level group was chaired (in name) by the treasury secretary and vice chaired by the secretary of state, the executive secretary tended to run the body. The first executive secretary was Norman Bailey, followed by Roger Robinson.[49] The body reported through Bill Clark to the president, enabling Clark to personally ensure that all government agencies executed their policies in a way that was consistent with Reagan's priorities.[50] This meant that SIG-IEP ensured that national security would trump commercial interests and profits, when it came to economic policy related to the USSR and Soviet bloc.[51]

As executive secretary of SIG-IEP, it was now clearer than ever to Norm Bailey that things were changing quickly, and Reagan was playing to win. Bailey stated: "The fact is that the first Reagan administration adopted, designed, and successfully implemented an integrated set of policies, strategies, and tactics specifically directed toward the eventual destruction (without war) of the Soviet empire and the successful ending of the Cold War with victory for the West."[52] This was not just an understanding by the man who ran SIG-IEP but also a mandate. Under Reagan, stated Bailey, "the policy of containment was changed from maintenance of the status quo to the goal of eventually ending the war through victory. Thus the strategic posture changed from defense to rollback." "In other words," continued the SIG-IEP executive secretary, "the Reagan administration designed and carried out a radical paradigm shift in the way the United States pursued the Cold War, from the war aims themselves to the strategies and tactics used to implement the new objectives."[53]

In SIG-IEP, with Bailey and Robinson and Clark, the Reagan team now had an actual vehicle for identifying and marshalling economic means of destruction. SIG-IEP could carry the water, making it especially difficult for other departments and agencies, like State, to set up roadblocks to try to stop Reagan's economic war.

GOING AFTER EASTERN EUROPE

Two months later, in September 1982, Ronald Reagan pondered: "How long can the Russians keep on being so belligerent and spending so much on arms when they can't even feed their own people?"[54] With each passing month, it seemed that he came to realize the acuity of the Soviet crisis more keenly, exciting him more than ever. This excitement in early fall led to the release of another major directive by Reagan's team. Titled "United States Policy Toward Eastern Europe" and issued September 2, 1982, NSDD-54 remains one of the most classified Reagan NSDDs. Still heavily redacted, there are roughly thirty-plus lines of text that continue to be blackened out of the directive.

Seeking the seemingly unattainable, the second line of NSDD-54 states: "I [Reagan] have determined that the primary long-term U.S. goal in Eastern Europe is to . . . facilitate its eventual reintegration into the European community of nations." Within this quote, the ". . ." unfortunately substitutes for almost a full line of text that remains secret to this day. "The United States," the NSDD insists, "can have an important impact on the region."

While one of the administration's goals for NSDD-54 remains classified, there were five that have been subsequently released:

- "Encouraging more liberal trends in the region."

- "Furthering human and civil rights in East European countries."

- "Reinforcing the pro-Western orientation of their peoples."

- "Lessening their economic and political dependence on the USSR and facilitating their association with the free nations of Western Europe."

- "Encouraging more private market-oriented development of their economies, free trade union activity, etc."[55]

Ambitious by any standard, in 1982 these goals were downright implausible as they sought a shift by the Communist bloc away from the USSR and toward the free-market democracies of the West. Anyone reading these objectives at the time, outside the Reagan administration, would have judged them utter fantasy. Their significance is self-evident, though the directive itself was short on elaboration—not that much else needed to be said.

Finally, NSDD-54 stated that the administration "will employ commercial, financial, exchange, informational, and diplomatic instruments in implementing its policy toward Eastern Europe." The directive listed ten such instruments, five of which—probably the most important—remain partly or fully redacted.[56]

Though much of the directive's contents remain unclear at best, the language clearly set out broad goals for eliciting dramatic change in Eastern Europe. Fittingly, it was at this time that some members of the mainstream media began catching on to this plan for Eastern Europe. On September 23, 1982, the *New York Times* ran an extraordinary article under the self-explanatory headline, "After Détente, the Goal is to Prevail." In this giant wake-up call to its readership, the *Times* noted that from President Truman to President Johnson, the watchword on policy toward the USSR was "containment." From President Nixon through President Carter, it was "détente." Now under Reagan, the watchword was "to prevail." The article quoted Tom Reed: "We believe the free world can prevail." Speaking for Reagan, Reed said that "not since John Kennedy left the world stage have we dared to dream such dreams." At the same time, Reed tried to reassure the typical *Times* reader: "there's nothing wrong with winning."[57]

The *New York Times* piece was quite detailed, going through the precise logistics over what exactly it meant to prevail. As Reagan officials explained, "It means pushing Russian influence back inside the borders of the Soviet Union." The *Times* added: "To prevail . . . means to reverse the geographic expansion of Soviet political influence and military power. It means to loosen Russian controls over Eastern Europe, Cuba, Vietnam, and North Korea." How would this be done?—"with the combined pressure of a military buildup along with diplomatic, economic, and propaganda measures."

Prevailing meant rearming America with increased military spending, impeding the Siberian pipeline, and reducing Soviet access to U.S. technology. The "concept of prevailing," said the *Times*, was also "at the heart" of Reagan's policies in Poland, Afghanistan, China, and elsewhere. According to the *Times*, Reagan officials believed that such pressures meant that "in the long run," victory would be achieved without armed conflict. The article concluded with a prophetic timeline, noting that Reagan officials believed that Soviet economic weaknesses would "intensify toward the end of this decade," as America regained economic and military strength.[58]

All of this was exactly right—a testimony to fine journalism and to the fact that what Reagan officials said after the presidency perfectly matched what they said during the presidency. The *Times* published these observations only three weeks after the release of NSDD-54.

THE ECONOMIC CAMPAIGN FALLS INTO PLACE: NSDD-66

While the *Times* article stirred up some domestic controversy, to the administration everything remained business as usual, and just before Leonid Brezhnev's death in November 1982, Ronald Reagan decided to impose limited trade and credit restrictions on the Kremlin. Reagan said that this meant that "none of us"—an overly idealistic reference to the G-7 economic partners—"would subsidize the Soviet economy or the Soviet military expansion by offering preferential trading terms or easy credits," nor would they "restrict the flow of products and technology that would increase Soviet military capabilities."[59]

This thinking, which was manifest in September's NSDD-54, was also evident in November's NSDD-66—another product of Bill Clark's NSC, where it was written chiefly by Roger Robinson.[60] In the strongest form to date, NSDD-66 formally authorized comprehensive economic warfare against the USSR. Robinson called the directive the "centerpiece" of the economic strategy.[61] In the words of the CIA's senior economic warfare strategist, Henry Rowen, the intent of NSDD-66 was to "cause such stress on the [Soviet] system that it will implode."[62] Championed by Clark, Casey, and Weinberger, the directive found warm reception with the president.

The directive was declassified in 1995, sooner than expected given its aggressive nature and the fact that its circulation was restricted to a small group of individuals.[63] (This also means that only a few Reagan officials knew how all the components fit together.[64]) Reading the document, however, it is not surprising that it was declassified so soon: It is written vaguely enough to seem almost innocuous.

Not at all innocuous were the destructive implications unleashed by the document's subtle language. What Reagan authorized in signing NSDD-66 was nothing less than a full search for means to destroy vital elements of the Soviet economy, authorizing an all-out search-and-destroy mission. Until NSDD-66 was declassified with the harmless title, "East-West Economic

Relations and Poland-Related Sanctions," the working title of the NSDD was the more provocative (and accurate) "Protracted Economic Warfare Against the USSR."[65]

Beginning auspiciously with the statement: "This framework agreement will govern East-West economic relations for the remainder of this decade and beyond," it was clear from the onset of the document that within the far-sighted text of NSDD-66 lay the potential to win the Cold War.[66] The directive sought formal commitment from leading Western European nations "not to sign or approve any new contracts for the purchase of Soviet gas" during a period specified in the NSDD. More specific language in NSDD-66 talked of reaching an agreement to "not commit to any incremental deliveries of Soviet gas beyond the amounts contracted for from the first strand of the Siberian pipeline."

In addition, NSDD-66 set forth three other objectives with Western European allies, the first of which was to gain agreement to add as many critical technologies as possible to the COCOM list. COCOM, the Coordinating Committee for Multilateral Export Controls, was created early in the Cold War to restrict certain Western exports to the Soviet bloc, particularly high-tech products. NSDD-66 looked to harmonize allied coordination in enforcing the COCOM product list and to produce an improved, effective list of results. The second objective of NSDD-66 was to secure an agreement on limits to advanced technology and equipment that went "beyond the expanded COCOM list," with, again, more emphasis on the oil and gas sector. Finally, the directive sought agreement among OECD countries (largely Western nations) to "substantially raise interest rates to the USSR to achieve further restraints on officially backed credits such as higher down payments, shortened maturities and an established framework to monitor this process."[67]

NSDD-66 had two tracks, one concerning action by the Reagan administration and the other involving Western European allies. Unfortunately for the Reagan team, Western Europe resisted much of this ideological-economic warfare. And, in fact, the NSDD acknowledged that the governments of the UK, West Germany, France, and Italy, not to mention Canada and Japan, had stated explicitly, "That it is not their purpose to engage in economic warfare against the USSR."[68] Still, this resistance never diminished NSDD-66's success. Overall, it was quite effective as a license to Reagan staff to identify Soviet economic vulnerabilities, even if the United States had to act alone.

Indeed, the administration was gung ho in its endeavor to act in its own best interest. Another section of NSDD-66, headed "Preparations within the

U.S. Government," identified SIG-IEP as the group "responsible for the attainment of U.S. objectives in the context of the work program." That dry language masked a malicious intent: so-called "interagency working groups" would be established under the supervision of SIG-IEP "to develop U.S. positions and strategies for the achievement of these objectives." In addition, "a working group will be established for an overall study of East-West economic relations in the context of political and strategic considerations." Through SIG-IEP, these working groups would submit to the president (for approval) various "strategies for attaining U.S. objectives."[69]

While the specific language of the directive may have been vague, its implications were not. In essence, the directive created an infrastructure to identify and exploit Soviet economic vulnerabilities—actual groups were established for that purpose. Though it was not as large scale, the directive was somewhat akin to the National Security Act of 1947, which authorized the creation of entire departments and groups like the NSC, CIA, and Joint Chiefs of Staff, to meet the new requirements demanded by the new war— the Cold War. Here, in late 1982, NSDD-66 spawned smaller groups for the new war silently declared by the Reagan team—the U.S. economic war against the USSR. Some of the strategies and efforts identified by these groups were pursued with the support of allies; others were not, and instead were achieved unilaterally by the administration or bilaterally with countries inside or outside Western Europe and the OECD.

Exploring where the most damage could be inflicted, Roger Robinson identified what he called the "strategic trade triad" of three critical Western resources that Moscow relied upon: bank credits, high-tech exports, and hard currency from energy exports. Among them, most vulnerable was the Soviet dependence on energy exports, which the administration was already exploiting and would continue to exploit in a number of ways.

As a result of these dependencies, there were many practical effects of NSDD-66; the Reagan administration proved to be quite successful at reducing Soviet access to Western technology. It pushed for much more vigorous prosecution of technology theft cases in order to disrupt sales of U.S. technology to Moscow; and this newly aggressive stance in fact caused a dramatic spike in the overall number of cases that were prosecuted. During the 1970s, only two or three theft cases were prosecuted by the federal government, but by January 1986, the Reagan Department of Justice was prosecuting more than 100 cases per month.

The administration also worked to limit the number of high-tech goods that could be legally shipped to the Soviet bloc, through the addition of twenty-six items to the U.S. Commodity Control List. Reagan's team also applied substantial pressure on allies and trade partners, including neutral countries like Switzerland and Sweden, in an attempt to limit their role as a transshipment point for high-tech materials headed to Moscow. COCOM, which included all NATO countries (except Iceland) and Japan, also tightened the controls on technology.[70]

The results were substantial: According to the U.S. Department of Commerce, in 1975 roughly 33% of all U.S. goods exported to the USSR were high-tech products, amounting to $219 million in sales. By 1983, the number dropped nearly seven-fold to 5%, dwindling to a paltry $39 million. By the fall of 1983, U.S. Customs agents seized some 1,400 illegal shipments headed to Moscow, valued at roughly $200 million.[71]

While these statistics were staggering, they were merely a sample of what was to come.[72] These were life-sustaining technologies at a fatal time for a Kremlin on the ropes, and depriving Moscow of them was a severe blow. Radomir Bogdanov, a former KGB official serving as deputy director of the USA and Canada Institute, had no doubt about Washington's mischief: The Reagan team was "trying to destroy our economy."[73]

MORE RESTRICTIONS ON TECHNOLOGY

On November 30, 1982, only a day after the release of NSDD-66, came NSDD-70, which restricted the transfer of "nuclear capable missile technology" to certain nations. It also limited "dual-use" goods—products that can have both a peaceful-commercial use and a belligerent-military use. In so doing, the directive restricted a broad category of high-tech exports to the USSR. NSDD-70 allowed the Reagan administration to, "Exempt on a case-by-case basis certain U.S. friends and allies from this policy." Soviet comrades were not among the "U.S. friends." The directive stated:

> All Executive Branch agencies having responsibilities or authorizations for export controls, including missile-related commodities, will adopt stringent export controls on technology and equipment which could make a direct or significant contribution to the design, development, production, inspection, testing or use of nuclear capable missile delivery systems and related components. At a minimum this will include guidance sub-systems and related software, propellants, propulsion systems, rocket nozzles and related

control systems, re-entry sub-systems, missile structure, and unique support equipment.

An expansive rubric of high-tech items fell under this designation, particularly through the use of ambiguous terms like "unique support equipment" and "at a minimum." Of course, the intention was to be expansive. The Reagan team was looking for a vehicle to restrict as many high-tech (and other) products as possible to the USSR—all the better to strangle the USSR. NSDD-70 offered yet another form of justification—on the nuclear side. With it, the administration could prevent even the most basic computer hardware and software from entering the USSR, plus more.[74]

While technology embargoes would prove to be central to the administration's efforts, these NSDDs from the fall of 1982 served a far greater purpose than merely restricting Soviet access to trade. Through their harsh language and unmistakable goals, these NSDDs became a clear articulation of Reagan's anti-Communist methodology for combating the Soviet economy. With the help of Clark and the NSC, the NSDDs became the frontline of Cold War combat, as Reagan demonstrated his unflinching desire to engage the Soviets on any economic front available.

As 1982 wound down, the administration was in overdrive as it sought to carry its economic assault into the new year. The NSDDs of the past year had given new focus to the administration's strategies, as Ronald Reagan saw his words turn into actionable policy. And yet there was no time for Reagan to rest on his laurels. He was sure that he had started to back the Soviets into a corner, and it was thus important for his administration to maintain its momentum as it pressed the foreign agenda forward. The year had been a busy one in the war on Communism, and as the world would soon find out, things were about to get busier.

12.

The Hottest Year in a Cold War: 1983

ON JANUARY 17, 1983, A TASS DISPATCH BY VLADIMIR SEROV fired off a warning concerning Ronald Reagan and his band of fellow "crusaders." Making reference to the previous summer of 1982, Serov reminded readers that Reagan had devised a "strategy" that constituted a "'crusade' against communism"; this was a "'roll-back communism' policy."[1] To Moscow, 1983 was going to be a year in which the Reagan administration would pursue more of the same. More of the same rhetoric. More of the same economic sanctions. More of the same crusade. Little did the Soviets know what was actually in store.

Long after the papers had hit the stands in Russia that January 17, the day was dawning in the United States. January had begun slowly for the administration, which was still reeling from its wealth of activity in 1982. The NSDDs of the previous year were poised to yield tremendous results, all of which would take their toll on the ever-weakening USSR.

Ronald Reagan had big things on his mind that day. The goal was to start the year off with a bang. The goal was to keep momentum strong and morale high. That objective for early 1983 was evident on that winter day, in the form of NSDD-75.

The embodiment of Reagan's thinking on U.S.–USSR coexistence, NSDD-75 was probably the most important document in Cold War strategy by the Reagan administration, and certainly the most significant and sweeping directive in terms of institutionalizing the Reagan intent and grand strategy. Predicated on Ronald Reagan's belief that the Soviet Union was rotten to the core and should be broken, the document was fully committed to pursuing this end rather than maintaining the status quo that accepted Soviet existence.[2]

Norm Bailey would dub NSDD-75, "the strategic plan that won the Cold War."[3] His NSC colleague, Tom Reed, called it "the blueprint for the endgame" and "a confidential declaration of economic and political war."[4] One of the longest NSDDs, the directive covered nine pages, and took quite a while to craft. Its chief author, Richard Pipes, had been working on it since the spring of 1981, first under Richard Allen and then with the backing of Bill Clark and contributions from the likes of Roger Robinson—and against heavy obstruction by the State Department.[5] Pipes called it "a clear break from the past. [NSDD-75] said our goal was no longer to coexist with the Soviet Union but to change the Soviet system. At its root was the belief that we had it in our power to alter the Soviet system through the use of external pressure."[6]

Indeed, NSDD-75 was revolutionary, turning on its head the doctrine of containment that had formed the cornerstone of American foreign policy since George Kennan sent his famous "Long Telegram" from Moscow in February 1946.

As Bill Clark put it, NSDD-75's search for "internal pressure" to bring to bear on the USSR represented a "new objective of U.S. policy." "We worked hard," said Clark, "on that new policy element of trying to turn the Soviet Union inside itself." He notes that, under Reagan, for the first time U.S. policy went beyond containment and negotiations and toward encouraging "antitotalitarian changes within the USSR." America, said Clark, would "seek to weaken Moscow's hold on its empire."[7]

Partly based on previous NSDDs like 32, 45, 54, and 66, NSDD-75 was tamely titled, "U.S. Relations with the USSR." In the first paragraph, it declared that U.S. policy would focus on "external resistance to Soviet imperialism" and "internal pressure on the USSR to weaken the sources of Soviet imperialism." Within that, it stated two core "U.S. tasks:" First, "To contain and over time reverse Soviet expansionism. . . . This will remain the primary focus of U.S. policy toward the USSR." And, second, "To promote, within the narrow limits available to us, the process of change in the Soviet Union toward a more pluralistic political and economic system in which the power of the privileged ruling elite is gradually reduced."[8]

It was this front-page language that reflected Pipes' principle contribution.

He wrote and fought for this language, insisting that the document articulate the central aim of striving to reform the Soviet Union. "The State Department vehemently objected to that," recalled Pipes. "They saw it as meddling in Soviet internal affairs, as dangerous and futile in any event. We persisted and we got that in."[9]

In the end, the inclusion of those lines which were at once impossible but prophetic proved to be the defining language of NSDD-75. And yet those lines, whose prescience is chilling, whose historical significance cannot be overstated, were nearly removed by the State Department, which urged they be struck from the text. In spite of the diplomatic obstacles, the language remained intact, a testament to Reagan who, said Pipes, "insisted" on the language; indeed, this was the core of everything Reagan had always wanted.[10] It was the manifestation of his forty-year crusade and it would become the centerpiece of the flourishing effort to defeat Communism once and for all. It quietly signaled a new era in both presidential power and American foreign policy.

THE OBJECTIVES OF NSDD-75

The document repeatedly expressed the "U.S. objective"—which it alternately called "task" or "goal" or "policy"—of "promoting positive evolutionary change within the Soviet system" (p. 6), of "containing and reversing Soviet expansion and promoting evolutionary change within the Soviet Union itself " (p. 6), of "containing the expansion of Soviet power" (p. 7), and more. NSDD-75 reiterated these goals again three separate times on page eight.

After the introduction, a three-part analysis followed, the first part revealingly titled, "Shaping the Soviet Environment," wherein the administration laid out how it intended to affect that environment. Militarily, NSDD-75 called for a U.S. capability to resist the USSR around the world. Politically, and very much in keeping with Reagan's views and rhetoric, the directive said that U.S. policy "must have an ideological thrust which clearly affirms the superiority of U.S. and Western values . . . over the repressive features of Soviet Communism." The directive advised that these differences be pointed out and broadcast over international airwaves, and suggested increased U.S. government efforts to highlight Soviet human-rights violations.

NSDD-75 emphasized that U.S. economic relations with the USSR must serve strategic goals. "Above all," U.S. objectives must "ensure that East-West economic relations do not facilitate the Soviet military buildup." Here, Pipes said he had in mind past cases of U.S. and Western technology that had aided critical Soviet missile-related industries, such as the sale (by a U.S. company) in the early 1970s of equipment to manufacture miniature ball bearings, which ended up in Soviet missile-guidance systems.[11] This meant a close eye on technology transfer. Hence, NSDD-75 reiterated key economic-warfare aspects of NSDD-66, notably in regard to the gas pipeline, energy exports, bank credits, and the tightening of technologies on the COCOM list.

Reiterating and expanding on many of the previous NSDDs, NSDD-75 proved to be the most comprehensive policy example of the administration's mindset. Years later, Bill Clark elaborated on the rationale that the document expressed:

> The basic premise behind this new approach was that it made little sense to seek to stop Soviet imperialism externally while helping to strengthen the regime internally. This objective was to be attained by a combination of economic and ideological instrumentalities. Thus it became United States policy to avoid subsidizing the Soviet economy or unduly easing the burden of Soviet resource allocation decisions, so as not to dilute pressures for structural change within the Soviet system.[12]

Clark's words matched the language of NSDD-75.[13] NSDD-75 acknowledged: "The U.S. recognizes that Soviet aggressiveness has deep roots in the internal system, and that relations with the USSR should therefore take into account whether or not they help to strengthen this system and its capacity to engage in aggression."

Additionally, NSDD-75 addressed the following specifics:

ALLIES. While stating that the support of overall U.S. strategy by allies was "essential," NSDD-75 conceded that America "may on occasion" be forced to go it alone, sometimes "even in the face of Allied opposition." (p. 4) This applied not just economically but militarily, and was a harbinger of what was to come later in the year in Grenada.

THE THIRD WORLD. The directive said the United States must "resist Soviet encroachment" upon the Third World and "support effectively those Third World states that are willing to resist Soviet pressures or oppose Soviet initiatives hostile to the United States." American efforts in the Third World must involve "an important role for security assistance and foreign military sales, as well as readiness to use U.S. military forces where necessary." (p. 4)

SOVIET ALLIES WITHIN THE SOVIET EMPIRE. NSDD-75 said there were "a number of important weaknesses and vulnerabilities within the Soviet empire which the U.S. should exploit." U.S. policies, said the NSDD, "should seek wherever possible to encourage Soviet allies to distance themselves from Moscow in foreign policy and to move toward democratization domestically." (p. 4)

EASTERN EUROPE. "The primary U.S. objective in Eastern Europe," NSDD-75 made clear, "is to loosen Moscow's hold on the region" while also promoting human rights in the region. The directive said the Reagan administration could advance this objective by "carefully discriminating" in favor of countries that "show relative independence" from the USSR in their foreign policy, or those that show "a greater degree of internal liberalization." (p. 4) The White House had in mind Yugoslavia and Poland, respectively.

AFGHANISTAN. The document affirmed that in Afghanistan, "The U.S. objective is to keep maximum pressure on Moscow for withdrawal and to ensure that the Soviets' political, military, and other costs remain high while the occupation continues." (p. 4)

SOVIET THIRD WORLD ALLIANCES. The Reagan team listed an added objective "to weaken and, where possible, undermine the existing links" between the USSR and its Third World allies. U.S. policy "will include active efforts to encourage democratic movements and forces to bring about political change inside these countries." (p. 5)

CHINA. NSDD-75 happily reported that China "continues to support" U.S. efforts to "strengthen the world's defenses against Soviet expansionism." The directive said the United States should seek enhanced cooperation with China and reduce the possibility of a Sino-Soviet rapprochement. (p. 5)

ARMS CONTROL. The directive advised that the White House only enter into arms-control talks when they serve U.S. objectives. Speaking Reagan's own language, NSDD-75 insisted that arms-control agreements were "not an end in themselves." (p. 5)

OFFICIAL DIALOGUE. The directive said the White House "should insist" that Moscow address the "full range of U.S. concerns about Soviet internal behavior and human-rights violations." The administration should resist a U.S.–Soviet agenda focused on arms control at the expense of human rights. (p. 5) This reflected Reagan's long-held belief that it was wrong to ignore Soviet brutalities merely for the sake of gaining arms-control agreements; that was what détente had produced, and that was why it was wrong.

BREZHNEV'S SUCCESSOR. The directive even addressed the issue of Soviet General Secretary Leonid Brezhnev's successor. On that, NSDD-75 said the administration would "try to create incentives (positive and negative) for the new leadership to adopt policies less detrimental to U.S. interests." (p. 7) Even that would later be fulfilled in the form of a man named Gorbachev.

Spanning almost every continent, NSDD-75 proved to be the most ambitious assault on Soviet interests in decades, maybe ever. In fact, it held no recognition of Soviet interests. The document assumed that not only should the United States encourage support for democratic forces throughout the world but that it should discourage the USSR's ability to support Communist forces, which were repressive and undemocratic and thus illegitimate.

NSDD-75 revealed not only an intention to deter Soviet aggression but also to roll back the empire when possible. In a calculated move to avoid overt conflict, NSDD-75 did not advocate taking on the Soviets at every point of incursion, but instead focused merely on the areas where the Soviets were most vulnerable and the United States most capable of inflicting damage. Clark's deputy, Robert McFarlane, emphasized: "NSDD-75 did not say we should confront the Soviets at every point. It said we should look for vulnerabilities and try to beat them."[14]

Though this language concerning the reversal of Soviet expansionism was explicit, NSDD-75 did not explicitly predict a disintegration of the USSR. It did, however, seek to make overt attempts to change the nation internally or to make life miserable for the Soviet system. Unlike Reagan, the document made no predictions of victory or of placing Soviet Communism on the ash heap of history, as such a declaration would have been out of character for an official document.

Other similar distinctions are in order: NSDD-75 sought to reverse not only future but *past* Soviet expansion, and stated such a goal more explicitly than any previous Reagan administration document. "Past" expansion was crucial: That reflects an intention to roll back territory already taken by the Soviet Union. Tellingly, the name applied to NSDD by some administration insiders was "Operation Rollback." One can rightly say that the intent of NSDD-75 was to alter both the Soviet empire and the USSR.

Further, NSDD-75 endeavored to "change" and "gradually reduce" the Marxist system within the USSR. By seeking political pluralism, it also hoped to repudiate the Communist Party monopoly on power. Alas, because the Communist system and USSR were one in the same, the intent of NSDD-75 was to transform the USSR itself.

Somehow the Soviets were able to procure a copy of the highly classified document. The Moscow Domestic Service released two statements on the directive, dubbing the "plan" a "subversive" attempt "to try to influence the internal situation" within the USSR. "[T]he task," said Moscow, was "to exhaust the Soviet economy." The Reagan administration had "drawn up aggressive plans" for "mass political, economic, and ideological pressure against the Soviet Union in an attempt to undermine the socioeconomic system and international position of the Soviet state."[15] This interpretation was correct on all counts.

Throughout the Soviet media, the directive resonated, as the propaganda machine sought to make sense of the confrontational wording. A piece in *Sotsialisticheskaya Industriya* stated, "Directive 75 speaks of changing the Soviet Union's domestic policy. In other words, the powers that be in Washington are threatening the course of world history, neither more nor less."[16] In doubting such Washington ability, the publication quoted editorials in the *Toronto Star* and *Los Angeles Times*. "Our country," it quoted assurances from the *Los Angeles Times*, "simply has no means of exerting pressure of this sort (on the Soviet Union), and it was staggering to hear that the Reagan administration thinks otherwise." Going by such accounts, *Sotsialisticheskaya Industriya* confidently told Soviet readers that the general opinion of Western observers was that the grandiose "ideas of Reagan and Pipes" were "staggeringly naïve."[17]

LATE JANUARY TO FEBRUARY 1983

Despite the domestic and Soviet blowback, it was clear that NSDD-75 was a bold declaration of the administration's vision—one that Reagan sought to embrace publicly. A week after signing NSDD-75 in private, Reagan spoke of this vision openly in his January 25 State of the Union, in which he called for a "comprehensive strategy for peace with freedom." He referred to his

Westminster Address in London the previous June and reiterated his commitment to the development of an "infrastructure of democracy" throughout the world. "We intend to pursue this democratic initiative vigorously," said Reagan. "The future belongs not to governments and ideologies which oppress their peoples, but to democratic systems of self-government which encourage individual initiative and guarantee personal freedom." Though he did not outline these initiatives in his speech, they would take the form of the National Endowment for Democracy, the modernization of the Voice of America and other broadcast facilities, and the launch of Radio Marti, a station designed to broadcast inside Cuba.[18]

In addition to discussing his plans for foreign policy, Reagan's State of the Union also touched on the domestic front, making it clear that the economic turnaround he had desired in 1981 as his first priority was now underway, setting the stage for the strength he and his country needed to take on the Russians: "[O]ur strategy for peace with freedom must also be based on strength—economic strength and military strength," said the president. "A strong American economy is essential to the well-being and security of our friends and allies. The restoration of a strong, healthy American economy has been and remains one of the central pillars of our foreign policy."

This feeling of success continued beyond the State of the Union, and a month later, on February 22, he began heralding a full economic recovery. In a foreign-policy speech to the American Legion, in which he spoke of the need to rebuild defenses and take on the Soviets, he reminded, "Our first and highest priority was to restore a sound economic base here at home."[19] It was the beginning of a surge in prosperity on the home front, and Reagan enthusiastically welcomed the shift with open arms. The domestic foundation of his foreign policy was indeed becoming stronger by the day.

Feeling that such had been achieved, he was ready to move on to the larger battle, a fact made evident in another line in the speech: "History is not a darkening path twisting inevitably toward tyranny, as the forces of totalitarianism would have us believe. Indeed, the one clear pattern in world events—a pattern that's grown with each passing year of this century—is in the opposite direction."[20]

The sum total of these events was that *Pravda* now really had something to crow about. "The anticommunist 'crusade' program outlined by the President has begun to acquire concrete form," said the party organ. The "Washington 'Crusaders,' " declared *Pravda*, were "on the march."[21] Now more than ever, "Crusade" was the operative word in the Soviet press.[22] The

Reagan administration, said TASS, the official Soviet news agency, was "bent on pushing forward its 'crusade against communism.' "[23]

MARCH 8, 1983: "EVIL EMPIRE"

Just when it seemed that the Soviets could not be more apoplectic over what they had been hearing from the Oval Office, on March 8, 1983 Ronald Reagan stepped before a group of evangelical Christians at the Sheraton Twin Towers Hotel in Orlando, Florida and made quite a claim. The USSR, the president told the National Association of Evangelicals, was the "focus of evil in the modern world"; it was an "evil empire." The speech was polarizing, as was its intention: to draw a line of demarcation between the two superpowers. Said the Crusader to his fellow Christian soldiers:

> I urge you to speak out against those who would place the United States in a position of military and moral inferiority. . . . I urge you to beware the . . . temptation of blithely declaring yourselves above it all and label both sides equally at fault, to ignore the facts of history and the aggressive impulses of an evil empire, to simply call the arms race a giant misunderstanding and thereby remove yourself from the struggle between right and wrong and good and evil.

To suggest the United States and USSR were morally equal, judged Reagan, was "rubbish." Some observers might try to declare themselves "above it all," haughtily asserting that "both sides are wrong." To do so, said Reagan, would be to ignore the facts. In his eyes, this Cold War battle was a struggle between good and evil, and there was no doubt over which side was which. By making this bold proclamation in this speech, Reagan hoped that others would likewise connect the dots.[24]

From the start of the speech, it was clear that the language was atypical; it was also evident that Reagan, who was always keenly aware of his audience, was playing up the religious imagery, as he spoke to the Christian listeners about the sinful, fallen nature of man: "We know that living in this world means dealing with what philosophers would call the phenomenology of evil or, as theologians would put it, the doctrine of sin." He then clarified his motivations for saying what he was saying: "There is sin and evil in the world," said the president of the United States, "and we're enjoined by Scripture and the Lord Jesus to oppose it with all our might."

There was certainly a largesse of sin and evil in one part of the world: in Moscow. Now, Reagan wanted to announce the fact loud and clear, lest on-lookers had any confusion regarding what the Cold War was all about—ditto its origins. Unlike America's founders, said Reagan, the godfather of the Bol-shevik state had a twisted conception of morality:

> [A]s good Marxist-Leninists, the Soviet leaders have openly and publicly de-clared that the only morality they recognize is that which will further their cause, which is world revolution. Lenin . . . said in 1920 that they repudiate all morality that proceeds from supernatural ideas—that's their name for religion—or ideas that are outside class conceptions. Morality is entirely subordinate to the interests of class war. And everything is moral that is nec-essary for the annihilation of the old, exploiting social order and for uniting the proletariat.

Yet, to think that this address was all fire and brimstone would be a mis-take. Immediately after dubbing the USSR evil, Reagan, employing a line he inserted into the text himself, served up a plea for prayer for the USSR: "[L]et us pray for the salvation of all of those who live in that totalitarian darkness—pray they will discover the joy of knowing God." However, "until they do," said Reagan, "let us be aware that while they preach the supremacy of the state, declare its omnipotence over individual man, and predict its eventual domination of all peoples on the Earth, they are the focus of evil in the modern world."

Then, in a moment that has been almost completely neglected by history, Reagan drew similar parallels to his own country, turning the spotlight on the manifold sins of which the United States had been guilty since its inception. "Our nation, too, has a legacy of evil with which it must deal," he said, citing slavery, racism, bigotry, ethnic hatred, anti-Semitism, and the "long struggle of minority citizens for equal rights." America had also betrayed God, partic-ularly the commandment, "Thou shalt love thy neighbor as thyself."

While America was hardly exempt from sin and did not escape criticism in this address, the thrust of Reagan's message was that America was facing not simply any old enemy; it was facing an evil enemy, an Evil Empire. And that empire, said Reagan, with some good news, was doomed: "I believe that communism is another sad, bizarre chapter in human history whose last pages even now are being written," he said again.[25]

Because of these dramatic pronouncements, certain Reagan staff loved the speech. Bill Clark called it "probably his greatest speech."[26] Fred Ikle, Reagan's undersecretary of defense for policy, appreciated the underlying logic and strategy in such bombastic statements. Ikle admired Reagan for having the "courage" to articulate what was then an "incendiary idea" but is now a "hackneyed truth"; namely, "that we had a Cold War because of the evil empire, and could not end the Cold War without undoing that empire."

For every Ikle or Clark, however, there was a David Gergen or State Department official who got quite nervous about such Reagan remarks. To them, such comments were more incendiary than helpful, destined to produce international anger instead of progress.

Two people who shared this view were Nancy Reagan and her friend Stuart Spencer, a Reagan campaign adviser: About a week after the speech, Spencer was having one of his private dinners for three with the Reagans. Nancy and Spencer both expressed reservations over the president's speech a week earlier. Reagan waved them off: "It *is* an Evil Empire," he responded. "It's time to close it down."[27]

SDI

While the "Evil Empire" speech had shocked people on both sides of the Atlantic and had burned yet another page in the annals of searing Reagan rhetoric against the Soviets, March 1983 still had another surprise in store. "My fellow Americans, tonight we're launching an effort which holds the promise of changing the course of human history," declared the president on March 23, 1983, in a nationally televised address that caught even his top advisers off guard. Reagan penciled a line into the speech text: "I call upon the scientific community which gave us nuclear weapons to turn their talents to the cause of mankind and world peace; to give us the means of rendering these weapons impotent and obsolete."

With this introduction, he announced his Strategic Defense Initiative (SDI), a vision for a space-based missile-defense system. Among other lines he added to the text of his speech was this harbinger of what was to come: "[L]et me just say that I am totally committed to this course."[28] That would soon be abundantly clear to the world. Coming only two weeks after the Evil Empire speech and amid his push to place intermediate-range nuclear missiles in Western Europe, Reagan's remarks left Moscow shell-shocked, not to mention his own staff. Four days before the speech, only Bill Clark, Robert McFarlane, John Poindexter, and science advisor George Keyworth knew what was to come.[29] Even physicist Edward Teller, who had been involved in Reagan's thinking on SDI longer than any individual, was completely surprised.[30]

The secrecy was necessary: Reagan had a terrible problem with leaks, especially from White House moderates and the State Department, the latter of which adamantly opposed SDI and anything that rankled the Soviets. Leaks had been a particularly big problem that spring of 1983, to the point where Reagan had Clark investigate the matter and even considered employing a polygraph.[31] He did not want SDI to be sabotaged by an in-house opponent as a lame-brain idea prior to its announcement.

Keyworth recalled a Monday meeting in the Oval Office with Secretary of State Shultz before the Wednesday evening speech. "Shultz called me a lunatic in front of the president," remembered Keyworth, "and said the implication of this new initiative was that it would destroy the NATO alliance. It would not work . . . and was the idea of a blooming madman."[32] What Shultz did not realize when he confronted Keyworth was that he had just called Reagan a madman in front of his top advisers—since the idea was completely Reagan's.

Resistance came not only from Shultz's State Department but also from Pentagon hardliners who usually supported Reagan. As the speech approached, Richard Perle telephoned Keyworth from Portugal and told him to fall on his sword, to go so far as to tell the president he would oppose the new idea publicly, to do anything to stop the speech. Keyworth said that even Fred Ikle was "violently opposed." "I guess I have never seen such opposition to anything," Keyworth estimated. "But [Reagan] was absolutely committed."[33]

Bill Clark notes that SDI offers one of the best examples of Reagan's full control over a decision and its fate. He says that even McFarlane, who helped Reagan write the SDI speech, was opposed: "My deputy, McFarlane, gave me a memo the night before the speech and said, 'Bill, can you talk him out of doing this?' No way, I said. . . . No one supported him on this announcement but a few of us." Clark said of McFarlane: "[He] helped put the words together, as instructed, for that speech. As a military officer he saluted and drafted for us. He came to me almost in tears to say, 'You've got to talk to the President out of giving this. It's not ready and violates certain treaties.' "[34]

What many of the principles did not realize at the time was that this Reagan fascination with a technological breakthrough that might allow for defense against missiles had begun long before his presidency. For years, he had called it "my dream," and having carefully thought out the concept for over a decade, it was beyond question his baby.[35] Reagan had a plaque in his office stating that there is no limit to what a man can achieve if he doesn't mind who gets the credit—yet, he was the first to admit SDI was his idea.[36]

An early seed was planted during a 1967 meeting between Reagan and Edward Teller, the father of the hydrogen bomb and one of the twentieth century's most influential physicists.[37] When Reagan moved in to the governor's mansion, Teller requested a meeting, and the new governor, eager to accommodate, responded right away. They met at the mansion—their first of numerous meetings over the years. Teller said he did "nothing more" than invite Reagan to Lawrence Livermore National Laboratory for a series of briefings, which in November the governor proceeded to do. Teller shared with Reagan his research on using explosives to defend against a nuclear attack. In Reagan's first year in public office, these two historic men met on what would become a historic idea.[38]

At the lab, Reagan asked Teller about a dozen questions, listening intently but without giving a clear indication of whether he was for or against the concept. The governor's questions, said Teller, "were by no means obvious questions, but in a field that must have been quite new to him he saw the salient points."[39] He said Reagan asked "very intelligent questions" and was keenly interested. Until his literal dying days, Teller insisted Reagan was a "very bright . . . exceptionally intelligent individual," who was always greatly underestimated by partisan foes.[40]

The two continued to meet and correspond right up until March 23.[41] In one exchange, Teller contacted the White House about a "potential advance" in nuclear weapons technology: the massive energy released in a nuclear explosion might be harnessed to "pump" a laser, which could in turn be directed "in a straight line over great distances to strike a target." It was an intriguing idea to Keyworth, one that he relayed to Reagan, saying that it represented a potential breakthrough which could "represent a means" enabling nuclear weapons to be used defensively rather than offensively. In Keyworth's memo briefing the president on this development, Reagan in turn scribbled a note to Bill Clark: "Dear Bill, We should take this seriously and have a real look. Remember our country once turned down the submarine. Ron."[42]

Another event pivotal in Reagan's early thinking was a July 1979 visit to the North American Aerospace Defense Command inside Cheyenne Mountain, Colorado. When Reagan asked what would happen to the enormously sophisticated, well-fortified complex if a Soviet SS-18 landed 100 yards outside, Commanding General James Hill snapped, "It would blow us away." According to Reagan adviser Martin Anderson, Reagan at that moment was deeply disturbed that America had no means to defend against such an attack. "There must be something better than this," he demanded.[43]

This concern was fueled even further when as a newly sworn-in president he was shocked by a classified Defense Department briefing which stated that at least 150 million Americans would be killed in a nuclear war with the Soviet Union, even if America "won." For Americans who survived such a war, said Reagan, he could not imagine what life would be like. Even if a nuclear war did not mean human extinction, it would mean the end of civilization as the world knew it. "*No one* could 'win' a nuclear war," insisted Reagan. "Yet as long as nuclear weapons were in existence, there would always be risks they would be used. . . . My dream, then, became a world free of nuclear weapons."[44]

Despite Reagan's hawkish reputation, he detested nuclear weapons.[45] "With every ounce of my being," he said in 1983, "I pray the day will come when nuclear weapons no longer exist anywhere on Earth."[46] He asserted that nuclear war would be the "greatest tragedy . . . ever experienced by mankind, in the history of mankind." He insisted that, "No room should be left for doubt about a nuclear exchange; no one would win."[47] Once, he rhetorically asked a British television correspondent: "Where do we live after we have poisoned the Earth?"[48] These were points that he echoed routinely, and yet they have largely been forgotten.[49] Indeed, among Reagan's chief second-guessers, no less than *Time*'s Strobe Talbott described him as a "nuclear abolitionist."[50]

For these reasons, Reagan was uncomfortable with Mutually Assured Destruction, or "MAD," a doctrine devised by Secretary of Defense Robert McNamara in the 1960s. MAD posited that the frightening specter of tens of thousands of American and Soviet nuclear warheads was, strangely, a good thing: it ensured that neither side would launch missiles because doing so would mutually assure global destruction. As a result, said McNamara, not unreasonably, leaders on both sides would be extremely careful to never push the button. In that ironic way, these vast arsenals were a stabilizing presence.

Reagan was appalled by this thinking. As George Shultz notes, he believed that relying on MAD for deterrence was "morally abhorrent."[51] "MAD spells what it is—it's really mad," Reagan insisted.[52] A grim example of "the madness of the MAD policy," said Reagan, was a situation in which a president had "six minutes to decide how to respond to a blip on a radar scope and decide whether to unleash Armageddon!"[53] Reagan said that, "People who put their trust in MAD must trust it to work 100 percent—forever, no slip-ups, no mistakes." It also depended on no madmen and no unmanageable crisis. He often noted that if even one missile was accidentally fired at the United States, the president had no way to prevent the "wholesale destruction" of American lives and the only recourse would be a retaliation that would wipe out millions of lives on the other side. This position was "simply morally untenable."[54]

Instead Reagan preferred to rely on what he called "mutual assured survival"—which he hoped might be possible via missile defense.[55] SDI, then, would be a "noble enterprise to find an alternative to nuclear terror."[56] As Ed Meese noted, Reagan's idea was a shrewd way of responding to the fears of nuclear war—which the Soviets and Western leftists were blaming on Reagan—with an option to diminish nuclear war.[57] Reagan drew leftists into a corner, daring them to violate their own slogans. As arms-control director Ken Adelman put it, it was ironic that liberals, who prided themselves on their humanity, advocated such a blood-curdling approach as MAD, which essentially said that all was well so long as both superpowers held the potential to vaporize hundreds of millions of people.[58]

For those not reassured by such a prospect, said Reagan, "we must ask: isn't it time to begin curing the world of this nuclear threat? If we have the medicine, can we in good conscience hold out on the patients? I believe that, given the gravity of the nuclear threat to humanity, any unnecessary delay in the development and deployment of SDI is unconscionable."[59] SDI, he assured, provided hope—a potential lifesaving alternative.

Aside from MAD, there was another crucial but neglected motivation in Reagan's pursuit of SDI: The Soviets had invested in their own defense system, a point Reagan made directly to the Soviets. He called the Soviet missile-defense effort the "Red Shield":

> It's no longer a secret that the Soviet Union has spent billions upon billions of dollars developing and deploying their own antiballistic missile defenses. Research and development in some parts of the Soviet strategic defense program—we call it the Red Shield—began more than 15 years ago. Today Soviet capabilities include everything from killer satellites to the modernized ABM defenses that ring Moscow. More than 10,000 Soviet scientists and engineers are working on military lasers alone, with thousands more developing other advanced technologies, such as particle beam and kinetic energy weapons.
>
> The Red Shield program actually dwarfs our SDI.[60]

This was not some Reagan fantasy. The Soviets had in fact been investing in missile defense for years, a point Reagan made often, but with no impact on the American media. Reagan hoped the "Red Shield" phrase would stick; it never did.[61] The press failed to make an issue of the Soviet program, no matter how often Reagan raised it.

"IMPENETRABLE SHIELD"

While the "Red Shield" phrase did not enter the public consciousness, other, less helpful phrases did, proving detrimental to certain perceptions of the program. One such phrase was "impenetrable shield," an expression which became associated with the plan soon after the March 23 speech. Contrary to popular criticism, Reagan did not propose SDI as an impenetrable shield that could be rapidly installed. His diary entry from the evening of the SDI speech reads: "I made no optimistic forecasts—said it might take 20 years or more but we had to do it."[62] A comparison of Reagan's various statements, along with a careful read of his original speech proposing SDI, suggests that he believed SDI was technologically possible, but not in the near future. As he said that March 23, "Current technology has attained a level of sophistication where it's reasonable to begin this effort."

An SDI might not achieve a 100 percent kill rate against 10,000 Soviet missiles raining over America; yet, a system might be developed that could take down a nuclear warhead launched via a Middle East "madman" or a "trigger-happy general" or a "slip up." Reagan turned this criticism on liberals: "They say it won't be 100% effective, which is odd, since they don't ask for 100% effectiveness in their social experiments."[63] George Keyworth was also greatly troubled with the perception that the administration was seeking to create a perfect shield. That image, said Keyworth, was "the toughest piece of propaganda we had to deal with."[64] It was always the biggest obstacle to SDI.

Actually, the media was partly responsible for the perception. Reagan aide David Gergen remembers the colorful video animation used by TV networks when they reported on SDI: perfectly narrowed lasers emanating from space-based platforms flawlessly neutralized onslaughts of Soviet ICBMs. The White House communications office did not create those exaggerated images. Ironically, the fictional images augmented SDI's status in the eyes of the American public and quite possibly Soviet officials.[65]

This is not to suggest that Reagan bears no responsibility for the misperception. He stated that SDI could someday "render nuclear weapons obsolete." Reagan envisioned an advanced system decades down the road, but more importantly, he hoped that an SDI system that was shared with the Soviets would make offensive nuclear weapons pointless and hence "obsolete" in that sense. Still, Reagan did not always pause to make such key distinctions when he employed words like "shield" and "screen."[66]

On the other hand, he was often careful and not guilty of hyperbole. A reflection of his realism—formally so—was the foreword by Reagan on the administration's later "Report on the Strategic Defense Initiative." Therein, Reagan revisited his initial March 23 announcement. The words he used were cautious and realistic:

> On March 23, 1983, I announced my decision to take an important first step toward this goal by directing the establishment of a comprehensive and intensive research program. . . . SDI is a program of vigorous research focused on advancing defensive technologies to provide a better basis for deterring aggression. . . . The SDI research program will provide to a future president and a future Congress the technical knowledge required to support a decision on whether to develop and later deploy advanced defensive systems.[67]

It is clear here that Reagan merely announced a *research program* to begin working toward the development of a system.[68] He told Margaret Thatcher that he was "simply embarking on a long-term research effort," not a commitment to deploy SDI. "Obviously," he told her, "it would be some time before we knew it would work as we hoped."[69] As late as the early 1990s, in his memoirs, Reagan said that SDI "might take decades to develop."[70] He added in his memoirs that he "never viewed the SDI as an impenetrable shield," adding categorically that he believed that "no defense could ever be expected to be one hundred percent effective."[71]

"STAR WARS"

It was difficult to tell who was most upset by SDI: certain members of Reagan's own administration, liberal Democrats, or the Soviets. Liberal Democrats dubbed the initiative "Star Wars," a term popularized by Senator Ted Kennedy (D-MA), who lampooned Reagan's SDI speech the following morning as "misleading Red-scare tactics and reckless Star Wars schemes."[72] With those words, Kennedy, who in 1983 had his eyes on a presidential run in 1984, had started something.

"Star Wars" became a vehicle to ridicule SDI. In the 1980s, Reagan was often caricatured as a dawdling fool, a lazy man and nostalgic ex-actor, who spent his time watching movies, where he lost himself in a world of make-believe. Surely, suggested the ridiculers, Reagan must have gotten the idea for SDI from the blockbuster movie "Star Wars," envisioning himself as kind of presidential Luke Skywalker combating the forces of darkness of Darth Vader's Evil Empire.[73] As a *New York Times* news story put it a week later, the SDI proposal was "Mr. Reagan's answer to the film 'Star Wars.' "[74]

If Kennedy had hoped to discredit the concept, he was making strides. Kennedy's "Star Wars" term became extremely damaging, especially once the partisan media at home and abroad delightfully ran with it. Reagan rightly feared that it suggested that he desired not a defensive system but an offensive war in space. It conjured "an image of destruction," he said, when, in fact, "I'm talking about a weapon, non-nuclear . . . [that] only destroys other weapons, doesn't kill people."[75] SDI "isn't about war, it's about peace."[76] He charitably allowed that the media probably did not envision such a deleterious effect, instead using "Star Wars" merely "to denigrate the whole idea."[77] However, privately Reagan told one friend that he "bristles" each time the media used the label.[78] Echoing these complaints, he confided to two other friends that the term was "never mine" but the media's, "and now they saddle me with it."[79]

In this endeavor, Reagan faced a huge PR problem begun by liberals in Congress and the press. This was evident in a later exchange with UPI's Helen Thomas:

THOMAS: Mr. President, if you are flexible, are you willing to trade off research on "Star Wars" . . . or are you against any negotiations on "Star Wars"?

REAGAN: Well, let me say, what has been called "Star Wars"—and, Helen, I wish whoever coined that expression would take it back again—

THOMAS: Well, Strategic Defense—

REAGAN:—because it gives a false impression of what it is we're talking about.[80]

Immediately after Reagan's plea, Thomas continued: "May I ask you, then, if 'Star Wars'—even if you don't like the term, it's quite popular. . . ."[81]

The term was popular because reporters used it. Reagan's request was reasonable: the program's name was the Strategic Defense Initiative. Objective reporters ought to be expected to use its proper name, not the name of derision used by partisan detractors. That did not matter. Reporter Chris Wallace followed Thomas: "I'm a little confused by your original answer on, if you'll forgive me, 'Star Wars'—if we can continue to use that term." Wallace then answered his own question: "The question is now, in the talks that are going to begin, would you consider setting limits on the deployment and the testing of 'Star Wars'?"[82] Clearly Reagan was fighting an uphill battle.

In Moscow, the Communist media loved Kennedy's term. To say that the Soviets embraced "Star Wars" is inadequate; they used the label in almost every story on SDI, rarely using the words Strategic Defense Initiative or the acronym. Tellingly, whereas the U.S. press typed "Star Wars" in uppercase to ridicule the idea as movie fiction, the Communists placed it in lowercase to suggest SDI was a vehicle for war amid the stars—"preparations for 'star wars,'" as the Moscow International Service put it.[83] The Kremlin seized upon the term with abandon to portray Reagan as a nuclear warmonger. The number of examples could fill this book.[84] Here are but a few:

In a commentary for the Moscow World Service, Viktor Olin claimed that "Preparations for star wars are under way in the United States."[85] The Red Army publication *Krasnaya Zvezda* published a lengthy, grave piece titled, "How the 'star wars' are being prepared."[86] Expressing similar sentiment on Moscow TV's "Studio 9" program, top propagandist Valentin Zorin, the KGB's Vitaly Kobysh, and academic Yevgeny Velikhov hammered Reagan's alleged "plans" to "fill the space around the entire planet with battle stations." "Only in his speech of 23 March 1983 did he formulate his idea, which became known as 'star wars,'" said Kobysh. This name, said Kobysh, was "very irritating" because of what it (allegedly) advocated. One group not fooled by Reagan, however, were liberal Democrats at home: "U.S. politicians," said Kobysh, "call it [SDI] the greatest deception of our time."[87]

While the Soviets did much of their own propagandizing, they admired certain American politicians and columnists, regularly employing their sentiments in their attempt to twist reality. The most quoted U.S. senator in the Soviet press in the 1980s was likely Ted Kennedy, whereas the most approved of columnist was James Reston of the *New York Times*; the two were cited so often that at times their names made their way into the same articles in *Pravda*.[88] For editorials, the *Washington Post* was a preferred source, particularly for *Izvestia* hatchet man Valentin Falin, who employed one *Post* editorial to attack Reagan and SDI as "evil," "cuckoo," and, according to the *Post* editorial, a "prospect of fabulous new riches for the [U.S.] military-industrial complex," a line that Falin underscored with special appreciation.[89]

One Moscow periodical, the English language *New Times*, noted that Reagan felt "hurt" by the term "star wars," and wished that the person who coined the expression would take it back because it misrepresented what he had in mind. Yet, the *New Times* was not fooled; it clarified Reagan's ruse: As evidence, it cited a *Los Angeles Times* piece by Thomas S. Powers. The *New Times* quoted Powers: "The secret of 'star wars' is that it is intended to defend weapons, not people. The purpose is not to keep the Soviets from threatening us, but to make sure we can threaten them."[90] TASS seconded the notion: "There is nothing defensive about SDI; an offensive system." Reagan's explanations were merely "clever tricks."[91]

For its part, the Moscow Domestic Service was grateful to the American media and to politicians like Senator Kennedy for "properly" labeling SDI:

> They christened it ["star wars"] with full justification, since this initiative envisages deploying strike weapons systems in space aimed at targets not only in earth orbit, but also on the ground. All the while, the White House has convinced itself that they have been misunderstood, that they have goodwill toward all mankind. . . . [The White House believes that] certain forces, it seems, have distorted the essence of the Strategic Defense Initiative by labeling it the "star wars" program. . . . However, Washington is resorting to mediocre verbal balancing acts in vain. There is nothing defensive about it.[92]

Reagan was left alone to deal with the consequences of how SDI was mislabeled and misreported, including against hostile Soviet reporters, to whom he protested: "We're not talking about star wars at all! We're talking about seeing if there isn't a defensive weapon that does not kill people." The Soviet reporters were incredulous; after all, they had gotten the term from Reagan's own American media which, the Soviets surely surmised, was certainly more objective on the matter than Reagan.[93]

THE WILLIAMSBURG SUMMIT

While the "Evil Empire" speech and SDI had created a furor in the spring of 1983, Reagan stayed the course on each of his crusade's many fronts. Despite the turmoil and indignation from inside and outside his administration, stubborn old Reagan still refused to back down on the Siberian gas pipeline. In the face of mounting criticism from around the world, he refused to give up his unrelenting attack on the project.

In May, the issue finally came to a head at a G-7 economic summit held at the College of William and Mary in Williamsburg, Virginia. Accompanying Reagan were Secretary of State Shultz and Treasury Secretary Don Regan. The other attending leaders were President Mitterand of France, Prime Minister Thatcher of Britain, Chancellor Helmut Kohl of West Germany, Prime Minister Yasuhiro Nakasone of Japan, Prime Minister Amintore Fanfani of Italy, Prime Minister Pierre Trudeau of Canada, and Gaston Thorn, president of the Commission of the European Communities. During the summit, the allies reached a compromise, settling for a one-strand pipeline— instead of the two-strand initially planned—in exchange for a pledge from Western Europe to impose tighter restrictions on technology exports and low-interest loans to the Soviet bloc. It was a victory for Reagan, or, at the very least, a notable half-victory.

During a Politburo meeting the following day, an angry General Secretary Yuri Andropov blamed everything squarely on Reagan. It was Ronald Reagan alone, said Andropov, who was the "bearer and creator of all anti-Soviet ideas." The American president was seeking to "put together a bloc against the USSR." To Andropov, the problem was not so much the Reagan administration and those around the president but Reagan himself.[94]

Andropov's resentment was understandable. Reagan had slowed construction of the pipeline by nearly two years, and, in the end, prevailed in stopping the construction of the second strand altogether—a remarkable personal achievement in light of the opposition he had faced in Western Europe and in his own cabinet.

All of this caused considerable damage to the Soviet economy. At the time, Soviet economist Abel Aganbegyan figured that "each month's delay" in the construction of the pipeline "costs us millions."[95] His estimates were correct. The Reagan NSC estimated that the Soviet Union after 1982 was deprived of an astonishing $10–15 billion in annual revenue—out of a total hard-currency income of $32 billion.[96] By 1990, when the pipelines would have been fully operational, Russian natural gas would have accounted for an estimated 23 percent of Western Europe's consumption.[97]

That loss was a massive blow to the Soviet economy, from which it never recovered. Equally significant, the lost revenue also deprived the USSR of the resources it needed to spend money on armaments and to subsidize client states. By 1990, a Communist despot like Fidel Castro suddenly found himself without his annual $6 billion check from Moscow, a sunken lifeline from which Havana likewise never recovered.

Williamsburg brought to a close a campaign that had begun more than a year earlier and had been one of Reagan's toughest stances on international policy toward the USSR. While the construction of the pipeline was something that Reagan had hoped to avoid altogether, this compromise proved to be the first decisive victory in his foreign policy against the Soviets, during which he had shown the world his refusal to acquiesce to international pressure.

JUNE–AUGUST 1983: POLAND

As the pipeline front was coming to a close, Poland appeared to be cooling off. Since the drama of December 1981, progress appeared to have stagnated, particularly with the leaders of Solidarity in prison or in hiding. Yet, beneath the placid surface, the water was simmering, and Reagan was hoping to make it boil so furiously as to knock the lid off.

By summer 1983, the covert aid that Reagan had authorized under earlier NSDDs was flowing to the Solidarity underground, as were a constant stream of words from the Oval Office.[98] These words constituted a powerful source of moral support for Poland's freedom fighters, and were so ubiquitous that the final index to Reagan's *Presidential Documents*—the official collection of all presidential statements—lists references to Solidarity or Poland on 216 pages, with multiple references on most pages, many of which came in 1983 alone.[99] In these, Reagan was an unflagging championing of Solidarity, and served up stinging rebukes of Soviet or Polish government actions.

While martial law was quietly lifted in July 1983—a big victory for Solidarity's man in the Oval Office—Reagan did not lift his verbal campaign on the subject. Poland, after all, still was not free. On June 23, speaking to Polish Americans in Chicago, the president said that Americans were bound to Poles, and that while time might pass, the American people would "never, never forget the brave people of Poland and their courageous struggle." Though martial law had descended like a dark cloud, the will of the Polish people had never been broken. "No one," he assured, "can crush the spirit of the Polish people."[100] He quoted Churchill, who once said Poland is like a rock: While it may occasionally be submerged by a tidal wave, it always remains a rock.[101]

A month later, on July 19, he commemorated free Poland in his observance of Captive Nations Week. The leader of the free world told all of those behind the Iron Curtain that he did not recognize their subjugation as a permanent condition. As for Poles in particular, the Crusader stated: "As Pope John Paul [II] told his beloved Poles, we are blessed by divine heritage. We are children of God and we cannot be slaves."[102]

Every chance he could, Reagan spoke at Polish events and even festivals, always using this kind of language, frequently quoting John Paul II and Lech Walesa—on and on, as if America had elected its first Polish president.[103] In fact, that thought occurred to Soviet hardliner Aleksander Bovin, who wrote in *Izvestia*: "One has the impression that it will not take much more for Reagan to start speaking fluent Polish!"[104]

These snipes regarding Reagan's remarks on Poland were quite common with the Soviet press. Propagandist Bovin maintained that Reagan "does not give a damn about Polish workers' rights." His talk about human rights was "hypocritical sympathy," a "cynical, dishonest, shameless farce and nothing more."[105]

If the likes of Bovin wanted genuine "hypocritical sympathy" concerning Poland, they needed to look no further than TASS, which called the members of Solidarity "anti-socialist scum"—striking language from an official government news agency.[106] Furthermore, TASS was irate that in his speeches Reagan "did not say a word about the liberatory mission of the Soviet Army and the Polish Army." By Polish Army, TASS meant the good work of Jaruzelski and his military group in regard to the martial law crackdown. By the Soviet Army, TASS was referring to the Red Army's role in driving the Nazis out of Poland in World War II. This role, said TASS, "apparently mean[t] nothing to Reagan."[107]

What TASS failed to mention was the Hitler-Stalin Pact that agreed to the September 1939 dual Nazi-Soviet invasion and division of Poland, and which launched World War II in the first place. TASS also conveniently overlooked the subsequent Katyn Woods massacre by the Soviet Army. These omissions were typical of the Soviet press, which not surprisingly turned a blind eye to those moments in history that did not prove its points. Picking up on these oversights, Reagan would frequently present accurate accounts of the Soviets' horrific treatment of the Polish people, ensuring that his audiences learned the unfiltered version of Poland's relationship to the USSR.

One unfiltered version of this treatment was witnessed by Solidarity member Wladyslaw Kaludzinski, who in 1983 was held in prison, where he and his buddies were unabashed supporters of the American president. One day, one of the more vicious guards, drawn from the secret police, kicked

Kaludzinski's friend in the stomach so hard that he vomited, prompting the guard to taunt him: "What? Reagan's sausage was not fresh?"[108]

As Reagan displayed his moral support for the Poles through actions and rhetoric, they in turn bolstered Reagan's conviction that he was doing the right thing for them. This cross-national support was evident when, in 1983, the organization *Paris Match* conducted a poll of 600 Poles traveling to the West. Upon being asked who was the "last hope" for Poland, these Poles placed Reagan third, behind only the pope and Virgin Mary, and ahead of Walesa.[109]

KAL 007 AND A "DEMOCRATIC" POLAND

As the summer came to a close, Reagan received a string of reports that General Secretary Andropov, the aged apparatchik, was suffering ongoing health ailments that were not improving. Yet, while the status of his condition remained uncertain, the status of U.S.–USSR relations was about to go into intensive care. On September 1, 1983, a South Korean commercial airliner, flight number 007, headed from New York City to Seoul, inadvertently strayed into Soviet airspace, where Soviet fighter pilots made a fateful decision to shoot it out of the sky. Instantly, 269 passengers, including 61 Americans, were dead.

Reagan was at his ranch in the hills north of Santa Barbara when he received a call from Bill Clark informing him of the tragedy, including the immediate suspicions—initially unconfirmed—that the Soviets had blasted it, contrary to official denials from the Kremlin. Reagan was horrified by the implications, and responded to his closest advisor: "Bill, let's pray it's not true."[110] Unfortunately, it did not take long for the truth to come to light, and upon confirmation of the deed, Reagan was furious. John Barletta, his riding companion at the ranch, overheard him: "Those were innocent people, those damned Russians! They knew that was a civilian aircraft."[111]

Though Reagan was steamed and blasted the Soviets in a statement, he reacted quite cautiously, and much more carefully than his critics would have imagined, belying his reputation as a stomping Cold Warrior.[112] He told Clark: "[L]et's be careful not to overreact to this. We have too much going on with the Soviets in arms control. We must not derail our progress." "Bill," he said, "we've got to protect against overreaction."[113]

Reagan did not want to start a war over this. Besides, he was already pounding the Soviets with the economic sledgehammer. Instead, he continued to deploy the verbal cruise missile, his nonlethal but brutally effective linguistical weapon. For the remainder of September 1983, he bashed the Soviets in harsh terms for KAL 007, simultaneously continuing his rhetorical and moral support for Poland. On September 25, for instance, Reagan spoke in New York City at the annual Pulaski Day Banquet. There, he linked the KAL 007 "crime" to the same Soviet totalitarian evil responsible for the World War II butchery in Poland's Katyn forest. "You know that downing a passenger airliner is totally consistent with a government that murdered 15,000 Polish officers in the Katyn forest," he averred. "We cannot let the world forget that crime, and we will not."[114]

Reagan maintained that while Poland had suffered so much, it had always given the world much more. He thanked "God for Pope John Paul II and all that he is doing." He asked the audience to pray that the pope's life be protected. America, he told his audience, remained a "mighty force for good" in the world. "Tonight, in your presence, I would like to reaffirm my commitment to a free and democratic Poland."[115] Those were key words. Until then, publicly at least, Reagan had used vague words like desiring a "reconciliation" and "renewal" for Poland—language he had used with Brezhnev in that December 1981 letter. Now, he openly spoke of desiring a free and democratic Poland.

This shift in his public rhetoric reflected the overall transformation that the administration had undergone over the course of that summer and early fall. As the pipeline deal had come to a close, Poland had moved to the front and center of the administration's agenda, a move that was palpable in Reagan's language and demeanor.

Emboldened by the success of the pipeline, the U.S. economy, and the economic war against Russia, Reagan was ready to tell the world that the return to greatness that his presidency had initiated was now, in his estimation, nearly complete. Morale in America was in fact higher than it had been in a decade, a trend that would only continue as Reagan and his administration turned their collective eye to the months ahead.

13.

Grenada and Winning: October to December 1983

ON OCTOBER 13, 1983, WASHINGTON WAS SHOCKED BY AN announcement that left true believers at the NSC reeling: Bill Clark was stepping down, leaving the NSC and Ronald Reagan's side. While the rationale for Clark's decision was complex, his choice was met by a mixture of disappointment from the anti-Communist stalwarts who were committed to undermining and triumph from White House moderates like Jim Baker, Richard Darman, David Gergen, Mike Deaver, and even Nancy Reagan, all of whom were accommodationists that did not like Clark's hard-line anti-Communism and wanted him out.

It was not the first time that Clark had tried to resign. As recently as December 1982, he handed Reagan a letter of resignation, which Reagan had refused. But this time, the president knew there was no dissuading his old friend, as Clark had given the decision careful consideration for some time. Reagan knew that Clark had never wanted to leave his California ranch for Washington in the first place, and despite the impassioned pleas of his loyal cadre inside the NSC—evident in the form of long, touching letters from the likes of Roger Robinson, John Lenczowski, Ken deGraffenreid, and Sven Kraemer[1]—Bill Clark would no longer be managing Ronald Reagan's national security.

During his tenure, Clark had overseen nearly a hundred of the most important NSDDs to be issued by any president. He had been a vital liaison to the Vatican. He stood aside Reagan when the president called the USSR an Evil Empire, when he pursued SDI, and for every fiery salvo directed at Moscow. NSC member Norm Bailey spoke for the stalwarts when he said that in Clark's two years as national security adviser, he "did more than any other individual to help the president change the course of history and put an end to an empire that was, indeed, the embodiment of 'evil.'" Bailey maintained that the nation owed a "very great debt" to the laconic rancher who embodied the image of the stoic, silent cowboy.[2]

A year that had been the NSC's loud call to arms suddenly seemed to be finishing with a whimper, or, perhaps in Bill Clark's case, with a characteristic whisper. But the letdown from Clark's departure was short-lived. There was no time for the hard-liners to mourn, as a storm was stirring in the Caribbean, and one of the Cold War's hottest years was about to get even hotter.

GRENADA

Since Ronald Reagan became president, there had been a lot of nasty words exchanged back and forth between Washington and Moscow. There had even been martial law in Poland, declarations of defensive technology initiatives, and the destruction of a civilian airliner. Neither the United States nor the USSR, however, had invaded a country since Reagan walked into the Oval Office on January 20, 1981. That was about to change.

On October 19, 1983, a radical Marxist group inside Grenada had murdered Prime Minister Maurice Bishop in a Cuba-inspired coup d'etat.[3] A violent Marxist military council trained by Cuba put itself in charge, shot and jailed Bishop supporters, enacted its own martial law, and imposed a shoot-on-sight, twenty-four-hour curfew. These edicts threatened not only those citizens of the island, but also the roughly 1,000 Americans present, most of which (some 700) were students at the St. George's School of Medicine.

The coup was not completely unexpected, and in actuality this was exactly the sort of Havana-backed deadly mischief the Reagan team feared for some time. Just a few months earlier, Maurice Bishop had come to Washington, where he met with Bill Clark in an attempt to explain the current situation in Grenada. After this meeting, Reagan had initiated a special task force to consider contingency plans for Grenada, placing Vice President Bush at its head.

This was not the first time that the president had been concerned about Grenada. Seven months earlier, on March 23, in no less than the SDI speech, Reagan had warned:

> On the small island of Grenada . . . the Cubans, with Soviet financing and backing, are in the process of building a 10,000-foot runway. Grenada doesn't even have an air force. Whom is it intended for? The Caribbean is a very important passageway for our international commerce and military lines of communication. More than half of all American oil imports pass through the Caribbean. The rapid buildup of Grenada's military potential is unrelated to any conceivable threat to this island country of 100,000 people and totally at odds with the pattern of other Eastern Caribbean states, most of which are unarmed.[4]

For the public, this statement was an early glimpse into Reagan's sense of the island's strategic importance. Though not mentioned in that particular speech, he was also apprehensive about how the turmoil might affect Panama Canal traffic.[5] Overall, what concerned Reagan most was how Grenada fit into the picture of global Communism. He dubbed the coup a "Communist power grab."[6] Intelligence told him that the USSR and Cuba were building military installations on the island, including the landing strip, and stockpiling materiel.

With this intelligence, Grenada posed the hazard of not only a joint military installation orchestrated by Moscow and Havana but also another full-fledged "Cuba" operating in the Western Hemisphere. He was already committed to ceding "not one inch" of territory to Communism anywhere, least of all in America's backyard, where he already dreaded Communism's existence in Nicaragua and its prospects in El Salvador.

After Bishop was shot, the situation deteriorated rapidly. Grenada's Caribbean neighbors were worried, and six of them sent a cable to Washington asking for help, which included a request to invade the island. That was the green light Reagan needed. Cap Weinberger's Pentagon had already drawn up plans, and was now awaiting orders.

"He was very unequivocal," recalled Reagan's new national security adviser, Robert McFarlane, who replaced Clark on October 17. "He couldn't wait."[7] When a White House staffer warned there could be a "harsh political reaction" to a U.S. invasion, Reagan replied: "I know that. I can accept that." He ordered simply: "Do it."[8]

They did. On October 25, 1983, some 5,000 U.S. troops charged the shores of Grenada in the largest U.S. military operation since Vietnam. There were remarkably few casualties, particularly when measured against what Americans had been tragically accustomed to only eight years earlier. In total, only 19 died, with a little over 100 wounded. By comparison, the United States lost 58,000 dead, or over 300 killed per month, to the Vietnam experience.[9] The commander of the task force in Grenada, Vice Admiral Joseph Metcalf III, boasted: "We blew them away."[10]

While Americans supported the attack, it was quickly denounced by the international community. Even Reagan's buddy Margaret Thatcher shouted at him on the telephone, in the harshest, most disapproving tone and language she ever directed at her friend: "As soon as I heard her voice," said Reagan of their phone conversation, "I knew she was very angry."[11] Her response was indicative of the larger international reaction, as Reagan received almost no support abroad. The vote at the U.N. Security Council was 11 to 1 against the United States, while the General Assembly vote was a staggering 108 to 9, with America joined only by El Salvador, Israel, and (tellingly) the six Caribbean neighbors that requested U.S. assistance in the first place.[12]

Yet, once the job was done, U.S. troops found an enormous cache of weapons, armored vehicles, and military patrol boats. This included 10,000 assault rifles, 4,500 submachine guns, 11.5 million rounds of ammunition, 294 portable rocket launchers with 16,000 rockets, 15,000 hand grenades, 7,000 land mines, 23,000 uniforms, and much more.[13] Also significant, during battle U.S. troops engaged 800 Cuban soldiers, making it clear that Grenada was not as isolated as many had believed.

Though the intervention later received some Pentagon criticism, with operational problems under the surface, it was a decisive victory and an emotional one. Only thirty hours after the start of the "rescue mission" (as Reagan called it), the first evacuated medical student to debark the airplane dropped to his knees and kissed the tarmac as he touched the soil of Charleston, South Carolina. It was the sort of smiling military triumph that had become frowningly unfamiliar to Americans. The student's gesture brought a lump to the throat and tear to the eye of many Americans, their president included.

Images like the student and of the triumphant American soldiers provided an additional success for the war in that it played a notable role in restoring military morale, a process which Reagan had implemented at the start of his presidency. Here the stakes were much larger than simply an island; Grenada held the possibility to exorcise some of Vietnam's ghosts, reminding Americans of their nation's military might and the just cause for which Reagan wanted them to fight.

In the 1970s, Reagan judged that the USSR had stepped into a void vacated by a defeatist America; to make a positive difference in the world again, the United States needed to reclaim its resolve. And such could not happen, he felt, under a president who produced "more Vietnams." Any real use of U.S. military power needed to be rare and quick and successful—meaning rapid victory with few lives lost. Despite his hawkish reputation, Reagan was quite restrained in using military force, and did so infrequently.[14] Grenada could be the anti-Vietnam. And it became the anti-Vietnam; it eroded the Vietnam Syndrome and reversed the defeatism.

Adding credence to that sentiment were the words from recent Republican commander-in-chief Richard Nixon, who commended Reagan and Grenada, noting that the operation "helped Reagan tremendously and lifted the spirit of the country. If you go in and nail the bastards without losing your men, you can go a long way politically. . . . [O]nce you make the commitment to do it, you've got to go in there and bomb the hell out of them with everything you've got." Nixon assessed: "There can be no hemming and hawing or hand wringing. Make the goddamned decision, and do it. . . . Keep it as short as possible; the short operations always work best." We saw that in Grenada, said Nixon.[15]

It was puzzling, then, that the Grenada operation was ridiculed by liberals because it appeared so easy. Future Clinton administration Secretary of State Madeleine Albright dismissively likened the operation to a football game pitting an NFL team against "The Little Sisters of the Poor."[16] Her sarcasm was mild compared to Democratic Party voices like former vice president Walter Mondale, Jesse Jackson, Senator Pat Leahy, Senator John Kerry, and the editorial pages of the *New York Times*.

This form of criticism was as disturbing as baffling: The Grenada operation was not simple, as is true for any military operation—a fact painfully obvious to those of the Vietnam generation, like Madeleine Albright and John Kerry. In truth, there are no easy military victories, and as demonstrated by U.S. interventions from the Bay of Pigs to Desert One, many seemingly straightforward operations had turned into fiascoes.

Despite criticism from the left, the need for a morale boost in the form of a sound military victory had been clear to Reagan for quite some time, a point that he reiterated during a speech at the Heritage Foundation shortly before the invasion. This was one of Reagan's most illuminating presidential talks, in which he again used the "sad, bizarre" line on Communism's doom as well as assuring his audience that "a democratic revolution is underway." He laid down his goal: "The goal of the free world must no longer be stated in the negative, that is: resistance to Soviet expansionism. The goal of the free world must instead be stated in the affirmative. We must go on the offensive with a forward strategy for freedom."[17]

Equally interesting, particularly in light of the Grenada operation soon to come, was what Reagan stated about morale:

> You can all remember the days of national malaise and international humiliation. Everywhere in the world freedom was in retreat and America's prestige and influence were at low ebb. In Afghanistan, the liberty of a proud people was crushed by brutal Soviet aggression; in Central America and Africa, Soviet-backed attempts to install Marxist dictatorships were successfully underway; in Iran, international law and common decency were mocked as 50 American citizens were held hostage; and in international forums, the United States was held up to abuse and ridicule by outlaw regimes and police state dictatorships. . . . All this is changing. While we cannot end decades of decay in only 1000 days, we have fundamentally reversed the ominous trends of a few years ago.[18]

Coming only weeks before Grenada, this statement is significant in two ways: First, Reagan sensed a turn around in the nation's spirits *before* Grenada. Second, he unknowingly sensed such a turnaround as a time approached when his Caribbean neighbors would request U.S. help in Grenada. As a result, he was especially prone to view Grenada as not just a liberation but a morale restorer once the request came in.

As the crisis unfolded three weeks later, Reagan saw a chance to chip away at the syndrome that had tied the hands of recent American leaders. He later wrote:

> Frankly, there was another reason I wanted secrecy [for the Grenada mission]. It was what I call the "post-Vietnam syndrome," the resistance of many in Congress to the use of military force abroad for any reason, because of our nation's experience in Vietnam. . . . We were already running into this phenomenon in our efforts to halt the spread of communism in Central America, and some congressmen were raising the issue of "another

Vietnam" in Lebanon while fighting to restrict the president's constitutional powers as commander in chief.

We couldn't say no to those six small countries [Caribbean neighbors of Grenada] who had asked us for help. We'd have no credibility or standing in the Americas if we did. If it ever became known, which I knew it would, that we had turned them down, few of our friends around the world would trust us completely as an ally again.[19]

Moreover, very importantly, Reagan knew that the Americans in Grenada were at risk as potential hostages, posing a repeat of the Iranian crisis that had already sunk American spirit. Hence, doing nothing in Grenada could hazard another loss to morale and become an even greater detriment to the American position.

Further, there was an additional, previously unexpected need to boost American morale: Just two days before the invasion, 241 U.S. Marines were killed by a radical Muslim suicide bomber at their barracks in Beirut, Lebanon. While the decision to invade Grenada was made before the Beirut disaster, the operation became a face-saving counterweight to the tragedy in Beirut.

MOSCOW'S GRENADA REACTION

Despite the high morale that Grenada produced in the United States, Moscow was predictably unhappy with the invasion's outcome. Reagan's Grenada "triumphalism" nauseated the USSR—a measure of the operation's success. The Kremlin knew that Reagan's run for a second term was only a year away. Moscow hoped upon hope that Reagan would lose next November. Thus, to the Soviets, the win in Grenada was bad news not only because it stemmed Communism's advance but because it boosted Reagan politically.

TASS decried how "the master of the White House" had "strived to convince his compatriots that they 'can be proud' of that operation." What was there to be proud of? America had flung a "mighty naval armada" and thousands of Marines at a "tiny island state" that did nothing wrong in "a policy of state-inspired terrorism."[20]

If that was not strong enough, TASS released this stinging satire of an imaginary conversation between Reagan and Secretary of Defense Weinberger:

The telephone rang in the U.S. President's bedroom in late evening on November the Second.

"My President," the familiar triumphant voice of the defense secretary was heard over the phone, "the island has been done with! The sky is cloudless over the whole of Grenada. . . ."

"Caspar, dear, you are a hero! And I will accept no objections from you on that score! You have razed to the ground the island's university and thus liberated a thousand of our guys studying there. Glory to you, Cap! Unless you object, at tomorrow's news conference I shall refer to the invasion not as an 'invasion' as I was rather rash to describe it previously, but as a 'remarkable operation,' or even as a 'rescue mission.'"

"Very much so, Mr. President."

"As to your guys, well, those who are . . . so quick to react with their guns and rifles to anything when you land them in other countries, I shall describe them as 'our best missionaries abroad.'"

"It will be a perfect description. Magnificent guys! Incidentally, Mr. President, I gave instructions to cook up photos depicting them giving a hand to doctors inoculating the locals . . ."

"Perfect of you, Cap! You are a big strategist and a great politician rolled into one. And in the future, in all countries we intend to conquer, we must inoculate all the locals against the ideological contagion brought in by the 'Red agents'. . . ."

"I would hate to hide from you, my President, that there are some killed and wounded."

"Let this not trouble you. I'll just say that it was the Americans who were dying and getting wounded while defending the lives of others and upholding freedom and peace . . ."

"You are a giant, Mr. President!"

"Thank you, my loyal minister, for this accurate comparison and [I'll] be sure [to] repeat it to Nancy. She will be very pleased . . ."[21]

The phrase "our best missionaries abroad" is a slap at Reagan's religiousness and anti-Communist crusading. Not stopping there, the TASS satire ended with a fanatical Reagan going nuclear a few days later:

Seated in his armchair in the Oval Office back from the news conference, Reagan thought he richly deserved praise for his firmness: "Well done for me to have told them that we can do any other country in like we did with Grenada." Then his thought strayed and he clearly saw himself in "Air Force One" on the morrow after a nuclear free-for-all. He is eating his breakfast

with gusto now and then glancing into the porthole. Enters the defense secretary who reports: "My President, the earth is over and done with," and
then proceeds to decorate him with the Pentagon medal "for the destruction
of life on the planet."

"Serves them right!" the President says. "Next time they will think
twice before interfering with my attempts to restructure the world the way
I like. . . ."

Yes, the nut from California had finally done it: he had pushed the button. The Grenada "cowboy," as the Soviet press dubbed him, had nuked the
world.

Like the American left, a popular tactic on the Soviet side was to down-
play the U.S. action as a petty operation directed at a tiny, and thereby unim-
portant, country—a line that contradicted Moscow's obsessive attention. To
buttress this viewpoint, the Soviets frequently borrowed from American
columnists. To cite just one example, TASS devoted an entire statement to an
article by *Washington Post* associate editor Robert Kaiser, titled, "Is This a Foreign Policy or a Recipe for Disaster?"[22] Kaiser's 3,000-word op-ed in the Sunday "Outlook" section excoriated Reagan policy, and the Soviets loved it:
"History?" Kaiser begged. "It has no apparent place in Ronald Reagan's view
of the world, except for the caricatured version he has carried around in his
head for years." In this passage, Kaiser inadvertently gave the Soviet propaganda machine a gem on Grenada, and TASS seized it, quoting it liberally and
circulating it around the world.[23]

Another line of Moscow's propaganda assault, which was prolific in
Pravda, was to refer to the Caribbean "adventure" as a "piratic attack."[24] Soviet commentator Valentin Zorin, who judged Reagan a "blockhead" who
"does not care to think,"[25] called the invasion "naked banditry." Americans,
he claimed, were "the most misinformed people on earth." This was the result, he said, curiously speaking of the United States and not of the USSR, of
"a gigantic propaganda machine. . . . When you leave the country, you get access to normal information." Yet, a frustrated Zorin reluctantly conceded that,
"A considerable number of Americans applauded Reagan."[26]

"WE'RE GOING TO WIN THE COLD WAR"

The radiance exuding from Reagan and his team was clear in an internal report only days after Grenada. That November, Reagan and his contingent of

underminers got an exciting glimpse of the depth of Soviet desperation, courtesy of Herb Meyer.

Meyer was special assistant to Bill Casey and vice chair of the National Intelligence Council, a prestigious seat at the CIA, where he observed the full scope and brunt of the Reagan strategy. That strategy, said Meyer, citing the tandem of Reagan and Casey, was "*very* dangerous . . . very gutsy. . . . And there were a lot of people who said, 'Oh dear, you're right, the bear is wounded. Don't poke sticks at a wounded bear.' But the Reagan-Casey approach was: 'Hey, my enemy is on his knees. It's a good time to break his head.' "[27]

It was in this spirit that Herb Meyer revealed dramatic intelligence findings in the weeks after Grenada in November 1983. This intelligence demonstrated that the Soviets now understood Reagan's rejection of their nation's very existence: "[Y]ou could see that in the intelligence. What they were saying was: 'Uh, oh, he caught on.' Because they knew they could be had. One thing about a bully: he knows exactly how strong he is and how weak he is. And the Soviets knew that Reagan knew that they could be had."[28]

This new development caused Meyer to write an extraordinary memo, read carefully up through the chain of command. In his memo, Meyer assessed that the USSR was entering a "terminal phase," but in spite of this good news, there was a danger: a cornered Kremlin might opt for war. On the other hand, wrote Meyer, "If present trends continue, we're going to win the Cold War."[29]

It was a view that mirrored Reagan's own precisely, only this time it was supported by concrete analysis from the intelligence community. This memo was the articulation of a point which Reagan had known long before he was president: that the Soviet system would weaken with each passing year, and at some point would collapse, giving rise to a wave of a freedom. It was a proclamation that he reiterated that December just before Christmas, telling French reporters from *Le Figaro* that contrary to what Communism professed, it was "freedom which is infectious and democracy which is the wave of the future. The tide of history is a freedom tide."[30]

His year of strong words and decisive action was helping to expedite the Soviet demise. Now, as the focus turned to the election year ahead, the strategy to undermine would need to rely on the mechanisms that the NSC had already put in motion. Next, it was time to let the people decide who would lead them forward in the four years ahead.

14.

Winning the Second Term: 1984

WITH THE COLD WAR STRATEGY NOW IN PLACE, RONALD REAgan had to focus on another political strategy, one that, if it failed, would never allow an opportunity for that "take-down" strategy that had consumed the fertile anti-Soviet minds at Bill Clark's National Security Council. The president needed to preserve himself before he could hasten the decomposition of the system that he believed was rotting as quickly as the occupant of Lenin's tomb. In short, Ronald Reagan had to get reelected. As such, 1984 was not so much about formulating policy to win the Cold War but, instead, devising a political campaign to win Reagan the Electoral College.

Because Reagan believed that Grenada reflected America's recovery and return to greatness, the operation continued to play a prominent role in his speeches and words well into 1984. In a January interview with the *Washington Post,* kicking off the election year, the president said he perceived a "new feeling," a "great change" in the "confidence" of Americans. "I think the reaction of our people to the success of our rescue mission in Grenada was an indication."[1] These were sentiments that he would continue to reiterate throughout 1984, as he claimed that Grenada had not only helped that region "but perhaps helped all Americans stand a little taller."[2]

Not only was Grenada in the win column, but the economy was exploding, as one-time double-digit unemployment, interest rates, and inflation were down to single digits, including an inflation rate near 2 percent, not to mention thriving growth rates not seen in decades. Buoyed by the good news, Ronald Reagan could now be heard everywhere making statements about the turn-around in the national psyche.[3]

While his 1980 campaign had focused on the need to *restore* morale, the 1984 campaign highlighted the *renewal* of morale. A bumper sticker circu-lated by the campaign in 1984 was already a bit behind the times when it stated simply: "PRESIDENT REAGAN: Bringing America Back." It was al-ready back; it was "Morning Again in America," as the warm and fuzzy Rea-gan TV commercials celebrated. These ads borrowed the then-new anthem of country music artist Lee Greenwood, which rang out, "I'm proud to be an American."

It was a strong message and one that was established early on in the year when Reagan's January 1984 State of the Union address set the tone for the months ahead. "There is renewed energy and optimism throughout the land," he proclaimed. "America is back, standing tall, looking to the eighties with courage, confidence, and hope." After that declaration in the fourth sentence of the speech text, nearly every other paragraph in his address contained a word like "revival," "spirit," "confidence," "credibility," "purpose," or phrases like "crusade for renewal," "restore pride," or "new strength."[4] It was a turn-about for which an emotional Reagan would later literally thank God: "We can be grateful to God that we have seen such a rebirth of it [patriotism] here in this country."[5] He often said that the change in morale was among his "proudest" achievements.[6]

Indeed, this achievement was such a success that it has gone undisputed by even the political left, including academic political scientists and histori-ans.[7] In other words, credit for the accomplishment came from those not ex-pected to credit Reagan.[8] Outside of academia, there were as many if not more journalists who credited Reagan with recovery.[9] *Time*'s dean of presi-dential correspondents, Hugh Sidey, said flatly: "No one can deny that Ronald Reagan restored morale to a country that needed it"—a view sec-onded by veteran CBS reporter Mike Wallace, among numerous others.[10] Edmund Morris goes so far to claim that Reagan changed the national mood "overnight." The change was so quick, said Morris, "that it can only be as-cribed to him."[11]

Foreigners were also impressed: Canadian Doug Gamble was so taken that he moved to the United States at this time, desiring a country that wore its patriotism on its sleeve. Fondly recalling the Reagan era, he spoke descriptively of "the good old lump-in-your-throat, tears-in-your-eyes, hair-standing-up-on-the-back-of-your-neck patriotism that was so thick in the air you could almost reach out and grab a fistful of it."[12]

It is easy to forget that by 1984 Ronald Reagan had achieved this renewal while critics tore at him and his policies. He was called stupid, uncaring, a warmonger—and had been especially vilified in the previous, intense Cold War year.

THE KREMLIN VS. REAGAN

Most impressive, similar assessments came even from the enemy's camp— from as far leftward as the USSR.[13] This was later captured by *Literaturnaya Gazeta*, which informed Soviet citizens: "The years of his presidency have seen an unprecedented surge in America's self-belief, and quite a marked recovery in the economy. . . . Reagan restored America's belief that it is capable of achieving a lot." It closed glowingly: "Reagan is giving America what it has been yearning for. Optimism. Self-belief. Heroes."[14]

As Reagan pursued a second term, this morale boost had the key dual, opposite effect of sapping Soviet confidence. Vladimir Kontorovich and Michael Ellman noted that the Soviet leadership had always been keenly aware of the need to "score successes" in the competition between the Soviet and American systems. However, by the mid-1980s, the Soviets noticed that market economies like that of the United States were full of confidence in the superiority of their system. This confidence was driven home via Reagan's constant declarations of that superiority. This, reported Kontorovich and Ellman, had a psychological effect, as it devastatingly reflected reality, particularly among elite government officials who traveled to the United States and other Western countries, where they discovered the stark contrast in the two systems.[15] Clearly, the restoration of the United States was playing a key role in the Cold War, cementing the role of Reagan's domestic policy in his broader foreign policy mentality. Here his focus on morale had paid off in that it led to broad popular support at home and came at the expense of Soviet morale.

Yet, while there may have been a grudging appreciation of what Reagan had accomplished, the Kremlin still wanted him defeated, and badly so. There was severe apprehension of that first Reagan administration and its prospects for a second term. Yevgenny Novikov recalled: "The Central Committee realized that they were facing a committed government in Washington. They saw activity on all fronts. . . . It frightened them to death."[16] At the start of 1984, the Soviet media was filled with examples of this siege mentality: TASS "economic writer" Vladimir Pirogov said it was "no secret" that Reagan was aiming to "exhaust" the USSR.[17]

According to a number of sources, such fears prompted a KGB "active measures campaign" that was underway by January 1984 and designed to disrupt Reagan's reelection prospects. None of these sources elaborate on the details, though such an operation would not be a surprise.[18] "The clear and widespread belief was anyone was preferable to Reagan," said Yevgenny Novikov.[19] If Reagan won a second term, the Soviets faced four more years in the crosshairs of the Crusader. Not surprisingly, they did not sit silent. They would go down swinging.

As Reagan kicked off his reelection campaign in January, sirens were sounded in *Pravda*. In his January 10 column, a perceptive Communist named Vitaliy Korionov linked Reagan's intent to undermine to his religious motivations:

> [I]t was the present White House incumbent, invoking God, who declared the "crusade" against socialism. . . . [T]he present U.S. administration has announced in official documents that its aim is to "destroy socialism as a sociopolitical system." U.S. political, economic, and ideological life is increasingly subordinated to that unreal task. . . .
>
> As we can see, the psychological warfare conducted by the United States and its allies against real socialism is organized, coordinated, and directed. . . .
>
> Washington is deeply involved in an exceptionally dangerous "crusade" against socialism as a social system. The most highly placed U.S. officials, headed by the president, are the spearhead of this spiritual aggression. . . . The U.S. President personally participates in the subversive actions. He does this in different ways.[20]

Korionov was joined by Vladimir Lomeyko, who two weeks later wrote in *Literaturnaya Gazeta* that in declaring his "crusade," the "incumbent White House master"—in this reelection year the Soviet press suddenly began referring to Reagan as the "incumbent"—was explicitly seeking to "overthrow" the Soviet empire.[21]

One of the strongest Soviet statements on the Reagan challenge, including the economic assault in particular, was offered in a January 27 article by two Soviet academics who held high posts in the military. Titled, "Imperialism's Economic Aggression," the article's coauthors were listed as "Doctor of

Economic Sciences professor Major General A. Gurov and Candidate of Economic Sciences Lieutenant Colonel V. Martynenko." It was published in *Krasnaya Zvezda*.[22]

The United States, assured the authors, had "in point of fact, already mounted an economic, ideological, and psychological war against the USSR and the other socialist community countries." Among other tactics, they cited "the unprecedented arms race mounted by Washington." "The economic war," they wrote, "is very closely linked to the arms race." "[T]the Soviet Union and the other socialist community countries," warned Gurov and Martynenko, "cannot close their eyes to the fact that Washington has declared a 'crusade' against socialism as a social system." They zeroed in on the economic assault:

> Economic warfare occupies a very important place in the "crusade" against socialism. Its strategic aims are directly linked, first with attempts to interfere in the socialist countries' internal affairs and with the desire to undermine their economies. Second, the schemes of economic warfare are very closely interwoven with plans to achieve military superiority, since it is precisely the economy that is the material basis of defense. Third, imperialism's aggressive foreign economic actions are accompanied by corresponding acts of subversion in the political and spiritual spheres aimed at liquidating the socialist social systems.
>
> The intensity of the economic war is increasing at a very high rate. . . .
>
> The United States has taken on the role of "commander in chief" in this war. The Reagan administration spent several years preparing its allies for it and directing the elaboration of a "united approach" to economic relations with the East. This was one of the chief issues at the conferences of leaders of the capitalist "Seven" in Ottawa (1981), Versailles (1982), and Williamsburg (1983).

This assessment by comrades Gurov and Martynenko was unerringly correct. With exactness, the authors spelled out precisely what the Reagan administration was doing, laying out details numbing in their surprising accuracy. Their only mistake was believing the USSR and Soviet bloc—amid its alleged "deepening fraternal cooperation" and "unbreakable cohesion," as they put it—would survive Reagan's assault.[23]

The reality was that Moscow was on its heels, and bitter at Reagan's success. The Soviet leadership was particularly enraged that Reagan, who had been low in the polls through much of 1982 and 1983 because of the recession, was now unexpectedly resurgent. The Soviets wanted Reagan to lose in November so badly that it hurt, and it seemed that they were about to be solicited from the most unlikely of places—from no less than the U.S. Senate.

THE TED KENNEDY FACTOR

It was during the 1984 campaign season that, according to a high-level Soviet document, Moscow almost got help from an unlikely corner in its attempt to defeat Reagan. If what Soviet documents allege actually transpired, it would no doubt prove to be the most fascinating aspect of the 1984 presidential race—one which was known by only a handful of people at the very top of the Soviet leadership and, as a result, never made America's newspapers or nightly news broadcasts. If it had, it would have been the major story of the 1984 campaign.

During his first three years in office, and particularly since the spring of 1983, Ronald Reagan had pushed a plan to deploy intermediate-range nuclear forces (INFs), also known as Pershing IIs, in Western Europe. His goal was to prompt the Soviets to remove their medium-range nuclear missiles from Eastern Europe. He told Yuri Andropov that if the Kremlin removed its missiles, there would be no need for the United States to deploy INFs. Reagan called this the zero-zero option: he wanted both sides to slash INFs to zero levels. If Andropov would not agree to do this, Reagan would ask NATO to deploy INFs.

Opposition to this policy was ferocious, with the Soviet propaganda machine dubbing Reagan a nuclear warmonger and labeling him a belligerent who wanted to plant missiles on the soil of Western nations that never hosted such weapons. In addition, the Western nuclear freeze movement gave the Kremlin a vocal ally extending from London to Bonn to New York. The nuclear freezers led massive protests and assailed Reagan in the harshest terms. The international left was convinced that Ronald Reagan was dragging the world to nuclear Armageddon. Freezers like Dr. Helen Caldicott of Physicians for Social Responsibility and certain Catholic bishops feared Reagan might blow up the world.[24]

Ultimately, Andropov refused Reagan's offer, which was actually rooted in proposals made in the late 1970s by West Germany's Helmut Schmidt and President Jimmy Carter. Defying all odds, Reagan's team persuaded leaders like West Germany's Helmut Kohl and Britain's Margaret Thatcher to accept

INF deployments. Through 1984, the missiles were deployed, and the left was incensed.

Reagan ideas like the INFs, as well as SDI and many others, infuriated most liberals, including Senator Edward "Ted" Kennedy (D-MA), who, according to a highly sensitive KGB document, was motivated to do something quite unusual. Indeed, the most intriguing opposition to Reagan's nuclear policies has sat for decades in the Soviet archives.[25] If the details of the document (see Appendix) are accurate, Ted Kennedy may have pursued an extraordinary partnership with Yuri Andropov.

Back in 1983, specifically on May 14, 1983, KGB head Viktor Chebrikov had sent a message of "Special Importance" with the highest classification to Yuri Andropov. The subject head to the letter read: "Regarding Senator Kennedy's request to the General Secretary of the Communist Party Y. V. Andropov." It began: "On 9–10 May of this year, Senator Edward Kennedy's close friend and trusted confidant J. Tunney was in Moscow. The senator charged Tunney to convey the following message, through confidential contacts, to . . . Andropov." The Tunney referred to in the letter was former Senator John V. Tunney (D-CA), who served in the U.S. Senate from 1971 to 1977. Defeated in 1976, Tunney was a private citizen in 1983.[26]

For Tunney to serve as Kennedy's liaison would not be a surprise, since Tunney and Kennedy were close. They had been law school roommates at the University of Virginia and often went sailing and mountain climbing together after law school. Tunney named his first child after Ted Kennedy, who is his son's godfather.[27] Along with their former wives, they rode on Kennedy's yacht together. They frequented the same social circles and shared many of the same friends. Remarkably, Tunney was said to somehow even share Kennedy's Hyannisport accent, even though he was not a native of the area. They were so close that Tunney once feared, "I didn't want to go through the rest of [life] being known as Teddy Kennedy's friend."[28]

After Tunney had been friends with the Kennedy family for a while, he even got the chance to meet Ted's famous big brother. When Tunney first considered running for Congress, John F. Kennedy himself—who, unlike Ted, was a fierce anti-Communist—advised Tunney to ditch his middle name, Varick, the name used by his friends and family since childhood. It sounded too Russian, said Jack the Cold Warrior—Tunney might strike the electorate as a Communist.[29]

According to the KGB's account of events, in May 1983, John Varick Tunney went to Moscow to tell the Communists that Senator Ted Kennedy was "very troubled" by the state of U.S.–Soviet relations. Kennedy believed the main reason for the dangerous situation to be "Reagan's belligerence"—particularly his commitment to deploy INFs in Western Europe. "According to Kennedy," reported Viktor Chebrikov in his letter to Andropov, "the current threat is due to the President's refusal to engage any modification to his politics." That refusal, said the letter, was made worse because of Reagan's political success, which made him even surer of his course, and obstinate.

Chebrikov's view was that Kennedy held out hope that Reagan's 1984 reelection bid could be thwarted. But how? Where was the president vulnerable? The Soviet interpretation of what transpired concluded that Kennedy had provided a possible answer. "The only real threats to Reagan are problems of war and peace and Soviet-American relations," wrote Chebrikov. "These issues, according to the senator [Kennedy], will without a doubt become the most important of the election campaign." Within the nuclear freeze movement and Congress there was resistance to Reagan. Yet, according to Chebrikov, Kennedy lamented that the opposition to Reagan was still weak; Kennedy regretted that Reagan was good at "propaganda," whereas statements from Soviet officials (*not* propaganda, presumably) were quoted "out of context" or "whimsically discounted."

Chebrikov then relayed Kennedy's alleged offer to Andropov: "Kennedy believes that, given the state of current affairs, and in the interest of peace, it would be prudent and timely to undertake the following steps to counter the militaristic politics of Reagan." The first step, according to the Soviet view of the Tunney meeting, was a recommendation by Kennedy that Andropov invite him to Moscow for a personal meeting. Chebrikov reported: "The main purpose of the meeting, according to the senator, would be to arm Soviet officials with explanations regarding problems of nuclear disarmament so they would be better prepared and more convincing during appearances in the USA." Kennedy, reported Chebrikov, offered to bring along liberal Republican Senator Mark Hatfield.

Second, wrote the KGB head, "Kennedy believes that in order to influence Americans it would be important to organize in August–September of this year [1983], televised interviews with Y. V. Andropov in the USA." By Chebrikov's account, the Massachusetts senator had suggested a "direct appeal" by Andropov to the American people. "Kennedy and his friends," stated Chebrikov, would hook up Andropov with television reporters like Walter Cronkite and Barbara Walters. According to Chebrikov, Kennedy suggested arranging interviews not merely for Andropov but also for "lower level Soviet officials, particularly from the military," who "would also have an opportunity to appeal directly to the American people about the peaceful intentions of the USSR." This was judged necessary because of so-called distortion by the Reagan administration.

In essence, Chebrikov's interpretation was that Kennedy offered to help organize a Soviet PR campaign, which would "root out the threat of nuclear war," "improve Soviet-American relations," and "define the safety for the world." "Kennedy is very impressed with the activities of Y. V. Andropov and other Soviet leaders," explained Chebrikov.

In the final paragraph of the letter, Chebrikov talked of Kennedy's political plans and prospects for 1984. "Kennedy does not discount that during the 1984 campaign, the Democratic Party may officially turn to him to lead the fight against the Republicans and elect their candidate president." "We await instructions," finished the KGB head to the head of the USSR: The senator "underscored that he eagerly awaits a reply to his appeal," the answer to which could be delivered through Teddy Kennedy's friend—John Tunney.

According to the KGB document, Kennedy's goal in reaching out to Andropov was to defeat Reagan on two fronts: He hoped to reverse the president's defense policies and foil his 1984 reelection bid. If the memo is in fact an accurate account of what transpired, it constitutes a remarkable example of the lengths to which some on the political left, including a sitting U.S. senator, were willing to go to stop Ronald Reagan—gestures not surprising to some who worked for the president.[30] Reagan's moves from the spring of 1983 into 1984, from the INFs to SDI, apparently sent some liberals over the edge. Reagan faced enormous opposition not just from Moscow but in Washington, and almost from Moscow and Washington (or at least certain elements of Washington) working together.

After Reagan left office, Tunney admitted to the *London Times* that he had made some fifteen trips to the USSR during the period, in which he acted as a "go-between" (his word) for a number of members of the U.S. Senate, as well as some unnamed others. He conceded: "I represented the views of some senators." He disputed certain portions of Chebrikov's memo, a few sentences of which were reported in the *Times* in February 1992. "I told them instead it would be good PR if they announced a cutback [in arms]," he said only.[31]

At one level, observers will decry Kennedy's actions as shameless political opportunism. Irate conservatives may try to label his overture "treason," charging that he sought to assist America's enemy during a time of "war," albeit an undeclared "cold" war, and will dub him everything from a useful idiot to

Benedict Kennedy to Red Ted for reaching out to the Kremlin at the height of the Cold War.

Kennedy defenders, on the other hand, will maintain that the senator did what he felt was best for world peace, even if that meant extending his hand to Yuri Andropov.[32] They will say that he was rightly concerned that the U.S.–USSR confrontation was spiraling out of control, edging closer to the precipice of nuclear oblivion. Andropov and other high-level Soviets actually feared the United States might launch a nuclear attack, as did many leftists in the West, including Ted Kennedy.

Yet, Democrats eager to defend Kennedy should know that this may not have been the first time he reportedly reached out to Moscow to undermine a sitting president's foreign policy in an election season: according to another KGB document, the previous target of the Massachusetts senator had been Democratic President Jimmy Carter, Kennedy's own political flesh and blood, whom Kennedy somehow believed was guilty of belligerence toward the USSR over Afghanistan. In early 1980, Kennedy, for whatever reason, seemed convinced that Leonid Brezhnev was committed to a peaceful settlement in Afghanistan—a country which the Soviets had just invaded and where they remained for the next decade. According to Vasiliy Mitrokhin's 2002 report for the prestigious Cold War International History Project of the Woodrow Wilson Center in Washington, DC, which also cites Soviet documents, John Tunney went to Moscow on March 5, 1980 to relay yet another message on Kennedy's behalf: Here, too, Kennedy reportedly blamed not the Kremlin and its dictator for aggression and escalating tensions but instead the American commander-in-chief that Kennedy was campaigning against for the presidency. Tunney said that the great Massachusetts liberal saw it as (in Mitrokhin's words) "his duty to take action himself."[33]

As the 1984 race heated up, Ted Kennedy continued to do extraordinary things to prevent Reagan's reelection. He wrote a March 1984 article for *Rolling Stone*, in which he again denounced Reagan's "Star Wars schemes" and called Reagan "the best pretender as president that we have had in modern history," before leveling the irresponsible allegation that Reagan officials were "talking peace in 1984 as a prelude to making war in 1985." Kennedy spoke of his "fears about an administration whose officials have spoken of winnable nuclear conflict."[34]

Needless to say, Kennedy's warnings did not materialize. In neither 1985 nor any point thereafter did Ronald Reagan fire a nuclear warhead. If the Kremlin never developed a formal axis with Kennedy, it at least sensed it had a blistering ally to assist its PR campaign against Ronald Reagan and his policies.

THE SDI DECEPTION

Despite the seeming potential for collaboration with Kennedy, Soviet fears of SDI's feasibility were worsened by a shadowy, extraordinary incident in June 1984. While the Kremlin was unsure whether or not the United States could execute a missile defense system, it remained uneasy over SDI. This uneasiness proved ripe for exploitation, and the Reagan administration sought to take advantage of Soviet anxiety. In the ensuing program of deception reportedly approved by Secretary of Defense Cap Weinberger, the United States appears to have rigged an SDI test.

To convey this hoax, military planners reportedly scheduled four attempts to down a missile launched from California with an interceptor missile launched from the Pacific. After three failed attempts in this test, a fourth attempt was a perfect success. The fourth test worked flawlessly, or so it seemed, as the missile struck and exploded the target—an apparent major victory for the SDI program. The test, however, was rigged—a beacon was inserted in the target and a receiver was placed on the interceptor.

In a revelatory August 1993 piece, the *New York Times* cited unnamed Reagan administration officials who confirmed the fake, explaining the ruse as necessary to fool the Soviets into diverting crucial resources. The *Times* tracked down retired Secretary Weinberger from his home in Maine, where, wrote reporter Tim Weiner, he "would not confirm or deny that he had approved the deception." Nonetheless, acknowledged Weinberger, in a possible concession: "You always work on deception. You're always trying to practice deception. You are obviously trying to mislead your opponents."[35] Deception, or counterintelligence, or black operations, is a natural component of any war, including the Cold War.

In September 2005, I sought definitive confirmation on the fake test from Weinberger and other officials. The 88-year-old Weinberger, in poor health (he died seven months later), said he did not recall the incident or even the 1993 *Times* article. Weinberger's close friend Bill Clark was also unaware of the incident, which would not have been surprising, since Clark was no longer national security adviser in June 1984. I posed the question to author Peter Schweizer, a friend of Weinberger who partnered with the former secretary of defense on a number of books and projects. Schweizer confirmed the incident.

He pointed to a 2001 conversation in which Weinberger told him that the "disinformation" had indeed taken place.[36]

Unfortunately, I was not able to confirm Ronald Reagan's precise role in the situation, but regardless the deception worked. The Soviets were further alarmed about the realistic possibilities of SDI and suddenly had a greater imperative to avoid a second term of the Reagan administration.

RHETORICAL BOMBS

Ironically, though Reagan kept silent on something of such magnitude as a faked SDI test, two months later he blurted out an unintended statement that handed the Soviets a campaign issue, and during the time of the Republican and Democratic national conventions. During an August microphone test prior to one of his weekly radio broadcasts, the president jokingly told Americans that he had just signed a law outlawing the USSR and the bombs were about to start flying. Reagan thought he was speaking off the air, but he was on live. To the Soviets, the joke was not funny.

A parade of subsequent articles in the Communist press claimed that the gaffe proved Reagan was not merely out to end socialism, but, as one East Berlin reporter put it, to "bomb it out of existence"; his *real* intentions had been exposed.[37] The Bulgarian press said that while Reagan's team had always been "aggressive and constant in its anti-Sovietism," and "the moral and physical negation of real socialism has been the ultimate goal of the United States," now, finally, the whole world could recognize Reagan's unmistakable objectives through his insensitive humor.[38]

On Moscow television, Genrikh Borovik linked Reagan to Hitler, noting the similarity between Reagan's regular rhetoric—here Borovik had in mind recent Reagan words on Grenada, not the radio gaffe—and that of the Fuhrer. "The very same words were heard by the world from Berlin just over 45 years ago when Germany was occupying Czechoslovakia," opined Borovik, with little sense of proportion.[39]

This Kremlin chatter failed to silence Reagan. That August, Reagan himself made reference to the Nazis. The occasion was the fortieth anniversary of the Warsaw uprising, where he again made clear his intentions regarding Eastern Europe. He declared:

[L]et me state emphatically that we reject any interpretation of the Yalta agreement that suggests American consent for the division of Europe into spheres of influence. On the contrary, we see that agreement as a pledge by the three great powers to restore full independence and to allow free and democratic elections in all countries liberated from the Nazis after World War II and there is no reason to absolve the Soviet Union or ourselves from this commitment. . . .

Passively accepting the permanent subjugation of the people of Eastern Europe is not an acceptable alternative.[40]

Unlike in the past, Ronald Reagan now remonstrated against Yalta not as a presidential candidate but as leader of the free world, where he could make things happen. And he refused to accept Eastern Europe's continued condition.

In response, TASS expressed disbelief at the president's insolence. Complaining that this was hardly the first time Reagan had spoken against Yalta, a displeased TASS properly interpreted this particular case as an example of Reagan "actually call[ing] for the revision of the decisions of the Yalta conference."[41] What audacity!

FALL OF 1984: MIXED MESSAGES BEHIND THE CURTAIN

As the November presidential vote approached, the Soviets remained strident in their war of words. On September 7 during Moscow TV's "The World Today" program, watched by millions of Soviet citizens, Valentin Zorin warned of the dangers of a reelected Reagan, a president who was "banking on undermining the economy of the Soviet Union," on "stifling" the USSR, on "causing considerable harm."[42] He had to be stopped.

And yet, the protests were falling on deaf ears, as Reagan continued to climb ahead in the polls, leaving Democratic challenger Walter Mondale in the dust. Perhaps out of desperation, the Soviets voiced their concerns directly to Reagan. In a September 28 meeting, Foreign Minister Andrei Gromyko told the president, "Behind all this lies the clear calculation that the USSR will exhaust its material resources before the USA and therefore will finally be forced to surrender."[43]

Gromyko did not record Reagan's response. He didn't need to.

Of course, not everyone from the Communist world was fighting against Reagan's reelection, especially those who had escaped and could cast a ballot for him. As the first Tuesday in November approached, the president received an October 26 letter from Father S. C. Rokicki of Assumption B.V.M. Church in Detroit, who said he was writing on behalf of "a few of the Bishops, many priests and very very many people" he had met in Poland while there for three-and-a-half weeks. "They wanted me to tell you that 80 percent of Poland is praying for you for re-election," wrote Rokicki. "The Polish people are very proud that you stand up to the Communists. Since the sufferings and prayers of Poland have meant so much to God over the centuries, I am sure that you will be re-elected. The Polish people know weeks in advance you will win." Father Rokicki concluded by thanking Reagan "for being such a good President and for taking a strong stand against Communism and abortion."

The letter meant a lot to Reagan, as he took time out of his extremely busy campaign schedule to write back to Rokicki on November 4, only two days before the election: "I'm also very proud, yet humbly so, to have the approval of the Polish people and their prayers. . . . I pray that I can be deserving of their support."[44]

On November 5, the eve of the election, Reagan reminded Americans that only four years earlier, the United States had been mired in recession, the nation had fallen, and the Soviets were advancing. Even the most devout Democrat had to admit that all of this had indisputably changed for the better under Reagan. He invoked Main Street and the Shining City and told his fellow Americans that their nation's best days lay ahead. "And you ain't seen nothin' yet," he promised.

The next day, he won reelection in a landslide, defeating former Carter vice president Walter Mondale by sweeping 49 of 50 states, losing only Mondale's home state of Minnesota, and even then by a narrow margin. He very nearly took all fifty states, running away with the Electoral College—525 to 13.

Despite their attempts, Ted Kennedy and the Soviets could not derail Reagan, nor his vision for a brighter America, which had helped him win the overwhelming support of the American people. Reinforcing the agenda he had laid out over his first four years, the election reaffirmed Reagan's calling and liberated him to continue a tough course against the USSR. What had begun four years earlier would proceed unabated, giving Reagan the opportunity to see his goals to their conclusion.

After a busy year of campaigning, Reagan was free to again make the Kremlin the center of his attention. With reenergized vigor, he headed into the new year, prepared to confront the Soviets on an entirely different scale. As he would soon find out, 1985 was to be a monumental year in the history of American-Soviet relations, one that would see not merely a renewed focus in Reagan's economic war, but a new leader for the USSR.

PART III

The Second Term

The Emergence of Mikhail Gorbachev: March 1985

WHILE RONALD REAGAN'S REELECTION PROMISED MORE OF the same, the big change in 1985 took place not in the Oval Office but instead with a sudden shift at the pinnacle of power in Moscow. On March 10, 1985, things began to unfold quickly in the USSR when the aged General Secretary Konstantin Chernenko died. Now, the totalitarian system moved hastily to pick a replacement. Fittingly, the Soviet people, accustomed to being kept in the dark, had no clue about the transition taking place behind closed doors; indeed, a day after the fact, they still had not even been informed that Chernenko had succumbed.

Behind the scenes, Mikhail Gorbachev was nominated for the nation's highest office by Stalinist Andrei Gromyko, with the eventual enthusiastic support of the entire Central Committee.[1] Gorbachev beat out Moscow Party boss Viktor Grishin and Prime Minister Nikolai Tikhonov. On March 11, the totalitarian press simultaneously announced to the public Chernenko's death and Gorbachev's ascension.

Upon assuming power, Gorbachev was fifty-four years old, a marked departure from his elderly predecessors, the last three of which (Brezhnev, Andropov, and now Chernenko) had died in office. The Soviet Union, once ruled by Stalin for parts of four decades, had now had four leaders in only three years. Reagan quipped that he wanted to negotiate with the Soviet leaders "but they keep dying on me."

And perhaps Reagan was sending a message to Gorbachev when only a few weeks later he headed to Europe to assure a West German audience on May 6, 1985, four years before the wall came down: "[Y]ou can create the new Europe—a Europe democratic, a Europe united east and west, a Europe at long last completely free."[2]

Reagan was exuding confidence. When he got back home, he told a Florida audience: "I firmly believe the tide of history is moving away from communism and into the warm sunlight of human freedom."[3] At the top of the other superpower, a man named Gorbachev was listening closely.

BORN MARCH 2, 1931 IN A SMALL VILLAGE IN SOUTHERN RUS-sia, Mikhail Gorbachev, like Ronald Reagan, came from humble origins. He was raised a peasant on a collective farm during Stalin's disastrous collectivization, where his father worked at a tractor station. Like Dutch's father, Mikhail's dad struggled financially. Both boys' mothers endured their troubles by consulting the Bible, which was a considerably greater risk in Russia than it was in Dixon, Illinois.[4] Mikhail was young enough to be spared from World War II, a conflict that took the lives of more men from Russia than any other nation.

Gorbachev was admitted to Moscow State University, where his academic focus was law and agriculture, the latter of which became his area of expertise in the Communist Party. At Moscow State, politics also captured his interest, and Gorbachev began his political career as leader of Komsomol, the country's Communist youth group. There, in 1951, he met his future wife, Raisa, a devout Marxist-Leninist-atheist, a soul-mate who, like Nancy Reagan to Ronald Reagan, supported his political ambitions.

It was with this marked background of education and ambition that enabled the charismatic Gorbachev to rise quickly up the ranks of the Communist Party, eventually becoming a full member of the Politburo in 1980 and positioning himself for the top spot in March 1985. Perhaps it was fitting that as Reagan's second term was in its infancy he suddenly had a new counterpoint in Moscow, an individual that he and the world hoped would stem the tide of oppression that had flowed from the Kremlin for so long.

REAGAN'S ROLE IN GORBACHEV'S SELECTION

Notably, it may in fact have been a direct intention of Reagan policy to influence the rise of a Soviet general secretary like Gorbachev. As possible evidence, a remarkable letter exists in the personal files of Bill Clark: In January 1984, Clark, who was by then secretary of the interior, wrote a letter to his old friend, giving Reagan advice on seeking a second term. Therein, Clark proffered a remarkable statement, assuring Reagan: "Another few months of 'standing tall' should restore the arms balance in Europe and very likely influence the rise of a less dangerous Soviet leader than the dying Andropov." Such a less dangerous Soviet leader was indeed in the works.

Regardless of whether this was Reagan's intention, a variety of sources, including former Soviet officials and even certain members of the American left, including the academy, claim that Ronald Reagan helped to "create" Mikhail Gorbachev by generating policies that led to a decisive need for change in Moscow's inner circle, and that demanded a new leader like Gorbachev.

While this argument has some credence, it is impossible for Reagan's policies to take all the credit for Gorbachev's appointment, since Gorbachev began his rapid ascension through the Soviet ranks before Ronald Reagan became president. It was October 21, 1980, two weeks before Reagan was elected in the United States, that Gorbachev had been promoted to full membership in the Politburo; he thus joined a small group of Soviet politicians who were both full members of the Politburo and of the Central Committee, and was by far the youngest member of that select group. He had become a senior secretary of the Central Committee two years earlier in November 1978.[5]

Archie Brown, the insightful biographer of Gorbachev, who is inclined to little sympathy for Ronald Reagan, maintains that there is "not a shred of evidence" to support the argument that Reagan's policies played a role in bringing Gorbachev to power. There may indeed be an absence of evidence, since, as Richard Pipes points out, the Communists always cloaked the elections of their leaders in deepest secrecy.[6] Still, says Brown, the two were "totally unconnected." He says that if not for the rapid deterioration in the health of Andropov and Chernenko, Gorbachev would not have come to power at all while Reagan was in the White House.[7]

Put that way, Brown is no doubt correct. However, while Gorbachev was already on a path to power, Reagan's policies dramatically worsened the Soviet crisis—economically, militarily, politically—and added a sense of urgency for a dynamic, energetic leader like Gorbachev. Andropov badly wanted Gorbachev as general secretary, particularly over Chernenko and advanced Gorbachev's career more so than any Soviet official.[8] Andropov was extremely worried about how to respond to Ronald Reagan, a man he compared to Hitler. The Soviet fear of Reagan was so intense that KGB officer Boris Yuzhin later claimed that there was talk of assassinating the president; at the least, it was abundantly clear that they desperately wanted Reagan gone.[9]

More than ever, Andropov saw Gorbachev as a future Soviet leader[10]—a situation made critical because of the Reagan factor. Richard Pipes remembers that Soviet trepidation over Reagan was so acute that Western academic Sovietologists incessantly warned that Reagan's anti-Communism would produce another Stalin.[11] Quite the contrary: Andropov undoubtedly saw Gorbachev as the best response to the incursions by America.

Gorbachev and the Central Committee that supported his general secretaryship were gravely concerned with the international situation and the threat posed from Washington. Though the full import of Reagan's policies remains unclear, it is unreasonable to presume that the then perilous U.S.–USSR relationship was not a significant factor in considering the next Soviet leader. Archie Brown acknowledged that Gorbachev was selected by the Central Committee "as a moderator who would give dynamism to Soviet policy," a moderator whose mediation may have been necessary because of Reagan. Surely, such dynamism was needed because of the radically new circumstances now alarming the USSR from the Oval Office. It was no coincidence that, as Brown said, "during the first phase of Gorbachev's leadership" the Soviet Union was "preoccupied with the relationship with the United States" more than any other country.[12]

For instance, one cannot ignore the influence of a Reagan statement like the following, issued only weeks before the Politburo sat down to pick Chernenko's successor: "There is one boundary which Yalta symbolizes that can never be made legitimate, and that is the dividing line between freedom and repression. I do not hesitate to say we wish to undo this boundary. . . . Our forty-year pledge is to the goal of a restored community of free European nations. To this work we recommit ourselves today."[13]

In other words, on the fortieth anniversary of Yalta, Reagan made clear his desire and intent to undo Europe's division. It would be naïve to think that such statements did not influence Soviet decision-making in choosing the next leader.

Despite Brown's contrary opinion, Reagan supporters at the time and today ardently believed that Reagan helped to create Gorbachev. Margaret Thatcher made such a statement, as did William E. Brock, who held many posts under Reagan, including secretary of labor.[14] In a talk at the University of Virginia's Miller Center, Brock predicted that future historians "will give a lot of credit to Ronald Reagan for the emergence of Mikhail Gorbachev." "I think Reagan deserves credit for Gorbachev," said Brock, calling Reagan a "parent" of Gorbachev. "I really don't believe that Gorbachev would be there or would have survived were it not for Ronald Reagan."[15]

Surprisingly, outside of Reagan boosters, the late historian Stephen E. Ambrose, who judged Reagan incompetent and a disaster for America, stated as early as 1988 that "history will remember Reagan as . . . the American president who helped make it possible for Mikhail Gorbachev to begin the process of restructuring Soviet society."[16] This view is not uncommon, and there are a number of respected Cold War and presidential scholars who agree.[17] One such scholar was historian John Lewis Gaddis, who, in his classic 1992 work published by Oxford University Press, pointed to SDI as one among a number of influences that "contributed to the rise of Gorbachev."[18]

In its later documentary on Reagan, done as part of its outstanding series on presidents, PBS's "The American Experience" called Gorbachev "the Soviet system's best response to the challenge of Ronald Reagan." In the documentary, Reagan biographer Edmund Morris, who is seen as generally uncharitable toward Reagan, made the same assertion. He maintained that Andropov began grooming Gorbachev in 1983 as, in Morris's words, "the only and most likely Soviet leader who would be able to handle this formidable, adamantine anti-Communist on the other side of the Atlantic"—a point echoed by Georgetown professor and Harvard fellow Derek Leebaert.[19] Morris said that Andropov, aware that he was dying, saw Gorbachev as the only Soviet politician who could deal with the "canny, determined" Reagan.[20]

An especially observant assessment of Reagan's influence on Gorbachev's selection is provided by Peter Schweizer, a fellow with Stanford's Hoover Institution and author of a number of books on Reagan and the Cold War. Schweizer notes that after Stalin's death, the Kremlin quietly divided into two camps with competing visions over which direction the USSR should go: hard-liners and reformers. Prior to Reagan, the hard-liners had been ascendant, as their aggressive approach had worked: the Soviets made magnificent gains throughout the 1970s, while the United States declined. As Soviet Ambassador Oleg Grinevsky put it, the Soviet Union was "on the march . . . had reached the peak of its power."[21]

Under Reagan, however, the tables turned completely: The USSR was in its worst crisis since World War II and the United States was suddenly soaring. This opened the door for the reformers, for men like Gorbachev. This was so clearly the case that, as noted, no less than hard-liner Andrei Gromyko championed Gorbachev's selection by the Central Committee; in so doing, Gromyko, more than any other high-level Soviet official, enabled the choice of Gorbachev.[22] And it was Gromyko who constantly complained about Reagan with repeated dire warnings of how Reagan and his aides "want to cause trouble. They want to weaken the Soviet system. They want to bring it down."[23] Schweizer points to Gromyko's nominating speech for Gorbachev in March 1985: "Perhaps because of my official responsibilities [as a long-time ambassador and foreign minister], it is rather clearer to me than to other comrades that he [Gorbachev] can grasp very well and very quickly the essence of those developments that are building up outside our country in the international arena"—a direct reference to the U.S.–USSR relationship, and perhaps a key fragment of that "shred of evidence" that Archie Brown feels was lacking.

Significantly, Schweizer adds that the three mighty Soviet institutions— the foreign ministry, the KGB, and the military—that forged the coalition which brought Gorbachev to power happened to be the three chiefly concerned with foreign policy and the Cold War struggle, and the three most troubled by Reagan. Schweizer concludes: "In a very real sense, Gorbachev owed his selection to the pressures Reagan was exerting on the Soviet system."[24]

There are Soviets who openly endorse this view. Ilya Zaslavsky, later a popular member of the democratically elected Russian parliament, declared: "Ronald Reagan was the father of perestroika."[25] Yevgenny Novikov, who served on the Central Committee senior staff, said that Reagan's policies "were a major factor in the demise of the Soviet system," and tied the birth of Gorbachev's trademark *perestroika* to Reagan military spending. "*Perestroika* was in many ways a military initiative," said Novikov. "They spoke about it as early as 1982, and saw it as necessary to preserve the Soviet military capability, particularly in light of the American buildup."[26]

Even the odious Valentin Falin, whose continuing dedication to the Stalinist motherland is matched only by his unbridled hatred of Ronald Reagan, twenty years later expressed his lingering bitterness over the Reagan team's "strategic operation" to deprive the USSR of hard currency, which, said Falin, was so crippling that it had "called for the appearance of Gorbachev."[27]

This view was so common in the Soviet Union that a visitor to post–Cold War Russia in the 1990s could stroll into a Moscow gift shop and encounter a classic collection of Matryoshka dolls of Soviet general secretaries with an unusual twist: Ronald Reagan was underneath Mikhail Gorbachev.[28]

Though many continue to debate the underlying rationale behind the election of Gorbachev, the fact is that Reagan's presidency at the very least played a role in his rise to the leadership position. Though the full extent of that role may never be known, it is clear that Reagan's policies created an environment in which the Soviet Communists needed a person who could try to cushion Reagan's hard-line positions. In Mikhail Gorbachev they found an individual at once patriotic and practical, who would initiate the broadest steps toward democracy that the Soviet Union had ever seen.

THE SOVIET CENTURY

While much has been made of the conditions that led to Gorbachev's election, what was of greater importance were Gorbachev's intentions upon entering office. It was clear that his central objective from the outset was to preserve the USSR, and, even, to keep intact the Eastern European Communist bloc that formed the cornerstone of the broader Soviet empire. This unflinching desire to preserve the USSR was immediately obvious, and shortly after arriving in power, he put together what became his best-selling manifesto, *Perestroika*, a 254-page book written to the world for the purpose of informing the world of the thinking and plans of this dynamic new leader.[29] The goal of his signature policy of perestroika, wrote Gorbachev, was to make the Soviet Union "richer," "stronger," "better" and to raise it "to a qualitatively new level"—it was of "tremendous importance for the future of the USSR." With perestroika, there was "no stopping" Soviet society. He envisioned a Soviet Union glistening into the twenty-first century—a "golden age" ahead.[30]

Confident that Reagan would eat his words about the Soviet Union being consigned to history's garbage dump, Gorbachev said that "those hoping to overstrain the Soviet Union" were "presumptuous." In an almost mocking tone, he said: "So do not rush to toss us on the 'ash heap of history'; the idea only makes Soviet people smile." Continuing on, he insisted that, "The idea that our country is an 'evil empire,' the October revolution a blunder of history and the postrevolutionary period a 'zigzag in history,' is coming apart at the seams."[31] He dared Reagan to think he could "exhaust the Soviet Union economically . . . we sincerely advise Americans: try to get rid of such an approach to our country. . . . Nothing will come of these plans."[32]

Gorbachev said that the United States was suffering "delusions" and "illusions"; it was "naïve" in "the belief that the economic system of the Soviet Union is about to crumble and that the USSR will not succeed in restructuring." Reagan's "requiem" about an "ash heap of history," assured Gorbachev, was "clearly premature."[33]

To the Communist and non-Communist world, Gorbachev framed Marxism as the indisputable wave of the future.[34] He viewed the world through the prism of class. "Since time immemorial," said Gorbachev in eternal language, "class interests were the cornerstone of both foreign and domestic policies."[35] He elevated perestroika as the next logical step in the dialectical materialism of Marxism-Leninism; it would "fulfill" Lenin.[36] The problem, explained Gorbachev, was not Communism. Marx merely required a periodic readjustment, to be pushed to the next natural level of development.[37]

As he summed up in his conclusion to *Perestroika*: "We are motivated by the ideas of the 1917 October Revolution, the ideas of Lenin." "As perestroika continues, we again and again study Lenin's works," wrote Gorbachev.[38] "The present course—is a direct sequel to the great accomplishments started by the Leninist Party in the October days of 1917."[39] He called the October Revolution, which led to the deaths of tens of millions of Russians, a "turning point in the thousand-year history of our state and unparalleled in force of impact on mankind's development."[40]

Gorbachev also made clear his rather disturbing adoration of Vladimir Lenin, which continued years after he left office.[41] He spoke of the Bolshevik godfather in a saintly, god-like manner, as the infallible source of "objective" truth. He wrote prayerfully in *Perestroika*:

The works of Lenin and his ideals of socialism remain for us an inexhaustible source of dialectal creative thought, theoretical wealth and political sagacity. His very image is an undying example of lofty moral strength, all-round spiritual culture and selfless devotion to the cause of the people and to socialism. Lenin lives on in the minds and hearts of millions of people.[42]

A spiritual Gorbachev professed not merely an "interest in Lenin's legacy" but a "thirst to know him." Speaking of Lenin in a way akin to

Christ's redemptive sacrifice on the cross, Gorbachev said that he and the Party had "conceived the new mentality [of perestroika] through suffering. And we draw inspiration from Lenin. Turning to him, and 'reading' his works each time in a new way. . . . Lenin could see further." Gorbachev viewed Lenin as an eternal font of Soviet wisdom: "We have always learned, and continue to learn, from Lenin's creative approach."[43]

On the whole, Gorbachev whitewashed the vicious Lenin, once claiming, "More than once he [Lenin] spoke about the priority of interests common to all humanity over class interests."[44] This was arrant nonsense, as Ronald Reagan himself often reminded people in quoting Lenin's own words.[45] Gorbachev not only sugarcoated Lenin but transformed him into a liberal, a champion of social justice and human rights, a peaceful soul.[46]

Like Lenin, Gorbachev approvingly compared the October Revolution with the bloody, horrific French Revolution, which he called "the classical bourgeois revolution," the "Great Revolution of 1789–1793." Gorbachev reminded readers that Lenin had said that "socialism would consist of many attempts." The current Soviet leader was about to do his part to bring it all the way home, with perestroika as his mechanism: "perestroika is . . . a jump forward in the development of socialism."[47]

Lest there be any confusion, Gorbachev clarified: "There are different interpretations of perestroika in the West, including the United States," he said rightly. Perestroika does not, he affirmed at the start of his book, "signify disenchantment with socialism. . . . Nothing could be further from the truth." He reiterated: "Those in the West who expect us to give up socialism will be disappointed. It is high time they understood this. . . . I would like to be clearly understood that . . . we, the Soviet people, are for socialism." Gorbachev asked rhetorically: "Why should the Soviet people, who have grown and gained in strength under socialism, abandon that system?"[48] He explained:

> [H]ow can we agree that 1917 was a mistake and all the seventy years of our life, work, effort and battles were also a complete mistake, that we were going in the "wrong direction?" . . . We have no reason to speak about the October Revolution and socialism in a low voice, as though ashamed of them. Our successes are immense and indisputable.[49]

Far from being an evil empire, Gorbachev saw the USSR as a grand achievement to be extolled. This sentiment carried over into a definitive sentence from *Perestroika*, in which Gorbachev quoted Karl Marx's best-known maxim, linking his own vision directly with Marx's: "The policy of restructuring (perestroika) puts everything in place. We are fully restoring the principle of socialism: 'From each according to his ability, to each according to his work (or needs).' "[50]

Similarly, *Perestroika* revealed a host of Gorbachev's beliefs regarding the status of Eastern Europe and how success in those vassal states would be achieved. These were beliefs that not only differed from Reagan's but were the antithesis of what the president believed. For instance, Gorbachev blindly asserted that "it was not socialism that was to blame for the difficulties and complexities of the socialist countries' development." Worse, speaking of the tragedies of Hungary in 1956, Poland in 1956, and Czechoslovakia in 1968, when thousands of freedom-seeking Eastern Europeans were murdered by Communist troops from the USSR or their own armies, Gorbachev wrote: "Through hard, and at times bitter, trials the socialist countries accumulated their experience in carrying out socialist transformations."[51] He portrayed these events not as cries for freedom crushed by Soviet tanks but, rather, heartwarming moments of perseverance in the glorious struggle for socialist victory. These socialist states, said Gorbachev, were engaged in "revolutionary creative work" to push Communism to that next dialectical stage on the heroic road to a classless society. They were all, together, proceeding to the next "crucial stage in world development."[52]

While Gorbachev displayed a fatally flawed interpretation of the Eastern bloc's past, his view of the region's present and future was equally in error. According to Gorbachev, "Now we can safely state that the socialist system has firmly established itself in a large group of nations, that the socialist countries economic potential has been steadily increasing, and that its cultural and spiritual values are profoundly moral and that they ennoble people."[53] This was Gorbachev's prognosis for a region that within just two years of the publication of his words would take to the streets to throw off the shackles of Communism, to smash the Berlin Wall, to jail and sometimes execute its Marxist leaders, and to toss every Communist out of office.

Importantly, it is difficult to say how much of this was illustrative of the real Gorbachev, the evolving Gorbachev, or the Gorbachev forced to frame the state of Soviet society in a positive light in order to mollify the hardliners that kept him in power.[54] Most likely, at this point in his shifting political-ideological life, it was a combination of the three, a combination that led to a refined Soviet ideology and a new outlook for the Crusader's enemy.

THE COMMUNIST AND THE CAPITALIST:
COMING TOGETHER

With this set of Communist Party line views, many in the Reagan camp feared that there could be no overlap or shared goals between their president and Mikhail Gorbachev. No doubt, there was much on which Reagan and Gorbachev disagreed, and which the president learned right away. Gorbachev, who characterized the United States as brimming with racists, Indian rapers, imperialists, robber barons, and the homeless, dubbed Reagan's image of America as a Shining City as "American propaganda—yes, propaganda."[55]

But while their policies had very little in common there was one ideal that remained consistent for both: Like Reagan, Gorbachev adamantly rejected the thought of attacking the United States or Western Europe, and vetoed ever entertaining a nuclear strike: "only a madman would unleash nuclear war." Between the two, there existed a mutual abhorrence for nuclear weapons, and Gorbachev made it a point to state this emphatically in *Perestroika* and elsewhere. Furthermore, he spurned the goal of global Communism—a fact Reagan noticed immediately and brought up repeatedly.[56] The notion that he would seek a one-world Marxist state, said Gorbachev, was "nonsense."[57] Most significant, he was committed to reforming the Evil Empire, the full extent of which remained to be seen.

Also, Gorbachev emphasized dialogue. Often, pleas for dialogue are a trite, veiled way of saying that all will be fine if you agree with *me*; for the Soviets, this had been the spirit, or at least their understanding, of détente. To his credit, Gorbachev sincerely sensed that he and Reagan could negotiate the life-death issues related to nuclear Armageddon and believed that they should do everything within their means to eliminate as many missiles as possible.[58] Gorbachev's instincts were correct, and in time, the dialogue between the two superpower leaders would lead to great mutual respect and genuine peace. Before this could happen, however, the two leaders would pass through some rough and tumble, and Gorbachev would need to experience the failure of his central task: saving the Soviet Union.

16.

Afghanistan, the Arms Race, and Gorbachev: April to November 1985

ALMOST FROM THE MOMENT THAT HE ASSUMED HIS NEW POSI-
tion, Mikhail Gorbachev had two overriding and dramatic interests: he
wanted to end the war in Afghanistan and he wanted to end the arms race.
The war in Afghanistan, which the Soviets had been waging since 1979, had
proven to be one of the costliest undertakings of any Soviet leader and was a
conflict that had outlived all three of Gorbachev's predecessors.

Gorbachev's initial dedication to the Soviet effort was evident in a very
disturbing section of his book *Perestroika*. Amazingly, he asserted that "progressive changes were charted" in Afghanistan under the brutal Marxist dictatorship that took over in the latter 1970s. These positive moves, Gorbachev
maintained, were halted by "imperialist quarters [that] began to pressure
Afghanistan from without. So, in keeping with the Soviet-Afghan treaty, its
leaders asked the Soviet Union for help." Speaking of Afghanistan's tyranni-
cal regime, Gorbachev insisted it be allowed "to decide which road to take,
what government to have, and what development programs to implement."

While Gorbachev justified the Soviet presence in Afghanistan with the
notion that the Red Army was resisting the "imperialist quarters" who stood
in the way of the happy Communist government, the truth was that the
Afghan rebels who had been combating Soviet intentions in the country since

the late 1970s were fighting a horrible Soviet-backed regime. Though Afghanistan had been a monarchy since 1953, the 1970s had seen the USSR support the concentrated buildup of a united Marxist party in the country. By April 1978, the party had gained sufficient strength to satistfy Moscow, which in turn sanctioned a Communist coup that overthrew the nation's monarch, King Daud.

From the onset, Afghanistan's experience with Communism was horrific, and human rights atrocities in the country were rampant. The Marxist trio that took over from 1978 to 1979 was vicious: Torture was a daily practice, prosecuted by appallingly innovative means. In March 1979 in the village of Kerala, 1,700 adults and children, including the entire male population, were machine-gunned in the village square. The dead and dying were bulldozed. Across the country there were thousands of horror stories like this, indelibly ingrained in the consciousness of millions of Afghans, all of which were deemed necessary for creating the Communist utopia. As Lenin and Trotsky had promised, Communism would not be ushered in with white gloves on a polished floor. Throughout Afghanistan, the floor was soaked red.

In the early stages following the Communist coup, the Mujahedin rebels formed the predominent resistance movement to the new Marxist regime, and as such served as an obstacle to Soviet hegemony. On December 25, 1979, the USSR invaded and installed a puppet government headed by Babrak Karmal, a former deputy prime minister of Afghanistan who had been living in exile in Eastern Europe. With their puppet show in place, the Kremlin then launched a nine-year total war against the entire population of the country, violating a multitude of international conventions and even going so far as to employ the use of chemical weapons in 1986 under Mikhail Gorbachev.[1]

In their attempt to win the hearts and minds of the Afghan people, chemicals were not the only weapons that the Soviets used. In addition, the Afghan Marxist regime also went after religious believers shortly after April 1979, not allowing themselves to be deterred by Afghanistan's deeply devout Muslim culture. Even there, like everywhere, Communists tried to purge religion. The Koran was burned in public and religious practices were banned. Imams and other religious leaders were arrested and shot. On January 6, 1979, at night, all 130 men of the Moaddedi clan, a leading Shiite group, were massacred.[2] The Communist regime's goal was nothing less than to transform human nature. For such to take place, the leading rival belief system—in this case, the Islamic faith—had to be purged. Ronald Reagan frequently complained of how "evil" Communists went after religion everywhere they took power. Afghanistan was no exception.

On the whole, the Soviet war destroyed the country. Out of a preinvasion population of 15.5 million, five to six million Afghans fled to neighboring Pakistan and India, where they lived in squalor for two full decades.[3] Countless Afghans were scattered among refugee camps, where they birthed over one million children who had never seen their homeland. In all, more than half the population was displaced. Observers estimate that one to two million Afghans were killed in the war with the Soviets, 90 percent of which were civilians; another two to four million were wounded.[4] And long after the last Soviet soldier left, Afghans were still being maimed by the five to ten million landmines buried in their soil.

REAGAN'S REACTION

From his first days in office, Reagan made sure that he was well-versed on the subject of Afghanistan. His team would brief him routinely on developments in the country, and he would often speak with dissidents and rebel leaders about how the United States could best proceed. To Reagan, the culprit was obvious: this was yet another predictable result of Communism. He used the bully pulpit of the presidency to denounce the Soviets and their intervention, and to ratchet up rhetorical support for the resistance. When he spoke of "freedom fighters" in Nicaragua or Poland, he often hailed them in Afghanistan as well. (More accurately, the Muj were fighting for freedom from the USSR, but were hardly Jeffersonian democrats.[5])

In every Captive Nations Day statement or human rights address or in most speeches blasting Communism, Reagan hit the Soviets on Afghanistan. Following the tone that he had set for Poland, Reagan issued proclamations and memorials, initiating Afghanistan Days and even releasing statements in support of the Observance of the Afghan New Year, where he lamented the "tragedy of Afghanistan," where the people were subject to "intolerable conditions," "devastation," "immense suffering," "total war," and were generally "being brutalized."[6] The USSR stood guilty of "indiscriminate Soviet attacks on civilians," who were "innocent victims of Soviet imperialism." "Massive Soviet military forces" had invaded the "sovereign country of Afghanistan."[7]

"Nowhere," claimed Reagan, were "basic human rights being more brutally violated than in Afghanistan." This was a result of the "brutal and unprovoked aggression by the Soviet Union." Moscow was employing "blanket bombing and chemical and biological weapons," a fact he noted frequently. He also noted that three million people had been driven into exile—one of every five Afghans. "The same proportion of Americans," he calculated, "would produce a staggering 50 million refugees."[8]

Reagan hoped to offer more than vocal support. As early as January 1980 he wanted to help the Muj fight and defeat the Soviets.[9] Once president, his administration worked hard to get weapons to the rebels, which included shipments not just from Washington but from sympathetic nations like Egypt, Saudi Arabia, China, and Pakistan. CIA director Bill Casey jetted around the world prodding nations for aid and arms; he reprimanded foreign leaders when the weapons were not up to par, once in April 1981 telling a stunned Anwar Sadat that the material being supplied by the Egyptians was "garbage."[10] The Chinese provided valuable AK-47s, plus other forms of aid—details that to this day are kept tightly under wraps. "We can see Beijing too in the trenches of this dirty war," said a reporter in an angry *Izvestia* article titled "Aggressors and Hypocrites."[11]

So concerned was Reagan about Afghanistan that he told Gorbachev directly at the Geneva summit, which occurred several months after Gorbachev took power, that "the continued Soviet occupation of Afghanistan remains an obstacle to overall improvement in our relationship."[12] At Geneva, he excoriated the Soviets, to Gorbachev's face, for human rights abuses in Afghanistan, including the dropping of booby-trapped toys from airplanes, which were picked up by Afghan children. Making this point in a heated reprimand of Gorbachev, Reagan then asked the general secretary angrily and pointedly: "Are you still trying to take over the world?!" Gorbachev was visibly shaken, staring at Reagan in silence, mouth agape, with a stunned expression.[13]

Reagan arms control director Ken Adelman, a witness, called Reagan's words in that exchange the most "harsh indictment of Soviet behavior ever delivered to the top Soviet man."[14] Edmund Morris reported that the only person who appeared more flabbergasted was the State Department note-taker.[15] Reagan was seething: like Poland, he had developed an emotional attachment to the Afghan experience. Even as it appeared that relations between the United States and the USSR were improving, Reagan continued to offer blistering assessments of the Soviet war, assessing the Soviet occupation as a "reign of terror" on the people of Afghanistan.[16]

Reagan's heated rhetoric enflamed the Soviet press. To Moscow, the Afghan rebels were not freedom fighters but rather, according to the standard over-the-top language of TASS and *Izvestia*, "a handful of corrupted and despotic parasites," "imperialist lackeys," "ruffians," "terrorists," the "worst enemies," or, on a good day, "barbarous and uncivilized bandits." To call them freedom fighters was a "malicious distortion." Hence, when Reagan spoke up for human rights in Afghanistan he was engaging in "hypocrisy," "slander," and "flagrant demagogy."[17]

Despite the linguistic counterpunch to Reagan, by spring 1985 it had become increasingly clear that the Soviets were going to do more than just talk about escalating the war. With Gorbachev at the helm and a renewed vigor to see Soviet success in Afghanistan, the Russian military initiated a new war plan under General Mikhail Zaitsev, making the Afghan war the highest priority. Under Zaitsev, who was transferred from the prestigious command of Soviet forces in East Germany, the USSR planned to shift one-third of total special forces, known as Spetsnaz, to Afghanistan. The very best paratroops and KGB operatives were sent in, along with top battlefield communications equipment which was deployed via sophisticated Omsk vans.[18] According to Aleksandr Lyakhovskii, a high-level military official who wrote an insider's account of the war, Gorbachev gave Zaitsev "a year or two" to win.[19]

NSDD-166 AND THE ESCALATION

This Soviet escalation by Gorbachev catalyzed the Reagan administration's response, which had been brewing in the mind of Ronald Reagan since 1980. For a long time, Reagan, Casey, and their team had searched for the most opportune moment to strike back at the Soviets on this new Asian front, and Gorbachev's heightening of the conflict provided just the impetus necessary.

Of all the overtures to the Afghan rebels, none was more important than NSDD-166, signed by Reagan in the spring of 1985.[20] NSDD-166 was not the first Reagan administration directive to voice support for the Mujahedin. In January 1983, NSDD-75 had said that the "U.S. objective" in Afghanistan was to "keep maximum pressure on Moscow for withdrawal" and to "ensure that the Soviets' political, military, and other costs remain high while the occupation continues."[21] Still, it was NSDD-166 that delivered the firepower.

Though nearly all of Reagan's NSDDs have been declassified, to this day NSDD-166 remains unavailable, even in redacted form.[22] Why would it remain so secretive a decade-plus after the end of the Cold War? The reason is its aggressive nature, which can be distilled from reports and interviews. In the words of Peter Schweizer, who interviewed those who crafted and implemented the directive, the objective of NSDD-166 was to provide covert assistance to enable the rebels to achieve "outright military victory" against the USSR[23]—a goal deemed impossible by all but the Reagan team. Another person who interviewed the principals, Steve Coll of the *Washington Post*, disclosed his findings in an explosive two-part, front-page series. He reported: "[T]he new, detailed Reagan directive used bold language to authorize stepped-up covert military aid to the Mujahedin, and it made clear that the secret Afghan war had a new goal: to defeat Soviet troops in Afghanistan through covert action and encourage a Soviet withdrawal."[24] According to Reagan official Leslie Gelb, academic researcher James M. Scott, and other knowledgeable sources, the directive looked to force the USSR out of Afghanistan "by all means possible," shifting the U.S. objective from "make Moscow pay a price" to "make Moscow get out."[25] Another informed source, Christopher Simpson, reported that the directive committed the United States "to support a significant escalation" in the war.[26] Reagan State Department official Peter W. Rodman said the directive stated "a clear policy of seeking to defeat the Soviet Union . . . and force a Soviet withdrawal."[27]

Ultimately, the directive committed U.S. security agencies to use "all means available" to assist the Muj in *defeating* the USSR and to prompt a Soviet pull-out.[28] It was a bold and overt initiative and one that was vastly different from President Carter's directive, which had the less ambitious goal of "harassment" of Soviet forces. Carter's classified 1980 directive did not speak of driving the Soviets from Afghanistan or beating them on the battlefield.[29] The contrast was noted by Milt Bearden, a career CIA official: "The CIA's covert action role in Afghanistan dating back to the Carter administration called for 'harassing' the Soviets, not driving them out," stated Bearden, who in 1985 became deputy director of the CIA's Soviet–East European Division. "Reagan was upping the ante, and now he actually believed he could win." Bearden said that the covert program "had taken a new turn"—"Reagan had rewritten the ground rules."[30]

THE RESULTS OF NSDD-166

The impact of NSDD-166 was immediately felt as it authorized a supply of highly advanced weapons and millions of dollars in covert aid—requesting from Congress over $450 million for 1986 alone. It reprogrammed an additional $200 million from an unspent DOD account and then sought reauthorization through congressional intelligence committees. Because of NSDD-166's floodgate, in 1985 the CIA delivered 10,000 rocket-propelled grenades and 200,000 rockets to the Afghan rebels, exceeding the total for the previous five years combined.[31]

By 1987, 65,000 tons of U.S. materiel was arriving annually, as was what Pakistani General Yousaf dubbed a "ceaseless stream" of Pentagon and CIA specialists. Also arriving were imagery specialists carrying satellite photos of Soviet targets in Afghanistan; communications specialists boasting sophisticated communications gear; experts armed with teachings on psychological warfare; demolitions specialists with explosives and timing devices for blowing up bridges, tunnels, fuel depots, and whatever else; and much more. Among these, the satellite photos were exhaustive. General Yousaf 's office was soon covered with maps of Soviet targets, along with carefully diagrammed approach and evacuation routes and even analysis of how Soviet troops could be expected to react once attacked.[32]

This was all provided courtesy of the Central Intelligence Agency. Milt Bearden never forgot his extraordinary order from maverick CIA director Casey: "go out and kill me 10,000 Soviets until they give up."[33] While the United States could not provide troops to do the killing, the CIA could train and arm Afghan soldiers to handle that task. Langley fielded operations officers who set up training schools to educate Afghan rebels in sabotage, guerrilla warfare, mine-laying, antitank attacks, secure communications, and more. This training ground was jokingly labeled "CIA U."[34]

Eventually the U.S. effort became so intense that Bill Casey reportedly sought to expand it to Soviet territory, gunning down Soviet troops on their own turf. He targeted Soviet factories, military installations, and storage depots in some of the most dangerous action of the entire Cold War. Intelligence officials became quite anxious over such ambitious lengths and how the Soviets might react. Already, the escalation had been "directed at killing Russian military officers," said one official. "That caused a lot of nervousness." And now, this was "an incredible escalation," said Graham Fuller, a senior intelligence official who was among those strongly opposed to raids on Soviet territory.[35]

As it turned out, extending the war within the USSR had been Casey's thinking before NSDD-166 had even been signed. Not surprisingly it was a prospect that also appealed to Reagan.[36] According to Peter Schweizer:

In early 1983, [Casey] met with the president and Bill Clark in the Oval Office to discuss the situation in Afghanistan. The conversation turned to raising the stakes for Moscow. The DCI suggested taking the war into the Soviet Union itself, and Reagan liked the idea. "The president and Bill Casey were determined that Moscow pay an even greater price for its brutal campaign in Afghanistan, including the possibility of taking the war into the Soviet Union itself," recalls Clark. . . . Soviet casualties were not running high enough as far as the Reagan administration was concerned. The Kremlin could sustain these losses, Reagan told aides and colleagues, because of the closed nature of its system. He wanted the numbers up, and he wanted the Soviet high command demoralized.[37]

Casey's audacious goal was also that of Reagan and Bill Clark, the latter of whom confirmed that he and Reagan personally gave "authorization to Afghanistan forces and their supporters 'to cross the river' if they were so inclined and sufficiently supported."[38] Though these instincts were first broached in 1983 among Reagan, Casey, and Clark, NSDD-166 turned them into formal administration policy, a bold initiative from a country that was looking to avoid direct confrontation with the Soviets. According to Milt Bearden, through NSDD-166 Reagan ordered that the Red Army be pushed back across the Amu Dar'ya, the river that marked the border between Afghanistan and the Soviet republic of Uzbekistan.[39]

This decision to extend the war into Soviet territory took off once Muj commanders and the Pakistani Intelligence Service (ISI), which was managing the flow of CIA aid to the Mujahedin, embraced the idea. Only then was it adopted and pursued by the White House. According to Schweizer, in 1985 and 1986 specially trained Muj units operating inside the USSR, equipped with high-tech explosives from the CIA and Chinese rocket launchers, sabotaged Soviet targets. They derailed trains, attacked border posts, and laid mines. On one occasion, thirty Muj fighters attacked two hydroelectric power stations in Tajikistan; in another, they orchestrated a rocket attack on a Soviet military airfield. Schweizer said there were dozens of ambushes.[40]

As if that were not explosive enough, the Reagan administration had a bitter internal debate over whether to ship sniper-rifle packages—rifles equipped with long-range, sophisticated sighting scopes—to the rebels. These would be employed to infiltrate the Afghan capital of Kabul and assassinate Soviet generals and senior military officials. American intelligence took the aggressive step of identifying the residences of Soviet generals in Kabul and tracking their regular movements.

The Reagan administration understood that providing sniper rifles might violate the 1977 presidential directive precluding assassinations—a question that some say was hotly debated internally. Ed Meese says that he was not privy to that debate. In a sense, however, he does not understand the fuss. "It doesn't strike me as unusual to send sniper rifles when you're already sending Stinger missiles to shoot down helicopters," said Meese. "Some of these helicopters were probably carrying Soviet officials."[41] In the end, the rifles were sent, but without the intelligence information, nor the night goggles necessary to carry out assassinations. General Yousaf recalled receiving a few dozen rifles, more than thirty but less than 100.[42]

Perhaps the most authoritative testimony on the subject of Washington's assistance came from General Yousaf, who supervised the covert war during the critical period of 1983–87, when Washington's weapons flowed like milk and honey. As he recalled: "[W]ithout the intelligence provided by the CIA, many battles would have been lost, and without the CIA training of our Pakistani instructors, the Mujahedin would have been fearfully ill-equipped to face—and ultimately defeat—a superpower."[43]

Altogether, the Reagan administration funneled over $2 billion in money and guns to the Mujahedin, compared to the $30 million sent under President Carter.[44] Steve Coll of the *Washington Post* was not alone in properly calling it the largest U.S. covert action program in the history of the CIA.[45] Moscow responded with reportedly $3 to $8 billion annually trying to win the war, an investment that far exceeded its means.[46]

In the Soviet press, the covert effort was reported as an "escalation of aggression" in the "undeclared war" by "militarists" in the Reagan administration and CIA in particular. The United States was "drilling" these "criminal Afghan dushmans" and "hired bandits" and "arming them to the teeth" as part of "American imperialism's dirty war in Afghanistan," a country that Soviet propaganda dishonestly insisted had become "democratic" Afghanistan as a result of the Communist "liberation." The Reagan administration was the "peddler" and the Muj its "puppets."[47]

These inflammatory accusations from the Soviet media failed to gain traction internationally, in spite of the fact that Reagan's moves in Afghanistan constituted the closest the administration ever came to fighting a direct war with the Soviets. Though actual troops were absent, Reagan ensured that the rebels had everything else necessary that they would need to fight the war and prevent a Soviet victory. By soliciting the help of Afghanistan's neighbors, Reagan had also made it easier for covert aid to flow, leaving fewer American fingerprints on the machinery of war.

But despite Reagan's efforts to end things decisively as quickly as possible, the battles continued as Gorbachev adhered to his two-year commitment to victory. Lives were being lost on both sides, but it was clear that the war was having a severe, negative impact on the Soviet Union. Forced to reconcile this military commitment to the myriad of other problems that Reagan had created, the new general secretary and his followers struggled to keep their heads above water, and Reagan's team searched for more means to make their lives miserable.

THE ROLE OF THE ARMS RACE

While Gorbachev hoped that victory might be possible in Afghanistan, he was not so convinced about the Soviet role in the arms race. For Mikhail Gorbachev, the Reagan arms challenge was the only issue that loomed larger than Afghanistan upon his arrival as general secretary, as he sought to stop what he called the "lethal," "costly and dangerous arms race," which he listed as his number one priority.[48] The arms race was his fixation upon entering office and absorbed the bulk of his focus during 1985.

To Gorbachev, this was an issue that was tearing apart the Soviet economy like no other, and it did not take long for him to convince his colleagues of the fact. Soon after his arrival, it became passé to flip on Moscow TV and watch Georgi Arbatov screech that "the most important contribution to the cause of true freedom"—Arbatov's "true freedom" was eternal totalitarian Communism—"would be the end of the arms race."[49]

This was the reaction Reagan had long desired to see, as he believed that the Soviet system and economic philosophy had created severe internal problems that would rip the USSR apart. Pondering how his administration "could use these cracks in the Soviet system to accelerate the process of collapse," he concluded that someone had to pry a crowbar between the cracks and twist. Communism had rotted the wood; the crowbar would tear it asunder.[50] In order to form this crowbar, Reagan focused on the internal contradictions, employing his staff to search out the cracks hidden below the surface and to engage them. As his aides pursued this strategy, one of the more prominent fissures was the continuous, detrimental impact that the arms race was having on the Soviet economy. From early on in Gorbachev's tenure, it became clear that his vehement opposition to the arms race stemmed in large part from his country's inability to sustain it.

This reality demonstrated the effectiveness of Reagan's arms policies over the previous several years. Despite appeals from Gorbachev abroad and Democrats at home, Reagan did not want to end an arms race that had proven to be one of his most effective weapons against the USSR. For every dollar the United States spent on a weapon, the Soviets increased spending by a corresponding amount. The Kremlin was struggling to get any bang for the ruble; an infusion of billions of U.S. dollars would markedly aggravate the task. The success of the arms race up until 1985 made the challenge at hand clear and the imperative immediate.

Despite the fact that many administration officials supported this interpretation of the arms race, the then Secretary of Defense, Cap Weinberger, interviewed nearly twenty years after the fact, said that he did not see his task as trying to spend Moscow into oblivion, and was never so instructed. "I actually did not hear [Reagan] talk about that at all," said Weinberger in an October 2002 interview. "I must say that I didn't see a deliberate attempt to bankrupt them." Rather, "I saw him [Reagan] trying to gain military strength for its own sake, for security, to counteract them, to make up for our weaknesses." According to Weinberger, Reagan desired military power "because not having it was dangerous for the United States." It was Weinberger's charge to lessen that danger by expanding the military.[51]

A month after the above interview, Weinberger was publicly asked pointedly if the Reagan administration pursued a military buildup to bankrupt the Soviet Union. "I don't think it was a conscious attempt to bankrupt them," he replied. "It was a conscious attempt to tell them they could never win or prevail in an arms race with us." In saying this, Weinberger quickly acknowledged that there does not appear to be a big difference in his two statements, though he ensured there is a difference.[52]

In Weinberger's judgment, the military had deteriorated to a "truly appalling" state under President Carter, an ill which he and Reagan believed they could remedy through a buildup that would bring the Soviets to the bargaining table. In his memoirs, Weinberger listed four goals for the buildup, none of which were to "bankrupt" the USSR. And yet, he believed the buildup had that effect. "Our military buildup had an economic impact [on the USSR]," wrote Weinberger. "What else, besides President Reagan's determination to win the Cold War, won it? First: Our military buildup."[53] In Weinberger's eyes, the arms race bankrupting the Soviets was an incidental effect of Reagan's policy, and an indirect result of the stated administration policy.

But while Weinberger did not view himself as tasked to spend the Soviets into the grave, other key administration players (and the Soviets) believed this to be precisely Reagan's goal for a military buildup. Richard Allen, the first national security adviser, and Reagan's foreign-policy adviser throughout the decisive latter 1970s (when Weinberger was not), states that Reagan decided from the start of his presidency that the United States would be dedicated to a plan of "spending it [the USSR] into oblivion."[54] He said Reagan saw an arms race as a means of shoving the enemy off the plank.[55]

This view of the arms race was one that Reagan seems to have advocated long before he became president. In discussing U.S.–USSR relations in that previously cited early 1960s speech, he said, "the only sure way to avoid war is to surrender without fighting." "The other way," he asserted, "is based on the belief (supported so far by all evidence) that in an all-out race our system is stronger, and eventually the enemy gives up the race as a hopeless cause."[56] In the 1970s, he explicitly stated that this "race" was an arms race. Similarly, during the 1980 campaign he told the *National Journal* and *Washington Post* that an arms race was "the last thing" the Soviets wanted to see from their American counterpart: "They know that if we turned our full industrial might into an arms race, they cannot keep pace with us. Why haven't we played that card?"[57]

Once in office, Reagan looked to play that card. He continued to reiterate the importance of engaging the Soviets in an arms race that forced them to spend money beyond their means. In an October 16, 1981 interview, Reagan said of the Soviets: "They cannot vastly increase their military productivity because they've already got their people on a starvation diet. . . . If we show them [we have] the will and determination to go forward with a military build-up . . . then they have to weigh, do they want to meet us realistically on a program of disarmament or do they want to face a legitimate arms race in which we're racing? . . . [N]ow they're going to be faced with [the fact] that we could go forward with an arms race and they can't keep up."[58]

During a January 20, 1983 press conference, Reagan said he was "hopeful and optimistic" that the Soviets "cannot go on down the road they're going in a perpetual arms race." "And so," he hinted, "this is one of the things in connection with our own arms race."[59] The next day he said that "if we ever hope to get disarmament, we will only get it by indicating to them that if they're going to keep on with that buildup, they're going to have to be able to match us, because we're going to build up."[60] He added this anecdote: "There was a cartoon that explained it all. Brezhnev, before he died, was supposed to be talking to a Russian general. And he said to the general, 'I liked the arms race better when we were the only ones in it.' "[61] Connecting the dots, Reagan summed up: "So, this is what we're doing. We want peace."[62]

Looking back, in his memoirs, Reagan said candidly, "we were going to

spend whatever it took to stay ahead of [the Soviets] in the arms race." He figured, "Someone in the Kremlin had to realize that in arming themselves to
the teeth, they were aggravating the desperate economic problems in the Soviet Union, which were the greatest evidence of the failure of Communism."
He noted, with irony that would have made Lenin cringe, that "the great dynamic success of capitalism had given us a powerful weapon in our battle
against Communism—*money.*" The Soviets, Reagan explained, "could never
win the arms race; we could outspend them forever. Moreover . . . we had the
capacity to maintain a technological edge over them forever."[63]

Although on the surface Cap Weinberger's caveat appeared to discount
the economic importance of the arms race, he may have been the lone dis-
senter who felt that Reagan's military buildup was for its own sake. In truth,
many of Reagan's own statements (particularly in regard to SDI) suggested
that he viewed the arms race as another front for the economic war. Indeed,
it would have been surprising if Reagan, for whom seemingly every policy
was an attempt to undo Communism, did not view the arms race as one of
the cornerstones to his crusade. After all, the arms race gave him the opportunity to exercise his deepest held beliefs concerning the inevitable triumph
of the capitalist system, since it pitted the two economies directly against one
another.

THE FISCAL IMPACT OF THE ARMS RACE

By 1985, Reagan was certain that the Soviets were feeling the pinch of the
arms race unlike ever before. His assessment was supported by high-level
members of the Soviet government, who spared no chance to voice their concern over Reagan's motivations. "The imperialists would like to exhaust us
economically," cried Soviet deputy R. N. Stakheyev.[64] *Izvestia* complained:
"They [the Reagan administration] want to impose on us an even more ru-
inous arms race. They calculate that the Soviet Union will not last the race. It
lacks the resources, it lacks the technical potential. They hope that our country's economy will be exhausted."[65]

That was Reagan's hope. At the time, CIA analysts struggled to pinpoint
precise data on the Soviet spending, a problem that Gorbachev himself faced,
since the totalitarian nature of Soviet society often left its own chieftains in
the dark. Outside estimates on Soviet military spending ranged from as low as
7 percent of GDP to as high as 73 percent.[66] Most estimates fell in the decidedly wide range of 20 to 50 percent.[67] Though he found it difficult to nail

down the specific numbers, Gorbachev knew that statements like this from the Moscow World Service in February 1984 were bald-faced lies: "We would like to remind you that in the Soviet Union defense spending has remained unchanged for several years. In 1984, it will amount to 4.66 percent of the national budget." This deliberately false statement was issued to pose a contrast to U.S. figures released by the Pentagon the day before, which, according to the Moscow Service, signaled American "preparations for war."[68]

When Gorbachev finally gained reasonably accurate figures, he was aghast to learn that military expenditures were not 16 percent of the budget, as he feared, but 40 percent, which he said was tantamount to 20 percent of GDP.[69] His foreign minister, Eduard Shevardnadze, believed the figure was 50 percent of GDP, which meant half of the country's wealth went to defense.[70] "For decades ours was probably the most militarized country," said Shevardnadze. "Gigantic proportions."[71] Genrikh Trofimenko estimated that Moscow spent 35 to 40 percent of its GDP on the military.[72] Today, the most authoritative estimates peg the figure at 32 percent of GDP.[73] Compare these figures to what the United States was spending during this peak period, which was usually less than 6 percent of GDP.[74]

How much did this equate to in rubles? Trofimenko says that Soviet GDP at the start of the decade was 619 billion rubles, or $103 billion—pathetically low for a supposed superpower.[75] The traditional official military budget figure for the USSR was 17 billion rubles per year, or about $3 billion, which was probably ten times less than what was actually spent.[76] A reliable projection is that the USSR spent roughly $30 billion annually on the military. By comparison, Reagan military spending hit $250 billion per year, eight times higher than Soviet levels.[77] Amazingly, the United States may have spent over twice on the military what the USSR possessed in total GDP. There was literally not enough GDP in the entire USSR to match Reagan defense levels—not even close.[78] Reagan no doubt had the numbers on his side.

While the percentage of Soviet military spending was staggering, there were other factors that contributed to the high domestic cost of the arms race for the USSR. Equally numbing were the resources of the country—industrial, technological, and human—that were going into military preparations.[79] The USSR was truly giving all it could but could not afford to step up competition even a notch, let alone to Reagan levels.

With these factors, Communism once again was proving to be its own worst enemy. From Moscow to Pyongyang to Havana, even if Marxists wanted to expand by force, their literally bankrupt ideology undercut revenues for the arsenals. In a supreme irony, the best way for them to generate revenue would have been to embrace the market systems of their adversaries.[80]

But while Soviet spending was bankrupting the economy of the USSR, Reagan's spending was providing him with his own set of domestic problems—by 1985, his military allocations had contributed to a 40 percent jump in the U.S. budget deficit.[81] Yet, he told Cap Weinberger repeatedly that he was willing to accept deficits as a necessary expense to aid his offensive against the USSR, which later prompted Lou Cannon to refer to these Reagan deficits as "wartime deficits."[82] Despite these deficits, Reagan was certain that the United States was more capable of sustaining high levels of military spending than the USSR. He knew that the American economy would be able to bounce back from deficits whereas Communism could not.

THE REACTION OF MIKHAIL GORBACHEV TO REAGAN'S ARMS RACE

The one Russian most fearful of his nation being bankrupted by the Reagan arms challenge was Mikhail Gorbachev. While Gorbachev had made it clear since his election that slowing the arms race was vital to him, the general secretary made the matter painfully evident to Reagan in his opening remarks at the Geneva summit in November 1985, in which he embarrassingly begged for Western aid. There, he made reference to the commercial and technological "weight" of America and the prosperity of Japan and West Germany, the two of which he said had the advantage of spending "so little on armaments." (He did not reconcile how the United States could spend so much on armaments and flourish economically.) He said that arms control was essential to enabling the USSR to reallocate funds from militaristic to peaceful purposes. "Mr. President," he pleaded, "the arms race is wiping us out. I'm starting my talk with this because it's crushing us and we can't afford it. Let's end it before there's nothing left of the USSR."[83]

Central to that arms race was Reagan's SDI challenge. In fact, SDI doubters in the United States were about to face a dilemma: If SDI was unworkable, as they insisted, why then did it scare the Soviets, especially Mikhail Gorbachev, so much? According to Foreign Minister Alexander Bessmyrtnykh, the Soviets were "enormously frightened" by SDI.[84] The initiative was "something very dangerous" that "made us realize we were in a very dangerous spot." He called SDI Gorbachev's "number-one preoccupation": "When we were talking about SDI, just the feeling that if we get involved in this SDI arms race, trying to do something like the U.S. was going to do with space-based programs, looked like a horror to Gorbachev."[85]

Within the USSR, there were mixed views concerning SDI's feasibility as a functional missile-defense system. Some articles in the Soviet press reacted as if the system were a certainty, whereas others were doubtful.[86] Gorbachev's regime seemed to believe that a so-called "impenetrable shield" was not in the cards.[87] The prevailing view within the USSR's inner sanctum (based on counsel from Soviet scientists) was that within fifteen to twenty years the United States might be able to produce a system that could take down a limited number of ICBMS, and thereby negate a Soviet return-strike capability after an initial U.S. nuclear strike—or, as Gorbachev put it: "the creation of a shield which would allow a first strike without fear of retaliation."[88] Gorbachev was sure of this, as had been General Secretary Chernenko and the Soviet press that parroted the line.[89] In Gorbachev's presentation at the second plenary meeting between him and Reagan at Geneva on November 19, 1985, he told the president that SDI "only makes sense if it is to defend against a retaliatory strike."[90]

FEARS OF THE "SDI ARMS RACE"

Despite the internal Soviet debate over SDI's practicality, Gorbachev concluded that SDI was yet another contributing factor to the downward spiral of the arms race. The Kremlin worried that responding to SDI in this form would further bankrupt the USSR. Ambassador Oleg Grinevsky recalls the dread of being "sucked into" a competition over dueling SDIs.[91] As Major-General Vladimir Slipchenko confirmed, "SDI did harm us. Our military industrial complex was able to obtain some money and we created our own SDI [research program] for Moscow, which nobody needed."[92] In addition, Gorbachev and his comrades feared SDI could be transformed into an offensive system equipped with sophisticated space-based lasers, with SDI research generating spin-off benefits for the U.S. military—that is, an entire line of new, cutting-edge technologies.

In the end, the Soviets made counter investments in response to SDI. Roald Sagdayev, head of the Soviet Space Research Institute, said Moscow spent "tens of billions of dollars" reacting to SDI. He confirms that Soviet generals insisted on measures to counter the missile-defense project, including an all-out effort to upgrade the USSR's land-based ICBMs. "This program became priority No. 1 after Mr. Reagan's announcement of the 'Star Wars' in

1983," said Sagdayev. Sagdayev said that this spending weakened the USSR and may have contributed to its demise.[93]

The importance of SDI was apparent throughout the Reagan-Gorbachev summits of 1985–86. Prior to the first summit in Geneva in November 1985, Reagan wrote a prophetic note to himself, predicting that if Gorbachev "really wants an arms control agreement, it will only be because he wants to reduce the burden of defense spending that is stagnating the Soviet economy." Furthermore, Reagan estimated that the technological challenge posed by SDI "could contribute" to Gorbachev's opposition to the program. The general secretary did not want "to face the cost of competing with us." Reagan realized that, "Any new move on our part, such as SDI, forces them to revamp, and change their plan at great cost."[94]

As recently declassified documents of conversations at the Geneva summit reveal, Reagan's assumptions regarding Gorbachev's negotiating tactics proved quite adept over the course of their meetings. In fact, during a conversation on November 20, Gorbachev even went so far as to insinuate that Reagan proceeded with SDI *solely* to hurt the USSR economically—the same thing he repeatedly told his Central Committee in tense sessions.[95] According to minutes from the meeting, Gorbachev told Reagan that he had observed to a Soviet scientist that he could see "no reason" for Reagan's commitment to SDI, but that the scientist had found the explanation: SDI would produce 600 billion to a trillion dollars in new military expenditures. In other words, SDI would be a tool in Reagan's economic war to bankrupt the USSR. "*That* was the reason" for SDI, said Gorbachev.[96]

At Geneva, Gorbachev urged that he and Reagan do everything possible to halt SDI and the arms race. So fearful was he of SDI that 90 percent of the official notes (declassified in 2000) from his second one-on-one encounter with Reagan at Geneva were consumed with jousting over SDI.[97] Anatoly Dobrynin confirms that Gorbachev's "principal goal" at Geneva was to stop SDI.[98] Similar sentiments were echoed at the Reykjavik summit in October 1986, where Reagan's Chief of Staff Don Regan said that Gorbachev was worried about SDI to the point of fixation.[99] Gorbachev was so concerned about SDI at Reykjavik that he actually proposed to eliminate all nuclear missiles if Reagan simply gave up missile defense.

At each summit, Reagan saw the panic in Gorbachev's eyes, and he liked what he saw.[100] Each time, Gorbachev demanded that Reagan abolish or at least curtail SDI.[101] So common were the Soviet general secretary's diatribes that Reagan officials began casually referring to Gorbachev's "usual tirade about SDI."[102] It was a tirade that was no doubt well-rehearsed: according to Reagan's director of the Arms Control and Disarmament Agency, three-quarters of the USSR's propaganda budget from 1984 to 1986 was directed solely against SDI.[103]

Despite stern warnings from Gorbachev, who repeatedly threatened to terminate all progress on cutting nuclear arms until Reagan gave up the program, Reagan held firm on SDI, refusing to negotiate it away. He remained true to two core convictions: his desire to reduce nuclear weapons and to continue research on missile defense, without compromising either principle in the face of Gorbachev's vigorous protests.

Reagan clearly understood SDI's persuasive power. And yet, when asked if SDI was "not just a bargaining chip," he responded flatly, "No. Oh, no. . . . No, no."—a point he reiterated on other occasions.[104] He wrote in his memoirs: "As the myths grew, one of them was that I had proposed the idea [of SDI] to produce a bargaining chip for use in getting the Soviets to reduce their weaponry. I've had to tell the Soviet leaders a hundred times that the SDI was not a bargaining chip."[105]

At the same time, he recognized SDI's leverage. "One thing is clear," he concluded. "SDI truly serves the purposes of offensive weapons reductions."[106] To Reagan, SDI was "not a bargaining chip in that sense—of being willing to trade off the research and stop what we're doing in order to get x number of missiles eliminated."[107] He pursued SDI on its own merits, as a missile-defense program intended to produce a missile-defense system, not as a bargaining chip to be negotiated away. Moreover, he told the press, if he suggested it might be a bargaining chip, it could indeed become one.[108] On the other hand, the power of SDI to bring the Soviets to the table became crystal clear to him.

Moreover, Reagan realized that the mere *pursuit* of SDI, even if the system was never built, served to bring the Soviets to the table, making it worthwhile for that reason alone. On one occasion, he mentioned this to Ed Harper, a fellow at the Brookings Institution and professor at Rutgers, who was Reagan's deputy director of OMB. "[I]n my role in the budget process," said Harper, "I assured him that we couldn't afford [SDI]. He responded, 'It doesn't make any difference. Out of whole cloth we are creating an important bargaining chip to deal with the Soviet Union in negotiating arms control. They don't know whether we are going to have it or not; we don't know whether we are going to have it or not. But it is something to negotiate with that we don't have today, so I'm going to do it.' "[109] This was a viewpoint that he mentioned frequently, sharing it with his first national security adviser, Richard Allen, his current national security adviser, John Poindexter, and Margaret Thatcher. To Thatcher, he stressed that SDI constituted an enormous economic burden on the USSR.[110]

At one point, Reagan's team put forth an NSDD on SDI. Written primarily by Ambassador Paul Nitze and signed on January 1, 1985, in anticipation of the Shultz-Gromyko meeting in Geneva, the directive formalized (administration-wide) the president's understanding of SDI's influence. "Another important factor influencing Soviet behavior," said NSDD-153, "especially in returning to nuclear arms reduction negotiations, is the Soviet desire to block our Strategic Defense Initiative." An "overriding importance of SDI" was that it provided "new and compelling incentives to the Soviet Union for seriously negotiating reductions in existing nuclear arsenals."[111]

And yet, it is important to understand the internal opposition that Reagan faced. By 1985–86, said one observer, nearly "every senior member" of the administration had recommended a "grand compromise" on SDI.[112] Among them, George Shultz never gave up, constantly pushing Reagan to relent on SDI, and was especially adamant at Reykjavik.[113] Along the same lines, Paul Nitze, a Reagan arms-control negotiator and respected, veteran figure of the Cold War, spoke of how his negotiating team was seeking a deal with the Soviets in which any movement toward deployment of SDI would be delayed at least ten years. All the Joint Chiefs of Staff were "wholly in agreement," said Nitze, "so that it was really just Reagan against most of the rest of the administration on this."[114]

Like the pipeline, SDI proved to be a scenario in which Reagan confronted the rest of the world, including his own advisors. It was an unrelenting resistence that came from his constant faith in the course he had chosen, and while the resistance would never subside, neither would Reagan.[115]

PUBLIC DENIAL, PRIVATE PURSUIT

While Reagan had long advocated an arms race as a means to weaken the USSR, he was considerably less forthcoming on the matter as president. During their November meeting in Geneva, he told Gorbachev, "our goal is not an arms race."[116] And yet, at the same summit, during their one-on-one "fireside chat" at the boathouse cottage, he told Gorbachev that the choice the two faced "was either an agreement to reduce arms or a continuation of the arms race, which," he taunted the general secretary, "I think you know you can't win."[117] On a separate occasion at Geneva, Gorbachev urged the president yet again: "I beg you to stop this arms race."[118] Reagan countered by again telling Gorbachev that he would "never let your country win [an] arms race."[119]

These statements were symptomatic of Reagan's delicate balancing game: He avoided publicly admitting, or acknowledging to the Soviets directly, that he was pursuing an arms race as a form of economic warfare, while he privately pursued that course. By its very nature, this political doublespeak led to situations in which Reagan was not truthful in his statements concerning an arms race or his intent to undercut the USSR. One such instance occurred in a May 8, 1985 speech to the European Parliament in Strasbourg, where Reagan pledged, "The United States does not seek to undermine or change the Soviet system."[120] The Soviets did not believe this for a minute. On May 11, a TASS report was published in *Pravda* under the headline, "What the U.S. President's Visit to Europe Showed." The piece stated categorically that the "aim" of Reagan's speech "was to rally his West European partners in the 'crusade' against Communism proclaimed by the head of the present U.S. Administration." TASS stated that Reagan's goal was no less than "overturning" the "postwar structure in Europe." As far as *Pravda* was concerned, the Reagan statement that the United States was not seeking to change or undermine the Soviet system was ludicrous; and this was one of the few times that *Pravda* was right.

Indeed, while Reagan admirers have long hailed his conviction and candor, on these issues Ronald Reagan was not forthright. Though the national security implications of the situation clearly influenced his words, the fact remained that he employed a deliberate and precise brand of diction in an attempt to avoid admitting to the world that an arms race was his goal. Sometimes he tried to avoid situations that would force him to twist his language, as he frequently denied or dodged the question, and not always adroitly. The *Presidential Documents*, which are a compilation of public statements made by Reagan, are replete with denials, dodges, and blanket statements like, "We don't seek an arms race."[121] He could usually get away with such a denial because his point seemed to be that he did not want to engage in an endless stockpile of nuclear-weapons manufacture with no intention of cutting the stockpile.

Though his statements on the arms race tended to be vague, on the issue of SDI specifically, the public Reagan was again circumspect. When Morton Kondracke and Richard H. Smith of *Newsweek* asked him pointedly if he was pursuing the initiative for the dual purpose of bankrupting the USSR, Reagan responded with an "oh" and a "well" and twice escaped into a rambling rejoinder, repeating at length something he said earlier in the interview.[122] There were other similar denials and dodges on this question. In one press conference only two days after the SDI announcement, he was confronted by reporter Sarah McClendon, who asked that in the spirit of "seeking better human relations" with the USSR, "Why don't we sell for cash some of the 190,000 tons of butter we're paying to store daily and daily adding to? The Soviets need butter desperately. . . . We have other surplus commodities." She pushed: "Why cannot we explore whether better living through sharing of food and consumer goods will make people turn from their warlords and bring about peace?"[123]

Reagan hemmed and hawed, evading the question with a three-paragraph ambiguous filibuster about disarmament, peace, food, agriculture, and commercial markets, not at all answering her question. McClendon, for her part, didn't get it, not realizing that Reagan did not want to help the Communist system that held captive ordinary Russians and the peoples of Eastern Europe.

While these two sides of Reagan appear to contradict one another, the reality is that they are easily reconcilable. As Bill Clark notes, "He didn't want to admit publicly that we were effectively at war [with the Soviets] or that we wanted to defeat them. He would [instead] say things like 'we will prevail as a free society,' and that 'the Soviet system is doomed to fail,' but he avoided using words like 'war' or a 'race to bankrupt them.' " Clark stressed that Reagan, despite perceptions, was "incredibly careful" about his words.[124] This "incredible care" was a game at which Clark himself was well-versed. During his time at the NSC, Clark frequently had to make similar public denials in the face of scrutiny from the press, while privately assembling a concerted campaign of economic warfare. When an interviewer from *U.S. News & World Report* pressed him on the issue, Clark was forced into a denial: "We are not engaging in economic warfare with the Soviets or with anyone else; we're seeking arrangements by which the Soviets . . . can shift their resources to peaceful ends. This does not require 'economic warfare.' "[125] Of course, Clark's NSC patented the product.

BUT DESPITE THE ADMINISTRATION'S OFT-REPEATED PUBLIC denials, internally there was no disputing the impact that the arms race was having on the Soviet economy. As 1985 was coming to a close, the Soviet economy was showing unprecedented signs of strain. Not only had the pace of the American military buildup become unmatchable, but the Soviet's very real fears of SDI were taking their toll as well.

Though they were experiencing great success with their efforts to date, Reagan and his stalwarts always focused on the horizon, ever in search of other more substantial ways to prevent the Soviets from gaining ground. The economic war was one that featured constantly shifting fronts, and Reagan, who now seemed a confirmed expert in the depths of Soviet desperation, sought to alter his efforts and correspond to evolving Soviet vulnerabilities. His was a battleground that produced constant motion; and while much of 1985 had seen him focus on arms, his team was about to turn attention to a resource that was a bit more crude.

17.

Conspiracies and Stingers: Late 1985 to 1986

SINCE RONALD REAGAN HAD BEGUN HIS HOMEWORK ON THE pipeline, which he shared in many a lecture to Western Europe, he had been well aware of the role that natural gas and oil played in the Soviet economy. In the face of overwhelming pressure to compete with America in the arms race, these two commodities had come to form the backbone of the Soviet economy, accounting for a remarkable 80 percent of Soviet hard currency earnings.[1] This currency was a life preserver that kept the USSR afloat, and as the world's largest producer of gas and oil, it was clear that these energy exports represented the centerpiece of Moscow's annual hard currency structure.[2]

Through the good work of the Reagan National Security Council, namely Roger Robinson, and Bill Casey's CIA, this weakness was identified. Bill Clark had left the NSC back in October 1983, but Casey stayed on at the CIA. Clark had shepherded the pipeline decision, which exploited the Soviet dependency on *gas* exports. Now, it would be up to Casey to carry the torch in regard to *oil* exports.

But still, a key question remained: How could the administration exploit the oil dependency? While a different CIA director might have merely raised an eyebrow at the Soviet economic inefficiency and moved on, Casey smelled blood. This was not just another Soviet internal contradiction to be recorded in the debit side of the ledger; to Casey, this Soviet vulnerability was an American opportunity. Casey's boss in the Oval Office agreed.

The answer came in the form of the most fundamental law of economics, supply and demand. After months of internal debate over the best way to take advantage of the situation, the underminers came to a solution: drastically lowering global oil prices and thereby undercutting the Soviet ability to earn vital hard currency. In charging a price for its oil exports, the USSR, like any supplier, was at the mercy of the market. If the world's major suppliers in the Middle East opted to drastically increase or decrease supply, they could alter prices. An increase in global supply would lead to an overall decrease in world prices and prove detrimental to the fragile Soviet economy.

The plan was perfect, but it contained a major flaw. Unfortunately, the Reagan administration could not quietly ask OPEC to boom supply simply to help Reagan hurt the Kremlin. The politics of the matter were too tenuous and rife with the potential for international backlash. Despite this apparent setback, the administration looked to persevere in its thinking, exploring the possibility that perhaps a certain lone, oil-rich Arab nation could be approached.

It was here that Saudi Arabia came to play a key role in Ronald Reagan's economic warfare campaign against the USSR. Reagan possessed an excellent rapport with Saudi King Fahd, and in an innovative, gutsy, and generally remarkable covert operation, Reagan and Casey convinced Fahd and his regime to help the administration bust the Soviets economically by shifting Saudi oil flows. The Saudi regime alone controlled so much supply that it held the power to dictate prices simply by virtue of how much it pumped.

TAPPING THE SAUDIS

From the onset of their covert relationship, the Reagan team would find the Saudis infinitely easier to deal with than the Western Europeans (certainly the French) in pursuing economic warfare schemes. The operation was undertaken in the strictest silence in Washington and Riyadh. After a lot of legwork by Casey, in late 1985 the Saudis agreed to increase oil production radically— ultimately raising output from less than two million barrels a day up to nine million barrels per day.[3] It was an unheard of boom, one that caused prices to plummet worldwide. In a matter of months the price plunged from $30 a barrel in the fall of 1985 to $10 in April 1986. There were even predictions it could hit $5 a barrel.[4]

When originally exploring the Soviet commodities cash flow, Roger Robinson calculated that every $1 rise in the per-barrel price of oil meant

$500 million to $1 billion in extra hard currency for the Kremlin, an estimate affirmed by the CIA. The reverse was true for dropping prices: Every $1 drop in oil meant a hard currency loss of $500 million to $1 billion for the Soviets.[5] So it came as no surprise to the Reagan administration that the massive price drop was absolutely devastating to the economy of the USSR; it destroyed Soviet hard currency earnings in 1985. Over the course of just a year, the Soviets shifted from a $700 million trade surplus with the West in 1984 to a deficit of $1.4 billion in 1985.

While those numbers were striking, they were merely the direct, calculable effect that the policy had on hard currency earnings. In addition, there was a ripple effect that had an impact which was harder to estimate, but nevertheless cost tens of billions of dollars more. Yevgenny Novikov, who served on the senior staff of the Central Committee, recalled: "The drop in oil prices was devastating, just devastating. It was a catastrophic event. Tens of billions were wiped away." Peter Schweizer, who first broke this incident in two separate books, *Victory* and *Reagan's War*, cites a classified May 1986 CIA study titled, "USSR: Facing the Dilemma of Hard Currency Shortages," which assessed the damage. Concluding that the sharp drop in world oil prices had "dramatically altered" Moscow's earnings, the report asserted that the total annual loss from the price drop was $13 billion—an enormous jolt to Soviet GDP.[6]

This Soviet crisis lingered into 1986, as the Saudis continued to maintain their high levels of output. Ultimately the crisis impacted not only the Soviet economy but Gorbachev's policies as well, a point that the May 1986 CIA report acknowledged, "The decline of Moscow's hard currency [capacity] . . . comes at a time when Gorbachev probably is counting on increased inputs from the West to assist his program of economic revitalization."[7] No, not *probably*—Gorbachev was *certainly* counting on such inputs. This made the Soviet crisis yet more acute, as if it could not get any worse.

Because of the earnings loss, the Kremlin was forced to cancel dozens of major industrial projects planned or underway.[8] For a country accustomed to parceling out large public works projects for employment, this was a sharp blow. The "Soviet way" was to divert massive funds from the alleged reptilian private sector into government projects to provide jobs to the smiling masses flourishing in the workers' paradise. The oil action halted this traditional Communist way of doing things.

By itself, the economic loss from the oil manipulation was crippling yet survivable, but when combined with Reagan's other efforts, the effect was paralyzing. It was an across-the-board assault that hit the Soviet body like a series of bullets. As Novikov noted: "The Central Committee realized that they were facing a committed government in Washington. They saw activity on all fronts. . . . It frightened them to death."[9]

OIL REVENUES: THE OLD SOVIET STANDBY

While the 1985–1986 oil shock would have caused big losses under any circumstance, what made the losses so extreme was that oil had been the chief stabilizing factor to the Soviet economy in the 1970s.[10] During the 1970s, according to Richard Allen, high oil prices had increased Soviet energy revenues more than tenfold—possibly fifteenfold, according to another source. Western energy dollars were an important component to the Soviet ability to "stay in the race."[11] Gorbachev made a similar comparison in his memoirs, saying that in the 1960s and 1970s the structural problems in the Soviet economy— that is, the systemic failure of Communism—were masked by high oil and gas prices.[12] In subsequent writings, Gorbachev would go even further saying that "in the final years before perestroika [our country] was able to exist only by virtue of oil and gas exports."[13] But now, without this valuable economic boom, the Soviet ship of state was sinking. The commodities that had long been the Soviets' principal assets were drowning them.[14]

In addition to the immediate impact on cash flow, the oil shock also proved detrimental in other ways. Because of the high prices in the 1970s, the USSR had begun the 1980s with a solid credit rating resulting from a long record of good payments. By the mid-1980s the USSR's trade surplus had all but evaporated as the country was hemorrhaging billions of dollars a year for lack of hard currency. From 1985 to 1986 alone, the budget deficit tripled. The Soviet bond rating on the international capital market was lousy, and foreign banks were unwilling to extend loans to the USSR.[15] Catalyzing the weaknesses of the already struggling economy, the oil shock proved essential to the eventual fall of the Soviet economy. The stellar credit rating with which the USSR began the decade had disappeared by end of the 1980s, due to an exorbitant trade deficit, a significant general budget deficit, large external debts, and payment delays on imports—much of which was a by-product of the damage from the depleted oil earnings.

In the end, the U.S.–Saudi oil shock was economic warfare of the highest degree, and when coupled with the lower than expected revenues generated by the lone strand of the natural gas pipeline, the losses were astronomical. Facing this unprecedented drop in currency earnings, aggrieved Politburo members sent a strongly worded letter to King Fahd in early 1986. The letter candidly warned Saudi Arabia against pushing prices lower and proposed secret meetings in Geneva to stabilize prices. The Soviet protests were joined by an angry Libya and Iran. Iran made public threats toward the Saudis, as the loss of oil revenue threatened an end to Tehran's planned wartime offensive against Iraq.[16]

Despite the open and repeated calls from the Soviets and other oil producers for price stabilization, the Saudis continued their massive output, keeping the source of their motives a secret from the world. Ultimately, it was not until September 2002, when the Saudi regime desperately needed a PR boost in the wake of the September 11, 2001 attacks, that Saudi officials finally gave even minor indications of the collusion. In a carefully worded September 17, 2002 op-ed in the *Washington Post*, Prince Turki al-Faisal, director of the Saudi General Intelligence Department from 1977 to 2001, sought to compensate for the fact that the 9/11 attacks were carried out by fifteen Saudi nationals by saying that the increase in Saudi oil production in the mid-1980s would not "have taken place without Saudi–U.S. cooperation." This, he said, "led to lower oil prices."[17] The prince's piece was far from a tell-all, but it nonetheless constituted the Saudis' biggest public admission of their role in the shock.

While this limited admission came a long time after the collusion had ended, the risks for both sides during the shock made it clear that disclosure of the truth would be far away. One of the most amazing aspects of this covert relationship was that it went forward in the face of manifold risks for both sides involved. The nervous Saudi leadership sought reassurances that the United States would provide verbal support against Soviet protests and physical support to ensure the security of Saudi oil shipped from the Persian Gulf, which was being threatened by Iran and Libya. Saudi Arabia was paying a price for assisting the U.S. economic war against the Kremlin. In Washington, Prince Bandar communicated his concerns to Weinberger and Casey, who reassured him and said that direct talks on the matter would take place during Vice President George Bush's visit to the Arab peninsula in April 1986.[18]

Although these talks tried to quell Saudi Arabia's fears, the United States also had to chance a potentially nightmarish situation in order to make the plan successful. On a domestic level, the action was harmful to the American oil industry, causing oil suppliers within the United States to lose millions of dollars in revenue. This negative was significant to Reagan, who was always a friend to big business, but, in the end, the president and his team estimated that the benefits of the situation would outweigh its costs to American industry. Also, while the domestic oil industry would be saddled, consumers would save billions of dollars in lower prices. Overall, the U.S. economy would benefit.

Internationally the United States would have to deal with the ramifications of potentially protecting Saudi ships should Iran, Libya, or the Soviets choose to take matters into their own hands. In addition, the United States bore the risk that should the world ever find out, the information could provoke a possibly violent confrontation with the Soviets.

Despite this multitude of risks on both sides, it was clear that Saudi Arabia had much more at stake in this gamble. Not only were the Saudis risking retribution from angry neighbors, but they were also risking their country's sole commodity and source of wealth. Why, then, in the face of so many risks, did the Saudis agree to the conspiracy? On the surface it seemed like a bad economic decision, but in truth it was not. While they lost money on a per-barrel basis, the Saudis made up the difference through the overall price of oil sold.

To the Saudis, the agreement was less of an economic arrangement and more of an investment in future relations with the United States, an intangible for which there could be no real price tag. Good relations with the United States had already paid off in October 1981 in the form of the Reagan administration's agreement to sell Airborne Warning and Control System (AWACS) aircraft and F-15 "enhancement items" to Saudi Arabia. The AWACS were capable of detecting aircraft within a perimeter of 350 miles, a wide range of space in the narrow Middle East. In the October 1, 1981 press conference explaining his decision, Reagan's choice of words was ironic in light of the U.S.–Saudi oil collaboration four years later:

> I have proposed this sale because it significantly enhances our own vital national security interests in the Middle East. By building confidence in the United States as a reliable security partner, the sale will greatly improve the chances of our working with Saudi Arabia. . . . As President, it's my duty to define and defend our broad national security objectives.[19]

Indeed, the AWACS sale was critical to such a working partnership later when oil, not planes, drew the two sides together. The administration was in fact able to work with Saudi Arabia to enhance its national security interests—against Moscow. In essence, the Reagan administration in 1985–86 cashed in its 1981 chips. Caspar Weinberger later said explicitly: "One of the reasons we were selling the Saudis those weapons is because of the hope that lower oil prices would result."[20]

On October 28, 1981, after intense lobbying from all corners, the AWACS sale was approved in the Senate by a small four-vote margin. Reagan's dedication to this issue was vital to its passage in Congress, a fact which was not lost on the Saudi regime. Indeed not only was Saudi cooperation in the oil shock repayment for Reagan's hard lobbying, but it also represented the potential for future military deals between the two countries. To this end, the administration did not hesitate to deliver, rewarding Saudi loyalty in May 1986 when Reagan vetoed a joint congressional resolution to prevent the sale of what Reagan dubbed "defensive missiles" to Saudi Arabia.[21] To the Saudis, this was significant support at an uneasy time, reinforcing the importance of their oil conspiracy with the United States and solidifying the connection between the two nations.

In the end, it was this strong connection between the two countries that produced an agreement that was conducive to success. The excellent rapport that began with Reagan and Fahd extended to include Casey and Prince Bandar as well. Though Casey was Irish Catholic and Bandar was a devout Muslim, both detested Soviet Communism and wanted to smash it any way they could. Both knew the USSR as a godless empire that suppressed and jailed religious believers, including Catholics and Muslims. Hatred of atheistic Communism was the glue that bound the two men and their camps together, making it possible for two very different nations to find common ground.

Ultimately, this 1985–1986 covert action was pure, unadulterated economic warfare, and by all accounts the joint oil manipulation was probably the most direct, damaging assault of the White House's economic warfare campaign. The Reagan team successfully carried out its plan without a major hitch and kept the secret through the life of the administration and beyond.[22]

To this day, researchers who venture through microfiche from 1985–1986 business weeklies will find no help in ascertaining the real reason for the drop in oil prices. Press accounts from the period simply profiled the merry befuddlement by the typical car owner over the cheaper prices at the tank—a pleasing polar reversal from the spikes, rationing, and gas lines of the previous decade. To understand the true reason for the 1980s price dive, researchers must look elsewhere.

A STINGING DEFEAT

With the severest months of the oil blow winding down in the fall of 1986, the conflict in Afghanistan thrust itself back onto center stage. As evening settled on September 26, 1986, three menacing Soviet Mi-24 helicopter gunships hovered over impoverished, war-ravaged Afghanistan. Lingering in the distance near the Jalalabad airport, the gunships were a frightening but common sight.[23] Hundreds of these aircraft had wreaked havoc since the Red Army invaded, and for almost seven years all the Muj rebels could do to combat them was hide in the brush and futilely fire bullets in the air as Soviet helicopters lit up the ground.

On this night, however, the entire war was about to change. A soldier named Mohammed Afzal emerged from the trees with a bazooka-looking weapon called a Stinger resting on his right shoulder. He aimed the barrel in the vicinity of one of the Soviet aircraft, fired, and watched the helicopter explode into thousands of pieces. The remaining pilots struggled to comprehend the calamity. Before they could come to their senses, Afzal fired at a second helicopter, which instantly burst into a fireball. The mortified third pilot was unable to escape. He too soon met his doom.

The Muj rebels were ecstatic. Not only had they struck down three Soviet gunships, but the weapon they now possessed would strike down the enemy's primary tactical advantage and thereby transform the war. "The shots were heard round the world," said reporter Fred Barnes.[24]

Of all the aid that NSDD-166 had authorized for the Afghan rebels, the crown jewel was the Stinger antiaircraft missile. It was a superb weapon with a range of roughly five kilometers, or 15,000 feet, racing upward 1,200 miles per hour to its target. Weighing only thirty-five pounds—astonishingly light for a tool whose cargo can obliterate a helicopter—it had less recoil than a shotgun. With an infrared heat-seeking mechanism that allows it to easily find its target, the Stinger missile enabled the operator to successfully destroy his target without needing to aim precisely at an aircraft's heat source. The infrared capability made the Stinger effective in all types of weather, more so than the Soviet-designed SA-7, also a shoulder-fired, heat-seeking, surface-to-air missile. The Stinger can distinguish between real targets and flares. Also unlike the SA-7, the operator of the Stinger did not need to sit in place and keep the target in his sights until impact, risking detection and counter fire.[25]

The Stinger kill rate was deadly. Of the first eleven missiles the rebels unleashed, ten took down Soviet helicopters. After that, the Muj averaged one destroyed plane or helicopter for each of the next 200 days.[26] A U.S. Army study conducted after the war found that of the 340 firings of Stingers in combat, 269 downed aircraft.[27] The Stingers were deadly against both the Soviet helicopter gunships that had proven so effective and the sophisticated, high-flying planes and helicopters that took off from the Turkistan Military District.

THE ORIGIN OF THE STINGERS

While attempts have been made to unearth who first broached the subject of Stingers in Afghanistan, the idea was long believed to have come up during internal administration discussions sometime in 1985 or 1986. But contrary to popular belief, such was not the case. In fact, the initial source of the idea came long before NSDD-166. Reagan himself cited the importance of anti-aircraft missiles in Afghanistan during a campaign speech in Pensacola, Florida on January 9, 1980. Coming just two weeks after the Soviet invasion, it was not clear who (if anyone) Reagan consulted, but the impact of the statement remained clear: the Stingers were a priority even before Reagan entered the Oval Office.

Reagan's recommendation was reported by Martin Schram of the *Washington Post*, who wrote that Reagan, "specifically urged the supplying of U.S. shoulder-launched, heat-seeking missiles that can shoot down Soviet helicopter gunships." Further, Reagan said the United States should supply weapons through Pakistan—another idea that later came to fruition. "There's nothing wrong with giving free people weapons to defend their freedom," said Reagan.[28] As early as 1980, he had apparently identified the remedy for turning the war against and ultimately defeating the USSR.[29]

Once he won the White House, Reagan began moving in this direction. In the first weeks of his administration, he started communication with the Afghan resistance. Two rebel leaders visited Washington in late February 1981. They held a press conference at the Capitol in which they expressed hope that they would receive not just rifles and ammunition from the administration but also, as the Moscow Domestic Service duly noted in a February 26 statement, "ground-to-air missiles."[30]

Though U.S. assistance came immediately, it took years to get Stingers to the Afghans. Reagan defense officials debated the wisdom of handing over one of the U.S. military's most precious weapons, allowing it to be not just used but also replicated. This was a source of opposition by the Joint Chiefs of Staff, some Pentagon officials, and even certain CIA members.[31] One Reagan official noted that some of the military brass reacted not like Cold Warriors, as they were typically portrayed, but like bureaucrats.[32]

While fears of compromising U.S. technology were legitimate, those who favored sending the Stingers saw no value in spending billions developing weapons simply to stockpile them. Moreover, it became apparent that the Stinger could be enormously productive in taking down the Soviets. Jack Wheeler, a conservative and friend of the Reagan administration, traveled to Afghanistan in 1983, where he witnessed the devastation wrought by Soviet helicopter gunships whose superiority the Muj could not counter. Upon his return, he was debriefed by officials from the White House, NSC, and CIA. "The Afghans control the ground," he told them. "The Soviets control the air. Take the Soviets out of the air and they lose."[33] He found little disagreement, particularly among Undersecretary of Defense Fred Ikle, Ikle's aide Michael Pillsbury, NSC intelligence official Vincent Cannistraro, and Ambassador Mort Abramowitz—all of whom championed the Stinger. At higher levels, Cap Weinberger joined Casey and Reagan in advocating the supplying of antiaircraft weapons.[34]

This top-level support notwithstanding, differences in opinion remained an obstacle. Reagan wanted the Stingers to be sent, but was blocked by bureaucratic rigidity. As State Department veteran Peter Rodman said of the Stingers: "[I]mplementing the policy in the U.S. government, as usual, was another matter. . . . [N]ever underestimate the power of working-level officials to obstruct a presidential decision they disagree with."[35]

In addition, Reagan also faced resistance from Congress, and it was not until 1985 that there was bipartisan support for sending Stingers. Nearly every Democrat on both intelligence committees opposed the idea. A key turning point was a June 1985 trip to Pakistan by Democratic Senators Bill Bradley and David Boren, members of the Senate Intelligence Committee. When they returned, they lobbied for increased support for the rebels.[36]

But much to the dismay of the Stinger proponents on Reagan's team, the support of the two Democratic senators was not enough and the obstruction continued. In April 1986, Reagan placed his signature on a decision memo that explicitly authorized Stingers to be delivered.[37] Even then, a maneuver by a CIA official left the first procurement of fifty Stingers sitting behind in a Virginia warehouse, as the official lamely argued that the Soviets were testing new antimissile defenses that had to be investigated before U.S. shipments could take place. It was a startling and stifling display, ironically emanating from the same agency that was training and arming the Muj and tearing up Soviet units in Afghanistan; from no less than William J. Casey's own organization. The disappointment could be seen by everyone. As National Security Adviser Bud McFarlane put it, "Everyone wanted to move fast, but he [Reagan] wanted to move even faster." Reagan instructed McFarlane and Casey: "Do whatever you have to do to help the Mujahedin not only to survive, but to *win*."[38]

The corner was finally turned when Pakistani officers traveled to the United States in June 1986 for special Stinger training—the same month that Reagan met with Afghan resistance leaders.[39] Days later, a clandestine training center was set up in Rawalpindi, Pakistan, featuring an advanced electronic simulator developed in the United States. Rebels fired mock missiles at a large screen that highlighted a hit or miss.[40] Shortly thereafter, the first shipment of Stingers finally arrived in Pakistan that summer.[41] The hard work of the administration would soon payoff as the missiles would reverse the course of the war.

Even before the first Stinger was fired in September 1986, Gorbachev had been frustrated by the war. In his opening speech to the twenty-seventh CPSU Congress, held February 25 to March 6, 1986, he called the war a "running sore."[42] Despite his efforts of the previous year, it was clear that his military initiatives with General Zaitsev and the KGB were not having the desired effect on the campaign.

Ultimately, the Stingers would prove to be a nail in the Soviet coffin, singlehandedly eliminating the air superiority which had accounted for much of the success of the Red Army. While the war continued to rage, the bells began tolling for the Soviet Union's military effort in Afghanistan on that day in September 1986 when the first helicopter exploded.

REYKJAVIK

As the fall of 1986 progressed, Reagan received yet more evidence that the Soviet economy was in dire shape. "It made me believe that, if nothing else, the Soviet economic tailspin would force Mikhail Gorbachev to come around on an arms reduction agreement we both could live with," said Reagan. "If we didn't deviate from our policies, I was convinced it would happen." Now more than ever, said Reagan, "the Soviet economy was a basket case."

He tried looking at the situation from Gorbachev's viewpoint: "I knew he had to be giving high priority to reducing the vast amounts of rubles the Soviets were spending on weapons. He had to be losing some sleep over the vitality of our economy . . . and he must have realized more than ever that we could outspend him as long as the Soviets insisted on prolonging the arms race."[43] More accurately, it was *Reagan* who was prolonging the arms race. Gorbachev was "prolonging" it by not tossing in the towel. When Reagan saw Gorbachev at the Reykjavik summit in October 1986, he warned him "to join in arms reductions or face an arms race he couldn't win."[44]

A few days after Reykjavik, Gorbachev shared his fears with Soviet citizens in a nationally televised address. "The United States wants to exhaust the Soviet Union economically through a race in the most up-to-date and expensive space weapons," he explained. "It wants to create various kinds of difficulties for the Soviet leadership to wreck its plans, including the social sphere, in the sphere of improving the standard of living of our people, thus arousing dissatisfaction among the people with their leadership."[45]

Two excellent high-level sources on Gorbachev's desperation at this moment were foreign ministers Eduard Shevardnadze and Alexander Bessmertnykh. Shevardnadze noted: "It was invariably argued that there was no reason why we should reduce our armaments, by, say, 10 or 16 items more than the other side. Yet the point was to stop the arms race. . . . Our country could not remain a militarized state."[46]

Bessmertnykh, who had been a member of the Soviet National Security Council and one of only five members of the Defense Council, acknowledged that Reagan's buildup, particularly the threat of SDI, did in fact prompt the Soviets to furiously counterspend: "We thought that the only way we could respond to the threat of SDI, especially, was to develop the ICBM program as much as possible. It was with that in mind . . . that we started to develop two modern ICBMs, which were the SS-24 and SS-25." And yet, the Soviets realized they had to relent. In long-range missiles in particular, Bessmertnykh said that Gorbachev "understood that we did not have a chance of catching up with the United States."[47]

It had been apparent to the Soviet government, said Bessmertnykh, that "Reagan decided to change the course of defense policy and start an enormous buildup," and "that the United States was serious about overwhelming the Soviet Union in one basic strategic effort." Quite significantly, Bessmertnykh states unequivocally that it was the "economic side of the arms race" that "was very much on Gorbachev's mind," and that, in turn, drove Gorbachev to propose arms reductions.[48] In other words, the Soviet leadership had reacted precisely the way that Ronald Reagan had long ago predicted.

But to what extent? From 1985 to 1986, the USSR may have spent an added $15–20 billion annually, strictly on military spending, in an attempt to match Reagan's hikes.[49] To the United States, an extra $15–20 billion would be exorbitant; to the USSR at this time, it was a silver bullet. According to a Soviet estimate, 62–63 percent of the money devoted to machine-building in the USSR in 1986 was for military purposes.[50] This was at a time when, as Marshal Akhromeev put it, "the USSR was not able to continue . . . the military confrontation with the USA and NATO. The economic possibilities for such a policy was exhausted."[51]

Nonetheless, Mikhail Gorbachev was trying. And, in responding to the Reagan challenge, the USSR was in the process of bankrupting itself.

18.

Calling for Liberation: 1987

ON JUNE 12, 1987, NINE YEARS AFTER HIS FIRST, FORMATIVE trip to East Berlin, Ronald Reagan returned to the Berlin Wall. Now, he voiced a number of opinions that he had wanted to say since his first visit, and this time with the power of the world's media behind him.

It was a clear but breezy day at the Brandenburg Gate, and likewise Reagan's unambiguous words flew off the page as he delivered one of the most notable speeches of his career, appealing directly to Mikhail Gorbachev to effect change in this city that was literally cut in half. "There is one sign the Soviets can make that would be unmistakable, that would advance dramatically the cause of freedom and peace," said the president. "General Secretary Gorbachev, if you seek peace, if you seek prosperity for the Soviet Union and Eastern Europe, if you seek liberalization: Come here to this gate. Mr. Gorbachev, open this gate. Mr. Gorbachev, tear down this wall!"[1]

The crowd roared in approval. Reagan's words placed the onus squarely on Mikhail Gorbachev, who indeed was the one man who held the power to tear down the wall. And if Gorbachev was truly the near saintly figure depicted by Western liberals, then he ought to do one simple, right thing: order the dismantling of the Berlin Wall.

It was a remarkable moment, one that rippled across the world. At once a call to arms and a direct challenge to the man in Moscow, the line shuddered with the intensity and hopes of millions that were trapped behind the Berlin Wall, none of whom dared say what Reagan could and did.

While historians have noted the significance of Reagan's command to Gorbachev, the words that came after that famous line, at the conclusion of the speech, have not been discussed with the same fervor, but nevertheless should be, since they contained a rather prophetic prediction. As he was finishing his speech, Reagan looked out on the crowd, and from the Reichstag, he noticed the words "This wall will fall" crudely spray painted upon the wall. Pondering them for a moment, Reagan faced the crowd and asserted: "Yes, across Europe, this wall will fall."[2]

That line did not appear by mere chance. While the phrasing was new, the sentiment had been there all along. Said Peter Robinson, the author of the speech: "Among us speechwriters, I don't think there was any doubt at all about what the Gipper was up to: He intended to win. I'll grant that I was surprised the wall came down when it did. But that Reagan wanted to defeat the Soviet Union—that he wanted us to win and them to lose—was clear enough in the speechwriting shop."[3] Robinson and Reagan on that day intended to call for the collapse of more than just the wall dividing Berlin.

Almost from the instant that he uttered those words, the Soviet press began to vilify Reagan's remarks, subjecting a captive audience behind the Iron Curtain to a lengthy discussion by hardline Marxists who denounced Reagan's declaration. Speaking on "Studio 9," three of Russia's top "political analysts"—Valentin Sorin, Georgi Arbatov, and Valentin Falin—explained that the speech was "plain blackmail, blackmail by an American cowboy . . . threatening and waving his hands and staging a show of strength near the Berlin Wall." Arbatov informed Soviet citizens that Reagan's "hypocritical" statement was merely an attempt to deflect Americans' attention from the Iran-Contra scandal and the impending economic "crash" about to befall the capitalist United States.[4]

Likewise, Falin complained that, "Reagan came to this city [West Berlin] without a good knowledge of the fact that West Berlin is less suitable than many other places on the planet for demonstrations of force, threats, and the lowest and most—if I may say so—frantic demagoguery." Reagan, said Falin, was yet again "increasing tension" and "increasing the temperature." Continuing in this vein, Falin proceeded to justify the Berlin Wall's 1961 construction and continued existence: "The measures our GDR [East Germany] friends were forced to take in 1961 were of strictly defensive character. Their Warsaw Pact allies also asked them to take these measures. . . . [T]hose measures at the border . . . steeply reduced possibilities of using West Berlin for subversive activities against the GDR and other socialist states."

He plowed on: "Today the President turns to the vocabulary of the forties, fifties, and sixties and tries to plunge West Berlin back at least twenty years. He tries to plunge both West Berlin and central Europe back into the period when very acute crises arose from time to time in and around West Berlin. I would like to remind you that in 1961, precisely in West Berlin at the end of 1961—when the U.S. Administration was toying with the idea of bringing down this wall—the world found itself within 200 meters of World War III."

As far as Falin and the group were concerned, Reagan was seeking to move the world backward, not forward, and the wall was a source of stability. In fact, Falin believed that a divided Germany was imperative, its presence providing an assurance that fascism could not once again take hold. Calling Berlin the "capital of bloody fascism," he complained: "For Reagan the whole German question is the opening of the Brandenburg Gate. For all Europe and for us, however, it is the question of not allowing World War III to be unleashed again by German imperialism."

Chomping at the bit, Arbatov jumped in: "What Reagan is saying is simply political vulgarity. Political vulgarity!"

Moderator Zorin invited the commentators to offer their objective assessments on Reagan's request that Gorbachev "prove" he was serious about "openness" by coming to Berlin to open the Brandenburg Gate. Falin replied by bemoaning Reagan's dirty pool: "I think that Reagan is not only imposing political rules of the game on the other countries and the Soviet Union, but he also wants the political figures and leaders of the states to think precisely the same way as he does. But the Soviet leaders do not come into the house of someone else to either close or open gates. Soviet leaders know and respect international laws, know and respect sovereign interests of other countries. . . ."

He was cut off by Zorin, "Unlike the U.S. President." Yes, Falin shot back, ". . . unlike the U.S. President."

To these Marxists-Leninists, who held the stage at the height of Mikhail Gorbachev's glasnost, it was clear that Reagan had acted rudely and without respect for Soviet sovereignty. For them, Reagan's call to tear down the wall had nothing to do with liberating people; it was a form of blackmail that cruelly exploited contented East Germany. This discussion on "Studio 9" was indicative of the larger mood that permeated the Soviet press and leadership, and it was clear that East Berliners' best friend in June 1987 sat in the Oval

Office, rather than behind a camera in Moscow. When Reagan made this speech, there was no one in the USSR—including Mikhail Gorbachev—who was making similar calls for the wall's dismantling, and despite all claims of the USSR's Western-looking government, the Soviet leadership and media remained rooted in the past.

In fact, at the time, Gorbachev was on record favoring a divided Germany and the very wall that separated it. Indeed, his shockingly insensitive euphemism for the wall in his 1987 bestseller *Perestroika* is worth highlighting: West and East Germany, wrote Gorbachev, casually, "are divided by an international border passing, in particular, through Berlin."[5]

It was certainly not Gorbachev's intent to knock down the Berlin Wall, which is why Ronald Reagan issued that challenge in front of the Brandenburg Gate. In fact, a month after Reagan's June 1987 challenge, Gorbachev visited with West German President Richard von Weizsacker. Weizsacker asked Gorbachev "just for the record" about his thoughts on the prospect of German unity. Gorbachev's response was hardly a ringing endorsement: He said that "history would decide what would happen in a hundred years."[6] Even Archie Brown says that when Gorbachev later famously told East German dictator Erich Honecker that "life punishes those who wait," he was speaking of and looking forward to the kind of liberalization of East German Communism that Gorbachev was pursuing in the USSR—he was not expecting an end to the regime in East Germany.[7] Gorbachev had not been helpful in prodding Honecker and unfettering East Germany. Two months after Gorbachev had come to power, when he first met privately with Honecker, Gorbachev told the East German despot: "There is only one model, Marxist-Leninist socialism."[8]

This was why Reagan challenged Gorbachev in Berlin in June 1987. The president was convinced that a challenge needed to be made, in spite of his advisers' protests and insistence that he remove the "provocative" line from the speech.

Despite the massive coverage that his call received, the Brandenburg Gate speech was hardly the first time that Reagan had made a call for the razing of the wall. On the contrary, he had made similar calls many times over the course of his crusade, a few in just the last year. The wall's removal had been a theme throughout much of his crusade, as the destruction of this symbol would mean the destruction of Communism. On three occasions prior to his presidency, May 1967, May 1968, and November 1978, Reagan called for the removal of the Berlin Wall. In each of these instances, the Crusader displayed his hatred of the symbol, seeking not only to end the division in Germany, but throughout the whole of the Eastern bloc. It was an idea that seemed far-fetched in 1967, and remained so twenty years later.

Equally striking was the fact that Reagan, as president, had called on Gorbachev to tear down the wall several times before his June 1987 speech. In August 1986 alone, he remarked on the destruction of the wall three times—on August 7, 12, and 13.[9] Ten days before the Brandenburg Gate address, Reagan reiterated his call in a June 2 interview with a West German newspaper, saying: "In a word, we want the Berlin Wall to come down." This time, Reagan added that he desired the "elimination of all barriers" between East and West Berlin as well as "the reintegration of all four sectors of the city into one unit again."[10] Significantly, this was an expressed desire for not merely the removal of the wall but for a unified Germany.

On June 12, Reagan made yet another "wall call" in a written response to interview questions submitted by the West German publication, *Die Welt*, where he said that if the "Soviet leadership" truly understood and acknowledged the "benefits of freedom," then "there is one step they could take that would be unmistakable: Tear down the wall, open the gates."[11] With that last line, he had now spoken almost the exact same language as he did in his dramatic, televised speech that same day at the Brandenburg Gate.

Not only did this dramatic language convey the gravity and importance of the wall, but it also succeeded in catching the attention of people around the world. With these clear and unmistakable words, Reagan reinforced his goals to the world, eliminating any lingering doubt over what his presidency was seeking to accomplish. To Reagan and to the world, the Berlin Wall represented more than a mere division between East and West; it represented the oppression and control of the Soviet government that had trapped the people of Eastern Europe. It was the embodiment of Yalta's failure to protect Eastern Europeans from Soviet incursion, and as such it had to go.

REAGAN ON POLAND

While 1987 saw Reagan focusing world attention on the Berlin Wall, he simultaneously returned his attention to Poland, as he continued to stress the plight of the Polish people and their containment inside the Soviet bloc. It was a subject that would remain a fixture for Reagan during his final two years in office, as he continued to use the power of the bully pulpit to show solidarity with the Polish people.[12]

On February 19, 1987, he claimed that, "the light of freedom continues to shine in Poland. The commitment and sacrifice of hundreds of thousands of Polish men and women have kept the flame alive, even amid the gloom." He cited the slogan of the nineteenth-century Polish independence movement: "For Your Freedom And Ours." "That," said Reagan, "is our slogan, too. And it is more than a slogan; it is a *program of action*." (Emphasis added.) He concluded by telling the Polish people that he continued to believe that Polish freedom and liberty remained a "possible dream." The "flame of justice and liberty," represented by those candles he asked Americans to light for Poles during Christmas 1981, would "never be extinguished."[13]

Reagan seemed especially inspired—spiritually so—in a memorable May 18, 1987 meeting with his speechwriters, one of which, Josh Gilder, was preparing remarks for the president to deliver at his next visit to the Vatican. The pope would soon thereafter be traveling to Poland again. Gilder asked Reagan if there was something in particular he wanted to say to the pope. Reagan replied with these exact words, recorded verbatim by Gilder:

> Our prayers will go with you [to Poland] in the profound hope that soon the hand of God will lighten the terrible burden of brave people everywhere who yearn for freedom, even as all men and women yearn for the freedom that God gave us all when He gave us a free will. We see the power of the spiritual force in that troubled land [Poland], uniting a people in hope. . . . Perhaps it's not too much to hope that true change will come to all countries that now deny or hinder the freedom to worship God. And perhaps we'll see that change come through the reemergence of faith, through the irresistible power of a religious renewal.[14]

This was not just a powerful spiritual statement aimed solely at Poles but to all believers suffering behind the Iron Curtain.

COMMUNISM AND LATIN AMERICA

As 1987 developed, it became clear that Reagan's prognostications on behalf of freedom were not limited to Europe. In actuality, his second term was filled with declarations to spread democracy throughout the world, in the Eastern bloc but also in Latin America. Having begun the term by affirming that one of his "consistent goals" in the Western Hemisphere was to "promote the development of democratic institutions," Reagan continued to voice these ideas throughout 1987 and 1988, saying that "American foreign policy is not simply focused on the prevention of war but the expansion of freedom."[15] It was an important idea for Latin America and one that he conveyed frequently during these last two years in office.[16]

This in fact was not a new concept for Reagan, who for much of his crusade had been acutely aware of democracy's absence in Latin America. In the 1970s, he routinely noted that over half of the region did not live under democracy, often promising to do something about it.[17] For Reagan, the fight against Communism in Central and South America was consistent with his promotion of freedom worldwide. He warned: "The transition to democracy, especially in Central America, has been accompanied by a concerted and well-financed effort by the Soviet bloc and Cuba to undermine democratic institutions and to seize power from those who believe in democracy." This "subversion" by Communists, said Reagan, had to be stopped by the forces of democracy.[18]

It was this desire to support the growth of democratic regimes and thwart the expansion of Communist ones which led to Reagan's aid of the Contra rebels in Nicaragua and a blemish on Reagan's presidency.

In late 1986, reports began circulating that the Reagan administration had secretly sold arms to Iran, a theocratic terror state that America did not publicly recognize, in exchange for hostages that had been held by the Iranians for what seemed liked an eternity to Reagan. While the United States had not had diplomatic relations with Iran since 1979's violent revolution, the sale of the weapons was deemed necessary in order to ensure the safe return of the American hostages. The money from these sales was then diverted to the Nicaraguan Contras, the anti-Communist rebels that Reagan hoped could supplant the Communist Sandinista government.

Attempting to support the Nicaraguan rebels was nothing new for Reagan, who had been trying for quite some time to provide them with aid but continuously ran into congressional opposition. A major roadblock came from a Democratic congressman from Massachusetts, Edward Boland, who had won the passage of several amendments designed to block the Reagan administration from sending military equipment to the Contras. Though the Iranian component of the deal sought to free American hostages, the affair generally raised questions in the press over the extent to which it sought a clandestine solution around a Congress that was unresponsive to the administration's pro-Contra agenda. As the story of the Iranian arms sale unfolded, the central question that began to emerge in the media concerned whether or not this arrangement violated Boland's amendments and thus United States law.

Once reporters smelled a scandal in the water, they began to confront the administration over the level of its involvement. Reagan's Attorney General and longtime aide and close friend Ed Meese learned of the situation and brought it to the president, and they together publicly disclosed it to the White House press corps on November 25.[19] Unsatisfied with the disclosure, the media wanted more, sensing what some journalists hoped might be "another Watergate" on the horizon. News anchors like CBS's Dan Rather provided continuous, wall-to-wall coverage in their broadcasts, while op-ed pages around the country overflowed with voices demanding answers.

Central to the discussion and a subject of great speculation was the level of Reagan's personal involvement in the Iran-Contra Affair, with reporters seeking to piece together the elusive timeline that explained what Reagan knew, when he knew it, and whether he was complicit in a cover-up. Quickly it became clear that no matter how much the administration wanted the issue to go away, it simply refused to do so. By December, critics of the administration had their voices heard, and Lawrence Walsh, a former federal judge, was appointed as a special Independent Counsel to investigate, which he proceeded to do intently.

While the investigation took place over the course of 1987, and failed to turn up any direct link between Reagan and the main questions of concern, the damage nevertheless had already been done. Occupying the administration's attention for the end of 1986 and much of 1987, the scandal became the focus of the news media and thus the country, forming an indelible stain on the administration's record. Perhaps the greatest tarnish of the Iran-Contra Affair was that it slowed Reagan's otherwise remarkable productivity. As the administration became increasingly focused on quelling the rising furor over Iran-Contra during 1987, Reagan was forced to spend less time focusing on his plans to fight Communism and more time centered on how to reduce public anger over the situation. While his prolific presidency had been marked by tremendous strides against the Communists, 1987 turned out to be a much different year, one in which Reagan was to some degree forced to shift focus away from the Communists and to domestic damage control.

But although this particular effort to get arms to the Contras landed Reagan in the midst of his presidency's largest scandal, it was consistent with his long-stated policy of promoting democracy and denying Communism wherever possible. "Our goal in Nicaragua must be to make democracy irreversible," he stated unambiguously. "Only the freedom fighters [Contras] can do that; only they can be our insurance policy for democracy in Central America." He insisted that "the freedom tide that has swept Latin America is pushing up against the borders of Nicaragua." That tide in that place, he warned, could "go either way."[20]

What Reagan recognized in his support of the Contras was that Communism lacked borders and would continue to spread as long as democracy was not there to counter it. In Nicaragua, Reagan feared the potential for much of Latin America to spin out of control and into the waiting arms of Communism, and in no less than America's backyard.

TALKING LIBERATION

Despite the limitations imposed by the Iran-Contra investigation, Reagan continued to employ the power of his voice whenever possible, using his words to compensate for the unfortunate distractions from his crusade. On August 26, 1987, Reagan gave a speech at the Century Plaza Hotel in Los Angeles, where his remarks were broadcast via satellite to a conference on U.S.–Soviet relations in Chautauqua, New York.[21] He received a warm reception as he approached the podium that day at 1:02 PM. The customary note cards he held in his suit jacket, wrapped tight by a rubber band and scribbled in bizarre shorthand coherent only to Reagan, contained some brash declarations. And yet, unlike the speech he had given at the Brandenburg Gate two months earlier, this one would somehow slip through the cracks of time.

Reagan began the speech by rejecting détente, declaring that his administration "could not gloss over" the differences between totalitarianism and democracy, nor could it be "content anymore with accepted spheres of influence, a world only half free." That was why his administration "sought to advance the cause of personal freedom wherever opportunities existed to do so. Sometimes this meant support for liberalization; sometimes, support for liberation." He spoke openly of the need to liberate that sphere controlled by the USSR, all the way to the remotest, jagged edges of the Iron Curtain:

> Our foreign policy . . . has been an attempt both to reassert the traditional elements of America's postwar strategy while at the same time moving beyond the doctrines of mutual assured destruction or containment. Our goal has been to break the deadlock of the past, to seek a forward strategy—a forward strategy for world peace, a forward strategy for world freedom. We

have not forsaken deterrence or containment, but working with our allies, we've sought something even beyond these doctrines. We have sought the elimination of the threat of nuclear weapons and an end to the threat of totalitarianism.

We know what real democracy constitutes; we understand its implications. . . . It means liberation of the captive people from the thralls of a ruling elite that presumes to know the people's good better than the people. . . . And that's why we know we must deal with the Soviet Union as it has been and as it is, and not as we would hope it to be. And yet we cannot rest with this. The opportunity before us is too great to let pass by.[22]

Reagan's intent was unmistakable: He wanted an "end to the threat of totalitarianism and liberation of the Communist peoples." He gazed intently at the audience: The opportunity was there—upon them, upon "us"—too momentous to miss. He concluded the speech by saying that in looking back over the previous six-and-a-half years, he could not help but reflect on "the most dramatic change to my own eyes: the exciting new prospects for the democratic cause." He hoped that "we may finally progress beyond the postwar standoff and fulfill the promises made at Yalta but never acted upon." To him, the promise of Yalta was a promise of freedom for those "captive people" in Eastern Europe.

Reagan continued the momentum after Los Angeles. In a September 16, 1987 speech in Washington, he likewise candidly stated that, "containment is not enough. Our goal has been to break the deadlock of the past, to seek a forward strategy—a forward strategy for world freedom."[23] This was a sentiment that Reagan would continue to echo throughout the remainder of that fall as he moved toward his final year in the White House. At once a restatement of containment's flaws and a forward-looking analysis of the current state of Communism, these words corralled the direction in which Reagan was headed as he entered the final year of his presidency.

Though the Iran-Contra scandal had marred another year of progress in Reagan's fight against the Soviets, the dramatic changes that would eventually culminate in the USSR's fall were already starting to show their faces. Once again the administration's attention would shift back to Afghanistan, where the final shots by the Mujahedin against the Soviets were being fired. This would prove to be just the beginning of Reagan's tumultuous final year, a year which helped set the stage for Communism's collapse.

19.

"Our formula for completing our crusade": 1988

ON FEBRUARY 8, 1988, THE GENERAL SECRETARY OF THE SOVIET Union made a dramatic announcement to the world: Soviet forces would begin withdrawal from Afghanistan effective immediately. To Washington, it was an exciting, thrilling announcement, one for which the administration had been hoping for some time.

While this news warmed Ronald Reagan's Cold War heart, it did not halt his aid to the Mujahedin, which Reagan continued until the Soviets were gone for good. A month after Mikhail Gorbachev's announcement, in March 1988, Reagan once again marked "Afghanistan Day," where he pledged his "constant support to the Afghans" and promised that "this support will continue . . . as long as is necessary. . . . Their struggle is our struggle."

As it turned out, Reagan didn't need to worry; Moscow's campaign was over for good.

Snapping Soviet morale, the introduction of the Stinger was the action most directly responsible for the Soviet retreat. By 1987, Soviet planes had almost entirely ceased flying over Afghanistan, and those that did fly, did so at extremely high altitudes and with anti-Stinger precautions. Senator Gordon Humphrey (R-NH) personally observed a Soviet military-escort plane arriving above Kabul, spiraling down toward the city from 30,000 feet as it launched heat-emitting decoys from all angles to divert any Stingers. "I'd never seen anything like that," said Humphrey of the spectacle. "They were scared to death of Stingers."[1] Soviet political officials recounted their terror as they sweated the entire trip inside Afghan territory, waiting to be blasted out of the sky. Sergei Tarasenko, then an assistant to Foreign Minister Eduard Shevardnadze, remembered:

> I went with Shevardnadze to Afghanistan six times, and when we were coming into Kabul airport, believe me, we were mindful of Stinger missiles. That's an unpleasant feeling. You were happy when you were crossing the border and the loudspeaker would say, "We are now in Soviet territory. Oh, my God. We made it!"[2]

When they did fly combat missions, Soviet pilots were forced to fly at such high altitude that their accuracy was diminished almost to the point of total ineffectiveness. Safer Soviet ground troops wryly began referring to their pilots as "cosmonauts."

On the whole, the Stinger antiaircraft missile achieved an 80 percent shoot-kill ratio, which demoralized the Soviets and eliminated their air superiority. Within twenty months of the Stinger's introduction, the Red Army, for whom surrender had been unthinkable, began its retreat in May 1988. At the June 1988 Moscow summit, Mikhail Gorbachev complained bitterly to Ronald Reagan that the rebels were firing upon withdrawing Soviet troops on their way home. His complaint was passed on to a contemptuous Muj leader, who responded: "They've killed a million of us. We've killed twenty thousand of them. It's important that they leave with the sound of bullets in their ears."[3] By August 1988, half the Soviet troops had left Afghanistan.

As the Soviets mourned defeat, champagne was uncorked at CIA headquarters in Langley, as operations officers celebrated their tactical victory. Joining in the celebration were unsung heroes like Charlie Wilson, the swaggering, hard-drinking Democratic congressman from East Texas who fought to fund the rebels from the House Appropriations Subcommittee, as well as a CIA officer from Aliquippa, Pennsylvania named Gust Avrakotos, and men like Frank Anderson and Jack Devine.[4] On all levels, the United States had emerged victorious in one of the largest covert operations ever undertaken, an operation that had helped David to beat Goliath. Against all odds, the mission impossible had been completed.

In the end, of one hundred twenty thousand troops deployed, the USSR suffered sixty thousand casualties—an extremely high percentage—including sixteen thousand killed.[5] To the contrary, the largest death toll endured by a Reagan mission in the 1980s was the nineteen who perished in Grenada, eight hundred times less than the Soviet loss in Afghanistan. Not only had Reagan avoided another Vietnam, he helped hand one to the Kremlin.

While Reagan himself may not have fired any shots at the Soviets, his decisions, money, and weapons did. Through middlemen, his administration armed and trained the Afghan people to fight and kill Soviet soldiers, which had a devastating impact on the Kremlin. In Afghanistan, Reagan abandoned the constraints of economic war, launching a campaign that involved live ammunition and real death. Here the Reagan attack was not simply economic; Washington was not merely murdering Soviet hard currency earnings. For the first time under his administration, United States dollars had translated into Soviet deaths. The stakes could not have been higher for Reagan.

This assistance allowed the Afghan rebels to pull off an enormous military upset, one that ran completely contrary to conventional wisdom and that the press had long viewed as impossible.[6] In June 1984, *Newsweek* had told its readers: "The Mujahedin can never be strong enough to drive the Soviets out of Afghanistan."[7] Three years later in the same weekly, reporters Russell Watson and John Barry assured subscribers that, "the anti-Communist insurgents can never hope to defeat their better equipped adversaries."[8] About the same time, Mikhail Gorbachev had concluded just the opposite. Similarly, in a January 1985 piece with the rhetorical title, "Why Aid Afghanistan?", Richard Cohen wrote with certainty in the *Washington Post*: "[W]e are covertly supplying arms to guerrillas who don't stand the slightest chance of winning. . . . Afghanistan is not the Soviet version of Vietnam." By April 1988, with Soviet defeat imminent, Cohen wrote a new piece titled, "The Soviets' Vietnam."[9]

GORBACHEV AND AFGHANISTAN

Mikhail Gorbachev later wrote in his memoirs that from the moment he took office he knew that, "We needed to withdraw from the damaging and costly war in Afghanistan." The conflict had "brought shame" on the USSR. By the late 1980s, approaching the end of the conflict, he called the war a "hopeless military adventure" and a "loss."[10] But despite these claims, he nevertheless undertook an expansion of the war that intensified fighting in an attempt to win, costing thousands of lives on both sides. According to Ken Adelman, the director of Reagan's Arms Control and Disarmament Agency, it was during the years that Gorbachev presided over the war that most of the dead Afghan civilians were killed. During Gorbachev's first year in power alone, a quarter to half of all Afghan peasants had their villages bombed and a quarter saw their water systems and livestock destroyed. Under Gorbachev, says Adelman, Communist forces repeatedly violated Pakistani airspace and perpetuated numerous "terrorist acts" within Pakistan, possibly even the 1988 plane crash which killed its president, Zia ul-Haq, and U.S. Ambassador Arnold Raphel.[11]

Ultimately, Gorbachev was forced to pull troops from Afghanistan because the USSR lost—a fate he sought to avoid. The proper metaphor for Gorbachev is not a dove flying out of Kabul with olive branch in mouth; the accurate image is an exasperated leader, who wanted what he felt was best for his country, who ordered his military to win—with subsequent brutal tactics and results—only to flee in frustration.

Moreover, unlike other causal factors in the downfall of the USSR, this one did *not* stem from a Soviet internal contradiction or from a failure inherent in Communism. The Reagan administration did not exploit a crack in the Soviet system, as it did with oil and energy exports, but instead it exploited a weakness in the Soviet military. Without U.S. aid, and most notably Stingers, the Soviets would have at worst stalemated in Afghanistan, but they would not have been humiliated and defeated in the theatre. Proactive Reagan administration policies, not Soviet internal inconsistencies, made the difference and proved fatal.

The end of the war in Afghanistan was a fitting start to 1988, as the loss proved symbolic of the larger Soviet defeat looming on the horizon. Military historians have argued that when a major dictatorial power—with varying designs on expansion—faces death, as the USSR did in the mid-1980s, that power typically makes a vital choice to turn outward or inward.[12] Strategists considered 1985—the year of Gorbachev's arrival—the arc of the Soviet crisis. Moscow struck outward by pouring resources into the Afghan effort, but to no avail. Thus, one of the incalculable benefits of the Reagan administration's victory in Afghanistan was that the Soviet leadership concluded that it could no longer turn outward. With that option closed off—and Gorbachev committed to internal reforms—the USSR was forced to turn inward, a move that was about to have enormous implications for the country's future.

That the Muj did the shooting and dying and experienced the literal pain of the front line should never be neglected. As one intimate observer put it, there are countless people breathing free today who have no idea of the contribution of a million "Afghan ghosts" who changed the world but have never been thanked.[13] And yet, the Mujahedin would not have persevered without help from the Reagan team.

THE MOSCOW SUMMIT

Whether or not Moscow knew it, the loss in Afghanistan meant that around the world the death toll for the Soviet Union had begun to ring. It was in this atmosphere that Reagan met with Gorbachev in a fourth summit that began in Moscow on Sunday, May 29, 1988. The two met privately, with no cameras, in a one-on-one negotiating session. Long classified as top secret, the federal government only recently approved the mintues of this session for release. Their contents reveal a Reagan who did not shy away from reiterating in private what he had said in public, as he called on the leader of the Soviet Union to tear down the Berlin Wall.

Since his dramatic call for the destruction of the Berlin Wall at the Brandenburg Gate a year earlier, Reagan had kept the rhetorical pressure on Moscow, as he routinely evoked the fall of the wall in many of his speeches—at least nine times since his address at the heart of divided Germany.[14] In February 1988, during his "Address to the Citizens of Western Europe," Reagan told the Soviet leadership that he had meant business, and was still waiting: "To the Soviets today I say: I made my Berlin proposals almost nine months ago. The people of Berlin and all of Europe deserve an answer." Importantly, Reagan did not stop there; he went further than his first call: "Make a start. Set a date, a specific date, when you will tear down the wall." He pressed: "And on that date, bring it down." This would, he rightly affirmed, "be an impressive demonstration of a true commitment to openness" ("openness" was a direct reference to Gorbachev's *glasnost*).[15] He was calling on Gorbachev yet again: if *glasnost* was truly *glasnost*, Gorbachev should prove it by dismantling the wall.

At the May 1988 summit, Reagan leaned hard on Gorbachev, telling him that Americans were encouraged by the reforms in the Soviet Union. With all those changes, he added, "wouldn't it be a good idea to tear down the Berlin Wall?" He noted that "nothing in the West symbolized the differences between it and the Soviet Union more than the wall." He told Gorbachev that the wall's removal "would be seen as a gesture symbolizing that the Soviet Union wanted to join the broader community of nations."

Reagan made his proposal in response to Gorbachev's request that the United States open up trade with the Soviet Union. Reagan responded to the general secretary by apparently suggesting a sort of linkage: increased trade might be possible if the Soviets bulldozed the wall.[16] Though desperate for cash, Gorbachev still would not budge on the Berlin Wall. In the words of his interpreter, Igor Korchilov, Gorbachev "said he could not agree with the president's view."[17]

SUMMER 1988

Despite the improvements in U.S.–Soviet relations that were made at the Moscow Summit, Reagan's verbal assault was unrelenting. He flew out of Moscow en route to London, where, on June 3, 1988, he spoke before an august audience at the dashing Royal Institute of International Affairs in London. Six years earlier, also in June, he had given the Westminster Address, where he prophesied Communism's doom. Now, he went further. He told his British audience about his pursuit of a U.S. policy that rejected the "permanency of totalitarian rule"—part of his "forward strategy of freedom." "Quite possibly," he said, "we're beginning to take down the barriers of the postwar era"—the "totalitarian" barriers to "freedom." "Quite possibly," he continued, "we are entering a new era in history, a time of lasting change in the Soviet Union."[18]

Standing within the Great Hall inside a venerable four-hundred-year-old Guildhall, Reagan articulated a future that looked beyond the fall of Communism. There he quoted those special words from the prophet Isaiah, as he did five years earlier in the Evil Empire speech, six years earlier in declaring a national day of prayer for Poland, and on many other occasions: "He giveth power to the faint, and to them that have no might, he increased their strength, but they that wait upon the Lord shall renew their strength. They shall mount up with wings as eagles. They shall run and not be weary."[19]

What Reagan did with that quote was eye-opening: "Here, then, is our formula for completing our crusade for freedom," he said, capping his oration by invoking the "rendezvous with destiny" which he first called upon in his October 1964 speech for Barry Goldwater. That rendezvous, "that crusade," said Reagan, was "well underway."

While this speech went far, a month later, on July 13, 1988, Reagan made another speech that went still further, declaring:

> [T]he tide has been turned. Despite decades of suffering, the will to freedom is alive. It has survived its tormentors. It will outlast the Communists.

And truly, I can think of no time in my adult life when the prospects for freedom were brighter than they are today. . . . The Communist idea is discredited and around the world new progressive forces are emerging as political change and liberation sweep the globe. America will continue to encourage the movement toward freedom, democracy, and reform by holding firm to our principles and speaking opening and truthfully about human rights and the fundamental moral difference between freedom and communism. And America shall light the path as the whole world climbs out of the dark abyss of tyranny to freedom.

And within the Soviet bloc there are hopeful signs.[20]

As the summer of 1988 progressed, Ronald Reagan could see that his policies had begun to bear fruit.[21] On August 15, he boasted that in the 2,765 days of his administration "not one inch of ground has fallen to the communists."[22] In the sixty-odd years prior to the start of the Reagan presidency, the Soviet Union had consistently advanced the frontiers of international Communism. There were dozens of countries that were part of the USSR, the Soviet Communist bloc, or were Soviet allies, satellites, or client states, a number that had increased each decade since the 1910s. The 1980s marked the first decade that halted that advance. Even more impressive was that the 1970s had been one of the busiest decades for Communist expansion, especially during the six-year period before Reagan's presidency.

Reagan saw Poland as a major part of this success. In a statement proclaiming Polish American Heritage Month, issued on September 28, 1988, Reagan saluted a handful of famous Poles, including recent names like Pope John Paul II and Lech Walesa. He asserted that the American people felt an "unwavering unity" with the Polish people, "now more than ever." He urged that "Poland's saga must be our own." The eternal optimist claimed that, "The freedom loved and advanced so much through the years by loyal Poles and Polish Americans is on the march in every continent today, because freedom is a universal and eternal cause."[23]

NSDD-320

While there was a clear consistency in Reagan's anti-Soviet policy throughout both of his terms, the second term witnessed decidedly fewer of his NSDDs. In truth, much of the mechanics articulated through the NSDDs were implemented under Bill Clark from 1982 to 1983, and thus there remained little in way of NSDDs for the administration to accomplish in the second term. Though there were some notable NSDDs during Reagan's second four years, few were as radical as those from the first term. Nevertheless, in November of 1988, two months before he would leave the Oval Office, Reagan still had one extraordinary NSDD left in him.

On November 20, Reagan signed NSDD-320, titled "National Policy on Strategic Trade Controls," which is today still heavily redacted. It began by acknowledging that, "The efforts of this administration to stem the flow of strategic technologies to the Soviet Union and its allies have been vigorous and the results substantial." The NSDD then listed results. Yet, NSDD-320 continued, "Despite recent improvements in U.S.–Soviet relations and the program of internal Soviet reform instituted by President Gorbachev, the military threat posed by the Soviet Union to the United States and its allies remains unchanged." As a result, "It therefore remains the policy of the United States to restrict the transfer of strategic technologies to the Soviet Union and its allies whenever such transfers would make a significant contribution to Soviet bloc military capabilities that would prove detrimental to the national security of the United States and its allies."

Thus, NSDD-320 pursued continued restrictions on key high-tech exports to the USSR. It enlisted greater export control laws and enforcement of laws by the federal government and COCOM.[24] The economic-warfare campaign would continue until the USSR's final gasps.

Although this move could be viewed as odd in light of the advances that had been made between the two countries, the reality was that this thinking was completely in line with Reagan's perception of the situation at the end of his presidency. Sure, Reagan had come to trust and even like Gorbachev, but he did not like the Soviet Union; he was trying to push the USSR over the edge, and nothing would stand in the way of that goal. Despite the fact that Gorbachev seemed to be a nice guy, Reagan was not about to alter his commitment to "reverse the wheels of history," as *Pravda* editor Tomas Kolesnichenko put it, in judging that the president still "views himself as some sort of provincial overthrower of Communism."[25] Or, as Georgi Arbatov put it, he was not about to give up his goal of dislodging the Bolsheviks from power, of "conquering" the "empire of evil." Reagan "wants to throw Communism and socialism on the dust heap of history."[26] Or, alas, as Manki Ponomarev laid out in the Red army publication *Krasnaya Zvezda* (*Red Star*), he was not about to retire his "crusade" to "wreck" the USSR.[27]

PAUL KENGOR   281

NSDD-320 proved to be the last significant piece of anti-Soviet policy that Ronald Reagan implemented before leaving the White House. It was fitting that the year would end in the same fashion in which his presidential crusade had begun, with a determined and hard NSDD. Like Reagan himself, the NSDDs which had done so much damage to the Soviets would now come to a close and the man who wreaked so much havoc on the Soviets would make his exit.

As the dust began to settle on 1988, there was little that members of the administration could do but watch and hope that the steps that they had taken over the previous eight years had made a difference. Over the course of the previous eight years, Reagan and his team had taken the most dramatic steps in forty years to bring about the fall of Communism and the USSR. Now, with all of their chips in place, there was little that they could do but wait for their plans to come to fruition, hoping for the realization of their goals, and for the freedom of those enslaved by the Soviet Communist state.

PART IV

The Fall of the Soviet Empire

20.

The March of Freedom: 1989

APPROACHING THE END OF HIS EIGHT YEARS IN THE WHITE House, Ronald Reagan was thrilled over the sudden resurgence of freedom around the world; it looked as though democracy was indeed on the rise. "More of the world's populace is today living in relative freedom than ever before in history," he triumphed in December 1988, "more and more nations are turning to freely elected democratic governments." He cited numbers from Freedom House, and noted that this democratic revolution was accompanied by a considerable change in economic thinking as well, as the world looked to free markets in addition to free elections.[1] And this was just the start; there was a "transformation" afoot alright, one "that many would have thought impossible only a decade ago"; yet, even the eternal optimist could not foresee the fast and furious shift in the months ahead.[2]

As he bade farewell, there was no doubt to Reagan about what he had committed America to in the 1980s: freedom at home and abroad. "[T]hat's what it was to be an American in the 1980s," he said with satisfaction to his fellow Americans in his Farewell Address on January 11, 1989. "We stood, again, for freedom." He summed up victoriously: "We meant to change a nation, and instead, we changed a world."[3] While clever, that line was not totally forthright: his intent to change both the nation and the world had been clear for a long time.

In that final speech, Reagan also could not resist one more poke at détente, even after it had lost currency: "The détente of the 1970s was based not on actions but promises," he reminded his fellow Americans. "They'd promise to treat their own people and the people of the world better. But the gulag was still the gulag, and the state was still expansionist, and they still waged proxy wars in Africa, Asia, and Latin America."[4] This desire to end détente, and the source and system that embraced it, never left him.

By the time of his Farewell Address, morale had long been restored at home, democracy was readying to spread abroad, and Communism had declined from its apogee of expansion in 1979. Reagan's popularity had grown substantially over the course of his presidency, even surviving setbacks like Iran-Contra. He was leaving office with the highest Gallup approval ratings of any president since Eisenhower, and there were many Americans who would have gladly voted for him for a third term if they had the option—as would many Poles. Andrew Nagorski, Bonn bureau chief for *Newsweek*, was in Poland in 1988, where he was asked wistfully by Poles: "Is it really true that Reagan can't run for a third term?"[5] Fortunately for the world, another Reagan term was not necessary for Communism to fall.

Alas, it was fitting that on the final day of the Reagan presidency, *Pravda* was sensing a similar shift. Looking back, columnist Gennady Vasilyev affirmed on January 20, 1989 that without question Reagan's objective had been "to roll back communism on a worldwide scale."[6] And that once unthinkable prospect now appeared quite possible.

POLAND UNRAVELS

In the spring of 1989, Reagan's beloved Poland was preparing for what in December 1981 would have been unimaginable: free and fair Parliamentary elections, open to candidates from any political party, Solidarity included. The inevitability of freedom was finally upon Poles, and former president Reagan eagerly anticipated the moment from his home in California.

A few weeks before the elections, Reagan received a visit from four gentlemen: two Solidarity members and the two Polish Americans hosting them. One of the hosts, Chris Zawitkowski, who today is head of the Polish-American Foundation for Economic Research and Education, asked the master campaigner if he had any words of wisdom or encouragement for the two Solidarity members as they prepared for the final stretch to the June elections. Zawitkowski was expecting to hear about political strategy, or even the invocation of Thomas Jefferson or the founders; instead, he was taken aback by what he heard from the seasoned candidate: "Listen to your conscience," said Reagan, "because that is where the Holy Spirit speaks to you."[7]

With this line, what began as a political appeal became a spiritual encounter. The ex-president pointed to a picture of Pope John Paul II, which hung on his office wall. "He is my best friend," explained Reagan. "Yes, you know I'm Protestant, but he's still my best friend." Antoni Macierewicz, a Solidarity member who had been imprisoned by the Communists, reciprocated by giving Reagan a Madonna that he had handcarved in a Polish gulag. Holding the gift in his hands, Reagan gladly accepted it and said that he and Nancy would be proud to have it in their home.[8]

Macierewicz and his friends soon had their day: In June 1989, unlike the Soviet broken promises at Yalta forty-four years earlier, Poland honored its promises and held genuine elections for a limited number of chambers in two houses of Parliament, which resulted in an overwhelming victory for Solidarity. The old chamber of the legislature opened 35 percent of its seats to balloting and a second newly created upper chamber was fully contested. In the old chamber, all 35 percent of the open seats were won by Solidarity candidates. Of the newly created upper chamber, which had 100 new seats, 99 were won by Solidarity candidates and the other by a millionaire capitalist. Not one seat was won by a Communist.

In short, Solidarity won over 99 percent of available seats in June 1989.

Lech Walesa capped the tumultuous decade by emerging as Poland's freely elected president, beating challenger Stanislaw Tyminski by a margin of 75 percent to 25 percent. (Tyminski was an émigré businessman from Canada. Communists were not to be found in this race.) Walesa refused to form a coalition government with the remaining Communist Party delegates in the old chamber, and eventually, Parliament accepted a Solidarity-led government. Antoni Macierewicz, the former prisoner, found new incarnation as Poland's minister of internal affairs.

A crucial component to this peaceful outcome came from Mikhail Gorbachev who, in 1989, spoke by phone with the head of the Polish Communist Party. During the phone call, he stated that the Soviet Union, which had tried to crush Solidarity, would accept the outcome of the free election—that is, a government with a Communist minority and non-Communist prime minister. "That phone call," wrote scholars Daniel Yergin and Joseph Stanislaw, "ended the Cold War."[9] That sentiment was shared by others, including Gorbachev himself, who later admitted that once Poland held elections, he and the Soviet foreign ministry knew the game was over. In his memoirs, he spoke of how the Soviet leadership realized that the emergence of Solidarity threatened not only "chaos in Poland" but the "ensuing breakup of the entire Socialist camp."[10]

THE REAGAN AID

It was only after the June 1989 elections that the full extent of Reagan's aid to Solidarity during the years after martial law was imposed became evident. As a result of NSDD-32 and other subsequent efforts, the Polish underground received tons of equipment—printing presses, photocopiers, cameras, computers, telephones, transmitters, short-wave radios, and fax machines (the first in Poland). Such equipment was smuggled into Poland by CIA agents, priests, and labor officials; among other things, it enabled the creation of underground newspapers and broadcasting operations. Zbigniew Brzezinski, national security adviser to President Carter and a Poland native, who played a key consulting role, emphasized: "This was about getting the message out and resisting: books, communications equipment, propaganda, ink and printing presses."[11]

While the actual amount that Reagan spent to keep Solidarity alive remains elusive, a commonly cited figure is roughly $8 million per year over a six- to seven-year period, or a total of approximately $50 million—a significant amount.[12] This aid was used in a variety of ways, among the most significant of which was to assist the crucial operation of *Tygodnik Mazowsze*, the underground Solidarity newspaper which was widely read and respected. The newspaper had a circulation of 70,000 to 80,000. Typically, it was a four-page paper, eight pages for a special issue. One source said that to produce a typical issue with a normal print run required sixteen tons of paper. He stressed that in a country in which everything had to be rationed, including paper, finding such a volume of material was a daunting task. "Some [of the paper] had to be stolen from government ministries and bureaucracy," said the source, who asked not to be identified. "It was quite a feat. The organizational effort that went into this was enormous." Hence, any support from the Reagan administration, particularly financial, was crucial.

Most vital to *Tygodnik Mazowsze*'s effectiveness was the wider, unstoppable means of mass communication that broadcast the newspaper's contents throughout Poland—namely, Radio Free Europe (RFE). In fact, most people digested the weekly by hearing its articles and opinion pieces read over RFE, a juggernaut unto itself, and among Ronald Reagan's most cherished projects.[13]

He never lost his love for radio, nor its inherent importance in communicating to huge swaths of people.

As Radek Sikorski put it, "everyone and his brother" listened to Radio Free Europe.[14] Reagan's team even added religious programming to RFE, including the Catholic Mass.[15] Poles were thrilled with this; in the USSR, priests were banned from radio and TV. Asked about RFE's import to Solidarity, Lech Walesa said, "Would there be the earth without the sun?"[16] He called RFE "the most precious gift of the West for me."[17]

Reagan administration financial and technical support of media was especially crucial because of the inability of internal sources of Polish freedom—such as Solidarity—to operate radio stations. Solidarity officials today ruefully remember those difficulties. "Inside [Poland], organizing underground radio was impossible on the regular basis," said one member in a 2003 interview.[18] At best, they settled for Pyrrhic victories: For instance, state-run stations employed many government technicians secretly allied with Solidarity; occasionally, the technicians cut in to state radio or television programs and inserted pro-Solidarity messages. Poles laughed as they listened or watched. The messages lasted only a few minutes until the military or police arrived at the station to shut down the malfeasance. These messages were programmed to emit automatically after the technicians left; the technicians did this at considerable personal risk. Try as they might, however, Communist authorities could not arrest RFE.

Richard Pipes, who, as a native of Poland, followed no other country with as much personal interest, said that the assistance that Ronald Reagan extended to Solidarity—in 1982 in particular—"later made it possible for Solidarity to survive and ultimately compel the communists to yield power." Specifically, Pipes cited the "strong moral support" that Solidarity received from the Reagan administration, which he says helped sustain Solidarity's morale and "by the end of the decade forced the communists to surrender power." He said that Ronald Reagan showed "a far deeper understanding" of the Poland situation and the stakes involved than anyone in the State Department, which said Pipes, had written off Poland as a total loss—as a "lost cause."[19]

In addition to Reagan, there were a number of other sources that contributed to Solidarity's ultimate success. Lane Kirkland, AFL-CIO president, funneled hundreds of thousands of dollars through the organization's headquarters. Similarly, the U.S. Congress also provided aid, sending $1 million in 1987 and 1988 during the buildup to Poland's June 1989 elections. And then there were individuals like Bill Casey, who, aside from his vital work at the CIA, personally donated $50,000 of his own money to purchase printing equipment that was smuggled into Poland through Vatican channels.[20]

But while this American aid was helpful, the role of John Paul II was monumental. On the Pope's role, Zbigniew Brzezinski, who himself was important, stated: "We involved the Pope directly; and I don't want to talk about it. I can't go into details. . . . [T]o sustain an underground effort takes a lot in terms of supplies, networks, etc. And this is why Solidarity wasn't crushed. . . . This is the first time that communist police suppression didn't succeed."[21] Russia historian and Librarian of Congress James Billington emphasized that in addition to Reagan's political support, John Paul II spiritually inspired the movement.[22]

Despite the involvement of many figures, few did more to help the Polish people than Reagan, and his impact was felt by Poles throughout the country. In early 1990, Arch Puddington, who worked for RFE, did a series of interviews with Eastern European émigrés and visitors. He asked their opinion of various U.S. presidents. He found that almost all had a highly favorable view of Reagan and his role in the fall of Communism—some even dubbed him "Uncle Reagan." He left a great impression on Poles.[23]

This was obvious to men like Colonel Henryk Piecuch, a high-level official in the Polish Interior Ministry in the 1980s, who said that, "Ronald Reagan was considered a god by some in our country. This pertains especially to the lower ranks of Solidarity."[24] This book could be filled with testimonies to Reagan's importance, whether from ordinary Polish workers to members of the Communist government and Solidarity to Polish academics now teaching in the United States, as well as men like Vladimir Bukovsky, who spent twelve years in the gulag.[25]

By the 1990s, free to speak, Lech Walesa spoke for many of these individuals, saying that he was thankful for Reagan's "very strong backing." He praised Reagan's devotion to "a few simple rules: human rights, democracy, freedom of speech" and his "conviction that it is not the people who are there for the sake of the state, but that the state is there for the sake of the citizens."[26] Reagan had indeed said exactly this many times, and apparently Walesa listened.

Once out of prison and in the presidency, and given a chance to thank Reagan directly, Walesa said to the American president: "We stood on the two sides of the artificially erected wall. Solidarity broke down this wall from the Eastern side and on the Western side it was you. . . . Your decisiveness and resolve were for us a hope and help in the most difficult moments."[27] He said that Reagan "emboldened" and "encouraged" him, and was an "inspiration."[28] Walesa saw Reagan as essential not just to Solidarity's survival but to the end of Communism, and went on to reference the Reagan radio gaffe that back in 1984 had enraged the Soviets and many liberals: "People thought it was unfunny, but I'm of the opposite opinion, that it was not only a good joke, but the words were also prophetic."[29]

Poles had needed a friend, said Walesa, "and such was Ronald Reagan." He said Reagan challenged rather than avoided problems and was "favored" by the "muse of history"; that muse liked Reagan "so much" that she whispered in his ear and told him what to do. "We owe so much to Ronald Reagan," concluded Walesa. "We Poles owe him freedom." Teary-eyed, he said Reaganesquely: "God bless America."[30]

POLAND WAS THE WEDGE

Ronald Reagan had long felt that Poland and Solidarity held the golden key. Recall that he had written in his diary on December 15, 1981, two days after martial law was declared, that at the NSC meeting that day he had taken a stand that "this may be the last chance in our lifetime to see a change in the Soviet empire's colonial policy re[garding] Eastern Europe."[31] Less than three weeks later, Bill Clark arrived to help ensure that the president's policies to make that happen were in fact carried out.

After his presidency, Reagan wrote in his memoirs: "The events in Poland were thrilling. . . . [T]he first break in the totalitarian dike of communism." He had "wanted to be sure we did nothing to impede this process and everything we could to spur it along. This was what we had been waiting for since World War II. What was happening in Poland might spread like a contagion throughout Eastern Europe."[32] The splinter could be a contagion to the body of the Soviet empire. As Reagan said at Eureka College in May 1982, the Soviets feared the "infectiousness" of the threat of freedom posed by Solidarity.[33]

In sum, Poland was the domino that catalyzed the collapse of Communism in Eastern Europe during the fall of 1989. And yet, despite its significance, the Poland story barely registers in the two preeminent biographies of Reagan; in fact, incredibly, Lech Walesa is not mentioned even once in either book.[34]

But while the biographers missed the importance of Solidarity (somehow even after the fact), Ronald Reagan never did. As it turned out, the labor movement—as well as the Polish pontiff—was every bit as influential as he had always estimated. Together, these factors inside Poland were the wedge, the splinter in the bloc. Once the wedge was hammered deep enough into the crack, the fissure spread, and the entire block (or bloc) came apart. As Bill Clark put it, in Poland, the Soviets lost an empire.[35]

NOVEMBER TO DECEMBER 1989

With Poland gone, it did not take long for the remainder of the Eastern bloc to break away, and less than six months later, many of them had started to fall. By November 1989, the Berlin Wall was rubble. The "March of Freedom" strolled into East Germany, Hungary, Czechoslovakia, Bulgaria, and, somehow, even to Romania, where, in December, Eastern Europe's most brutal dictator, Nicolai Ceausescu, was put on trial by the Romanian people, convicted, sentenced to death, lined up against a wall, and shot on Christmas Day—a day he tried to ban. That evening the news anchor on Romanian state television cheerfully announced his country's liberation: "Good news this Christmas Day: the Anti-Christ is dead!"[36]

Aside from Ceausescu's violent death, the end of Communism in Eastern Europe occurred with complete tranquility, without shooting, without a war, and certainly without World War III—the "DPs" were at long last free. As these events unfolded around the world, one could not help but think of Reagan's line from his Westminster Address, in which he had seven years earlier hoped that Eastern Europeans would "choose their own way to develop their own culture, to reconcile their own differences through *peaceful* means." (Emphasis added.) That was in fact surprisingly the way the revolution unfolded in Eastern Europe in the fall of 1989. Who would have guessed that if and when the Berlin Wall came down, it would fall peacefully as celebrants stood atop it cheering and toasting with champagne glasses? In Czechoslovakia, the revolution was so smooth that it was dubbed the Velvet Revolution.

MR. REAGAN GOES TO WARSAW

In September 1990, as most of Eastern Europe was approaching the first year anniversary of Communism's demise, Ronald Reagan visited Gdansk, home of the shipyard in which Solidarity was born.

There, the former president received what both UPI and Reuters described as a "hero's welcome." Despite a torrential rain and punishing hail, seven thousand braved the storm to greet Reagan in front of the shipyard gate where they sang "Sto Lat," which means "May He Live 100 Years," a Polish anthem sung only to honor the nation's heroes. They chanted "Thank you, thank you!" In a gesture Reagan must have loved, Lech Walesa's parish priest handed him a sword and explained: "I am giving you the saber for helping us to chop off the head of communism."[37] In turn, Reagan told them: "You have triggered fast changes in the political map of Central and Eastern Europe." Referring to the other dominoes that subsequently fell, Reagan said: "One might say that [this] was the shipyard that launched a half-dozen revolutions."[38]

Poles' gratitude to Reagan did not stop. They continue to look for ways to honor him. This includes a grassroots movement that looks to name things after Reagan. "I will not rest until there is a Ronald Reagan Square in Warsaw," says Radek Sikorski, who in the 1990s became deputy minister of foreign affairs and deputy minister of defense for free Poland. "We want some major form of commemoration," says Sikorski. "They [Poles] at least want a Reagan statue in a place of significance."[39] Precisely that was proposed by a committee of Poles called the Ronald Reagan Legacy Committee. Sikorski is chair of the committee, which includes Polish cabinet members, senators and members of Parliament, and major Solidarity figures.

Still, Poles did what they could. Reagan was made an honorary citizen of Gdansk and Krakow, the cities where the Solidarity tradition was strongest. Separately, a competition was held to choose a name for a square in front of a train station in Warsaw. No suggested names were listed. The public voted. Naturally, the winner was the architect who constructed the train station. Ronald Reagan, however, was runner up.

21.

The Coroner Comes to the Kremlin: 1990–1991

IT DID NOT TAKE LONG FOR THE JOYOUS DEVELOPMENTS OF 1990 in Eastern Europe to have a ripple effect on the Soviet Union, one that Mikhail Gorbachev did not desire, that he in fact dreaded, and that careened out of his control. For Gorbachev, Eastern European freedom was a train wreck screeching down the tracks and headed smack for the middle of Red Square.

It was not supposed to happen like this: Gorbachev had not initially favored what blossomed in Eastern Europe. "Gorbachev never foresaw that the whole of Eastern Europe would fly out of the Soviet orbit," said Ambassador Anatoly Dobrynin, stating the obvious. He became, said Dobrynin, a "helpless witness" to the inevitable.[1] He also did not foresee that at the start of 1990 Eastern Europe's next generation of leaders would move quickly toward free-market economies. Once given the freedom to chart their own course, they unwaveringly opted for the antithesis of the Leninism that Gorbachev publicly preached and even the milder socialism he privately pursued. With their first opportunity to choose their economic destiny, the former Soviet bloc nations looked toward the system championed not by Gorbachev but by Ronald Reagan, a man who in the 1980s was the face of free-market capitalism.[2]

Still, what was done was done, and to his immense credit, Gorbachev did not send in the tanks. He was unwilling to use force to try to hold onto Eastern

Europe, to the Soviet "ally" he had seen as central to the grandiose twenty-first century he envisioned for a better, stronger, kinder USSR.

When it came to the Soviet Union, Gorbachev was not willing to let go so easily, and as such the general secretary had planned a very different response. Oxford professor Archie Brown confirms that "it was no part of Gorbachev's intention to stimulate the breakup of the Soviet Union" and that "the last thing Gorbachev wanted was to lose any part of the Soviet Union following the loss . . . of Eastern Europe." He "wished to reform the Soviet system, not to destroy it."[3] Going further, Brown agrees that by introducing the democratizing elements of glasnost and perestroika, Gorbachev was unwittingly fostering the dissolution of the USSR.[4] Gorbachev himself came to understand this, but only as he scrambled to prevent a complete collapse.

DYING FOR THE MOTHERLAND

Approaching the end of 1990, Gorbachev, terrified by the hurricane he had generated, shuffled his inner cabinet, adding hardline Communist thugs dedicated to the Soviet motherland—men that Gorbachev thought would be needed to help him keep the USSR together—and immediately instructed them to draw up plans for emergency and full dictatorial power if need be. In response, his most influential adviser, Edward Shevardnadze, resigned, direly warning the legislature that "dictatorship is coming."[5]

A few weeks later, on January 12–13, 1991, Lithuania felt the brunt of this ugly turn, as Soviet special forces stormed the Baltic nation with orders to destroy the democratic opposition. Gorbachev, once again talking like a Marxist, spoke of the need for a "restoration of the bourgeois order" in Lithuania.[6] At least fourteen were killed and hundreds were injured. In an attempt to absolve himself of the situation, Gorbachev denied that the military acted on his orders but instead under the direction of his newly appointed lieutenants—a claim adamantly denied by those lieutenants. One of them, KGB chairman Vladimir Kryuchkov, later stated: "Everything was done with the agreement of Gorbachev. Absolutely. And I don't condemn Gorbachev for giving the command to intervene. I condemn him for his deception and his denial of this command. Can you imagine that in this country we would intervene without the president's permission?"[7]

Despite condemning sentiments such as these, some authorities believe that Gorbachev had little personal culpability in the matter. Archie Brown conceded that Gorbachev "is open to criticism" for his tougher line and sharp rhetoric toward Lithuanians in the days preceding the attack, as well as for his tardiness in condemning the killings; however, claims Brown, there is evidence that Gorbachev was angry over the attacks.[8] Anger that would lead some to conclude that Gorbachev did not have a direct role in the attacks themselves.

Regardless of Gorbachev's personal involvement, Boris Yeltsin responded by calling the actions in Lithuania "a powerful attack on democracy" and predicted that more force would follow in other republics, as it did only a week later in Riga, Latvia on the night of January 20–21, where four people were killed by Soviet troops.[9] Gorbachev countered by blasting Yeltsin for threatening the Soviet state and the socialist ideal it represented. He then again, as with Lithuania, publicly reaffirmed his devotion to Communism: "I am not ashamed to say anywhere in public that I am a Communist and believe in the socialist idea," said Gorbachev. "I will die believing this and will pass into the next world believing this."[10]

In the United States, two opposing ideological sources, the *New Republic* and *National Review*, the political bibles of the left and right, respectively, expressed outrage. In an editorial titled, "Gorbachev's Tanks," the *New Republic* stated: "It says something about the roots of glasnost that, in a twinkling of Gorbachev's eye, it can revert to levels of distortion even Brezhnev might envy." The editorial continued:

> [Gorbachev's] Western admirers are in acute discomfort, amazed that a man who has just amassed near-dictatorial powers in his revamped presidency should choose to use them . . . horrified that perestroika should seem now to mean the crushing of human beings with tanks. Yet no one who saw Gorbachev's complete lack of remorse, or witnessed his energetic verbal attack on the Lithuanians two days after the Soviet army assaulted their country, can doubt that he is in control of this strategy.[11]

This was not completely new. "Gorbachev's government has brutally repressed secessionist revolts before," noted the *New Republic*, citing Kazakhstan in 1986, Georgia and Uzbekistan in 1989, Armenia, Azerbaijan, and Tajikistan in 1990. According to the human rights group Helsinki Watch, over 200 people were killed in these incidents. The methods included not just tanks and machine guns, but, in the brutal case of Tbilisi, Georgia, chemical weapons.[12] To be fair, some of these were ethnic disputes and as such their violence was not Gorbachev's fault. Nonetheless, others were questionable and clearly do not stand as a shining star on Gorbachev's record.

Again, to the extent that Gorbachev backed these assaults, his purpose was to preserve the Soviet Union—his top priority.[13] Thus, Gorbachev aide Alexander Yakovlev could later justifiably argue that it was "unjust" to blame Gorbachev for the Soviet breakup: "He did everything possible to keep the country united."[14] He indeed did all he could, including resorting to violence on occasion, and continued to employ or consider coercion until his final moments in office.[15]

GORBACHEV'S HISTORIC CHANGES

Although these cruelties blemished Gorbachev's legacy, from 1990 to 1991, he implemented a series of crucial reforms that were fundamental to the burgeoning democratic movement in the USSR. The most critical of these changes began after the Moscow Summit in June 1988. Notably, Ronald Reagan's actions at this summit were extremely important. Gorbachev had been under fire from hard-liners. As U.S. Ambassador Jack Matlock recalls, Reagan's comments in Moscow at the summit that week "did more than any other single event to build support for Gorbachev's reforms."[16] When Reagan returned to Washington, an emboldened Gorbachev, who had struggled tirelessly for years, got back to work. The Communist Party Conference was scheduled. Soon, Gorbachev pushed the majority of his reform ideas through the Central Committee. Significantly, these ideas included granting real power to a legislature chosen by honestly contested elections, the establishment of an independent judiciary, due process for Soviet citizens, and an end to Party control of government institutions. Through these Gorbachev initiatives, the path to an ultimately pluralistic political system had begun.[17]

By 1990, the Soviet media was freer than ever, thus introducing a completely new concept in the USSR: a critical (though still cautious) press. Most significant, in February 1990, Gorbachev succeeded in banishing the Communist Party's guarantee as the USSR's sole, legitimate political party. Specifically, he backed a proposal by Alexander Yakovlev, a Politburo member and confidant, as well as other Soviet reformers, to repudiate Article 6 of the USSR Constitution, which had ensured the seventy-plus year Communist stranglehold on power. He also accepted a program that recommended the creation of a Western-style presidency and cabinet system. This historic shift toward political competition was greeted by a double-line top-of-the-fold headline that ran across the front page of the February 8 *New York Times*.[18]

At that moment, Mikhail Gorbachev formally ended the Communist monopoly in the USSR.[19] Yes, Ronald Reagan applied crucial pressure that set certain forces in motion. Yes, Boris Yeltsin later won two significant presidential elections that kept the Communists from taking back the executive branch. However, it was Mikhail Gorbachev who technically ended the unilateral rule by the Communist Party—an idea suggested to him (and which he contemplated) as early as 1985.[20]

Yet, as Archie Brown notes, despite this monumental move by Gorbachev that ended the unilateral rule of the Communists, "it was no part of Gorbachev's initial conception to introduce a fully-fledged political pluralism in which the Communist Party would become just one party competing with others." The reality of the situation was that once Gorbachev began making qualified statements on democracy, the concept of pluralism took on a life and momentum of its own, and though he eventually came to embrace this newfound pluralism, it was a far cry from his original conception.[21]

Equally significant, while this viral brand of pluralism was not Gorbachev's initial intent when he began the process of reform, it most certainly was Ronald Reagan's intent in many of the NSDDs. Indeed, the pluralism that swept the former Soviet Union was an objective of NSDD-32, which set the goal of encouraging democratic change within both the Soviet bloc and the USSR itself. Recall that shortly after the signing of NSDD-32, NSC member Tom Reed had spoken to the Armed Forces Communications and Electronics Association, where he shocked the audience by stating that it was the Reagan administration's "fondest hope" to "one day convince" the Soviet leadership to "seek the legitimacy that comes only from the consent of the governed." In addition, this goal was reiterated in the January 1983 directive NSDD-75, which advocated promoting "the process of change in the Soviet Union [and Eastern Europe] toward a more pluralistic political and economic system." What makes these Reagan administration objectives particularly resonant is that they took place two to three years before Gorbachev had even entered office. Encouraging democracy in the USSR had been a plan of Ronald Reagan.

Reagan watched this plan come to fruition, intuitively understanding the deeper forces at work in the hearts and souls of all people, meaning that each person, in Reagan's view, possesses a God-given yearning for freedom—including those living in the Evil Empire. As Gorbachev tried to get a grip on the situation, ex-president Reagan, speaking in Cambridge, England on December 5, 1990, demonstrated that he understood better than Gorbachev what Gorbachev's taste of freedom would bring: "As is always the case," said Reagan, "once people who have been deprived of basic freedom taste a little of it, they want all of it. It was as if Gorbachev had uncorked a magic bottle and a genie floated out, never to be put back in again. *Glasnost* was that genie."[22]

Now, everything was collapsing around Gorbachev, spinning beyond his control.[23] "Attempting to change society," said Valery Boldin, CPSU official and Gorbachev's chief of staff, he was "unintentionally" destroying "[our] statehood."[24]

THE END OF THE END

By 1991, Mikhail Gorbachev's star, and his beloved USSR, had been eclipsed by Boris Yeltsin's Russian Federation. Yeltsin was voted president of Russia on June 12, 1991; its first ever democratically elected leader. Gorbachev, too, had changed the title of his unelected post from general secretary to president. There were now literally two presidents in Moscow sharing the Kremlin— who, or what, would prevail?

The subsequent weeks were a blur, filled with dramatic events all beyond Gorbachev's ability to forestall, including a failed coup attempt in August. By September, nearly every remaining Soviet republic declared independence, a glorious event for just about everyone except Mikhail Gorbachev, who still hoped the USSR would persevere. Yet, there was no way Gorbachev could stop the flurry of freedom he had unleashed. Among the rebels, on September 6 a defiant Georgia severed all ties with its abusive parent. That same day, the USSR State Council recognized the independence of Estonia, Latvia, and Lithuania, and fed up residents of Leningrad restored the city's original name, St. Petersburg—a move that must have made Lenin howl from his tomb. On November 6, President Yeltsin—now the top dog after saving both Gorbachev and Russia from a coup—banned the Soviet and Russian Communist parties.

In a gesture of major symbolic importance, on December 18, 1991 the red Soviet hammer-and-sickle flag which had flown over the Kremlin for decades was replaced with the flag of the new Russian Federation. It then took Mikhail Gorbachev seven days to do the inevitable: to step down. He called President George H. W. Bush to say: "You can have a very quiet Christmas evening. I am saying good-bye and shaking your hand."[25]

That evening, Gorbachev went on television to announce he was leaving his post. He began his December 25 resignation speech by noting that he had stood "firmly . . . for the preservation of the union state, the unity of the country. Events went a different way. The policy prevailed of dismembering this country and disuniting the state, with which I cannot agree." He lamented the "breakup" of Soviet "statehood" and "the loss" of, curiously, "a great country." He also noted, rightly, that his "foremost achievement" was the political and spiritual freedom he brought to Soviet society, and highlighted the "historic significance" of eliminating the totalitarian system and creating a democratic society. He cited two principal reasons for the failure of the Soviet state: the Communist-command economy generally and the "terrible burden of the arms race."[26] The Reagan arms race was indeed devastating.

By resigning as head of the USSR that Christmas Day, Gorbachev also resigned the USSR and provided the time of death of the Communist empire. His peaceful departure reflected the fact that he had no other option, but it was also a sign of his gentleness and unique nature. He was that rarest of Soviet leaders: he walked out of office willingly rather than being carried out horizontally. As Ronald Reagan had frequently emphasized, Gorbachev was a decent man, certainly not a monster like his predecessors.

The symbolism of Gorbachev's resignation on that special day was rich: The Bolshevik dictatorship, born October 26, 1917, which declared war on Christians and other believers, was ended on the day the world celebrates the birth of Christ. From California, Reagan the Crusader must have relished the spiritual significance of the moment.

THE SDI FACTOR

Soon after Communism's fall, it became clear that while many of Reagan's initiatives had contributed to its unraveling, one stood out above them all: SDI. It was a point that a number of high-level Soviet officials have verified since the end of the Cold War. Gorbachev Foreign Minister Alexander Bessmertnykh said flatly that programs like SDI "accelerated the decline of the Soviet Union."[27] Ukrainian Alexander Donskiy, a Red Army veteran who served in the prestigious Strategic Rocket Forces division, says that he and many of those who served with him thought that SDI "was possible," that it could work, and that it "helped destroy the Soviet Union. The economic cost killed us."[28] Vladimir Lukhim, a high-ranking Soviet official, said: "It's clear that SDI accelerated our catastrophe by at least five years."[29] That is quite a claim to make in regard to a single research program.

Genrikh Trofimenko, head of the Institute for U.S.A. and Canada Studies of the Russian Academy of Sciences, says that with SDI Reagan merely reacted to a race the Soviets had started, and that Reagan "took up the gauntlet" thrown by Soviet leaders:

> So what did President Reagan do, being aware of all this? He said: "If you are that eager to compete, let us then go the whole hog; let's do whatever one likes best. The United States will continue competition not in offensive ballistic missiles, where you, Soviets, are on par with us. We will move into a realm of strategic missile defense, where we—Americans—have a little bit of something that you, probably, still do not have." And the whole business of competition, as you know, is to keep ahead. He actually says to Moscow: "You thought you've caught up with us in strategic gadgetry and that it is the end of the road. You are wrong—the road is endless. The United States is not defeated. So, let's have another go at it."[30]

Trofimenko says that Reagan responded to the challenge by one-upping the Soviets, with the intent of prompting the USSR to spend itself into oblivion. The USSR, says Trofimenko, did just that. Thus, he asserts, "it was Ronald Reagan who won the Cold War and brought it to an end. That is why the system's collapse is a clear and definite victory for the West—for President Ronald Reagan, in particular, who raised the cost of potential victory for Moscow so high that it collapsed from the strain."[31]

Of course, this was Reagan's intention. Trofimenko said that SDI "was the most effective single act to bring [Gorbachev] to his senses—to the understanding that he could not win. . . . [H]e had to cry 'uncle' and to vie for a peaceful interlude." It is interesting that Trofimenko used the phrase "cry uncle"—exactly the words Reagan had once written in his diary.[32] Trofimenko maintains that "ninety-nine percent of all Russians believe that Reagan won the Cold War because of [his] insistence on SDI."[33]

Outside the USSR, there is widespread agreement on SDI's impact, including from liberals like Strobe Talbott and the toughest Reagan biographers.[34] Edmund Morris insists: "And as we all know, the Strategic Defense Initiative was what brought about the final capitulation of the Soviet Union."[35] Agrees Lou Cannon: "SDI turned out to be very useful in getting the Soviets to the bargaining table. Reagan was right."[36] Even Secretary of

State George Shultz, who once deemed SDI a form of insanity, later called it "the propellant that would lead the Soviets to agree to deep reductions" in missiles. SDI "in fact proved to be the ultimate bargaining chip," Shultz writes in his memoirs. "And we played it for all it was worth."[37]

Shultz's words would have thrilled SDI progenitor Edward Teller. When I interviewed Teller as he laid on his deathbed, he kept bringing up the summit meeting in Reykjavik. He understood that SDI prompted Gorbachev's dramatic offers at Reykjavik, and that Reykjavik was crucial to ending the Cold War. He returned to the point again and again. "I believed then and I believe now that Reagan made a great contribution to stopping Communism in Russia," said Teller. The Soviets "could not match" the U.S. missile-defense effort; the pursuit of SDI contributed to the Soviet demise. "Without Reagan," contended Teller in July 2003, "there would be Communism now."[38] This was a sentiment that Margaret Thatcher reiterated when she maintained that SDI turned out to be the single most important decision of Reagan's presidency.[39]

Alas, despite the passions of liberals like Ted Kennedy, decisions like the pursuit of SDI and the deployment of INFs made the world a vastly better place—by ultimately cutting, not increasing, nuclear weapons and the risk of war. Eminent Cold War historian John Lewis Gaddis commends the impact of both SDI and the INFs on Soviet behavior and in eventually defusing the Cold War.[40] Even liberal Cold War scholar Raymond Garthoff grudgingly concedes that the Reagan military buildup and pursuit of SDI posed a military challenge that the Soviet Union was economically and technologically hard pressed to meet. He concedes that Gorbachev understood that the Soviet Union could not afford to match or overmatch the United States.[41]

MUTUAL PRAISE: REAGAN AND GORBACHEV

By the start of 1992, even the most recalcitrant Cold Warrior could no longer deny the obvious: the Cold War was over, as one of the two nations in the long battle no longer existed. Now, a second struggle commenced as historians and journalists sought to credit either Reagan or Gorbachev with the Cold War's demise.

While those partial to Reagan often fault Gorbachev's role, it is important to recall that Reagan himself repeatedly said that Gorbachev was a "different kind of leader." He came to view Gorbachev in an endearing way; indeed, as a "friend."[42] He said from the start that Gorbachev "has faith in" and "believes in" Communism and was "totally dedicated to their system."[43] Yet, he learned that Gorbachev was committed to reform and a better world. He nonetheless always insisted on a "trust but verify" relationship—translated into a Russian phrase he repeated so often to Gorbachev that it annoyed the general secretary: *dovorey no provorey*. Reagan captured Gorbachev quite well when he later summed up in his memoirs: "Whatever his reasons, Gorbachev had the intelligence to admit Communism was not working, the courage to battle for change, and, ultimately, the wisdom to introduce the beginnings of democracy, individual freedom, and free enterprise."[44]

Like Gorbachev, Reagan too had his detractors who were willing to gloss over the extent of Reagan's involvement and lavish Gorbachev with praise for his role. This position is one that Gorbachev himself disagrees with. In a letter to Reagan following the May–June 1988 Moscow summit, Gorbachev gave the president major credit for Soviet developments. "The Soviet people," he wrote in the letter, recently declassified, "have met you up close and have come to appreciate your goodwill, and your role in *everything* (emphasis added) that has been accomplished by our two countries."[45] Gorbachev wrote in his memoirs: "In my view, the fortieth President of the United States will go down in history for his rare perception."[46]

In an interview, he added: "He [Reagan] is really a very big person—a very great political leader."[47] When Reagan and Gorbachev met together for a seventh time, on September 17, 1990 in Moscow, when Reagan was no longer president, Gorbachev toasted Reagan "as a man who did a lot to make relations with our country the way they are now."[48] Over a decade later, at a dinner in Cambridge, England in 2001, a British academic called Reagan "rather an intellectual lightweight." Gorbachev would not tolerate the slight, and reprimanded his host: "You are wrong. President Reagan was a man of real insight, sound political judgment, and courage."[49]

In 2002, Gorbachev called Reagan "a major individual," and added: "If at that time someone else had been in his place, I don't know whether what happened would've happened."[50]

IN THE END, THE TRUE CREDIT IS DUE TO BOTH SIDES, AS EACH of the two leaders played vital and essential roles in the end of the Cold War. The dissolution of the Soviet Communist empire was something that they accomplished together, but the chief difference lay in the fact that of the two of them, only Reagan had intended that outcome from the start. Intentions are critical to bestowing credit and even greatness upon a leader. On the issue of

Cold War *intent,* Reagan stands on much firmer ground than Gorbachev—which is not to deny Gorbachev's magnificent reforms. In regard to the Soviet empire, Reagan achieved his central intent, impossible as it seemed; Gorbachev failed in his primary intent. This failure haunts Gorbachev to this day, having told *USA Today* in April 2006 that, "The Soviet Union could have been preserved and should have been preserved." Reagan wanted to undo the Soviet empire and the USSR itself; Gorbachev's central objective was to hold the USSR together. It was Reagan who accomplished what he aimed to do.

And yet, despite initial intentions, Mikhail Gorbachev's epitaph, like Ronald Reagan's, will rightfully read: he peacefully helped end the Cold War. For that, we should be forever grateful to both men.

22.

Drifting Back

"THIS IS WHERE I WAS A LIFEGUARD FOR SEVEN SUMMERS." SO said Ronald Reagan to his biographer on December 9, 1994, pointing to the painting of the Rock River hanging in his Los Angeles office. His longtime biographer sensed that his subject no longer recognized him. A month earlier Reagan had informed the world that he had Alzheimer's disease, which, he said in the words of only the most hopeful optimist, was now riding him into "the sunset" of his life. Though forgetful of much else, he clung to his memories of the Rock River. "I saved seventy-seven lives," he said, staring at that painting. "And you know, none of 'em ever thanked me."[1]

By the mid-1990s, the White House, the Soviets, and the great Cold War victory were vague, flickering memories to Reagan, sometimes accessible, but only rarely. The Rock River, however, seemed seared in his consciousness. One day in 1997, Michael Deaver paid a visit to Reagan at his office. Though he had spent thirty years at Reagan's side, he was not recognized by his old friend. Still, Reagan was cordial and polite, and somehow reflective. Trying to make conversation, he spoke to Deaver about the image on his wall, gazing longingly at those colorful brushstrokes of the spot where he patrolled the beach at Lowell Park.[2]

Ronald Reagan was not the only Dixonite who remembered his river days while battling the loss of other memories. On June 22, 2001, I sat with three elderly women at Heritage Square, a nursing home located among a line of cool shade trees standing along North Ottawa Avenue in Dixon, Illinois. The three, who had grown up with Dutch Reagan decades ago, reclined in the facility's

"Ronald Reagan Room," a roughly ten-by-ten-foot unimpressive space with a sink, refrigerator, bland cabinets, a desk with an enlargement apparatus for reading, and a simple table with a piece of paper taped to the surface that read in capital letters: "NO FOOD OR DRINK IN THE REAGAN ROOM DUE TO SENSITIVE EQUIPMENT THANK YOU." Pictures of the hometown boy adorned the walls. Marion Emmert Foster and sisters Olive and Savila Palmer squeezed around the small table and reminisced. Olive had been baptized with Ronald Reagan at the First Christian Church in June 1922 and was introduced by an administrator for the facility in this way: "This is Olive, she became a Christian the same day as the president."

Of all the topics they could have discussed, the three ladies brought up Reagan's lifeguarding and self-confidence, explicitly connecting the two. "Those lives he saved...really affected that," said Marion, the daughter of Reagan's beloved Sunday school teacher, Lloyd Emmert. "He was an extremely confident person," Savila Palmer chimed in, "all the way back to when he was a boy." "Yes," Marion summed up. "That was always true for him."[3]

Ronald Reagan often said that those he rescued from drowning never thanked him, but in truth their gratitude came forth in a more intangible way, one for which mere words could never do justice: Reagan was paid back through the self-confidence that those seventy-seven rescues brought to his various endeavors, particularly his can-do willingness to take on the Soviet Union. As his son Ron has pointed out, lifeguarding ingrained in Reagan a broader life-saving mentality, one that followed him through his life, guiding his decisions until his final days.

Pushing seventy years old when he arrived in Washington, DC, his new self-appointed rescue mission was directed at many more lives than just those swimming in the river. His was a self-appointed mission to save the world from the evil of atheistic, expansionary Soviet Communism—an ideology that took the lives of tens of millions in the USSR alone, and over 100 million worldwide throughout the twentieth century, twice the death toll of the first two world wars combined.[4] The lifeguard would lead the charge from the shores of the United States toward the "captive peoples" behind the Iron Curtain. He decided it was up to him to play the role of world saver.

Indeed, a rescuer rescues. By the end of the 1920s, Dutch Reagan had not yet become a crusader. A crusader is a rescuer driven by an ideology, a deeper, grander cause, sometimes even a religious mission. But a crusader needs something to crusade against. That target resided in Moscow.

Over the course of his crusade, Ronald Reagan had frequently cited data from Freedom House marking the number of free and unfree nations in the world. When he did so in the 1970s, he bemoaned the lack of freedom. As president, he dedicated himself to improving those numbers. Now, today, we can cite that same source to demonstrate the degree of his success: In 1980 there were 56 democracies in the world; by 1990, there were 76. The numbers continued an upward trajectory, hitting 91 in 1991, 99 in 1992, 108 in 1993, and 114 in 1994, a doubling since Reagan entered the Oval Office. By 1994, 60 percent of the world's nations were democracies. By contrast, when Reagan lamented the lack of freedom in the mid-1970s, the number was below 30 percent.[5] In the time he shifted from presidential candidate to ex-president, the number of democracies increased from under one-third to a strong majority.

Today, 120 of the world's 192 nations are democracies. Outside of Western Europe, 88 percent of Latin American and Caribbean nations are democracies, 91 percent of Pacific Island states, 92 percent of South American nations, and 93 percent of the nations of East Central Europe and the Baltic states—that is, the former Soviet region. There has been an explosion in freedom worldwide since the 1980s. This transformation is one of the great stories of modern humanity, and a momentous development that Ronald Reagan desired. Few presidents got so much of what they wanted.[6]

IN THE END, IT WAS THIS COLD WAR VICTORY—THIS REN-dezvous with destiny—which was remembered most when Ronald Wilson Reagan died at age ninety-three on June 5, 2004. That week, America witnessed an outpouring of emotion for a president not seen since the deaths of John F. Kennedy and Franklin Delano Roosevelt. The week was filled with eulogies. The world heard from Reagan's Cold War partner, Margaret Thatcher, from his vice president, George H. W. Bush, from the current President George W. Bush, from Reagan's children, and from numerous dignitaries and everyday Americans.

Mikhail Gorbachev immediately sent a letter to Nancy Reagan: "I shall always remember the years of working together with President Reagan, putting an end to confrontation between our two countries, and equally, our friendly rapport." He finished: "Your husband has earned a place in history and in people's hearts."[7] Gorbachev took a moment to tell reporters: "I take very hard the death of Ronald Reagan, a man whom by fate sat with me in perhaps the most difficult years at the end of the twentieth century."[8]

In the *Washington Post*, a newspaper hardly sympathetic to Reagan in the 1980s, the page one story on his death, written by respected Russia journalist David E. Hoffman, was titled simply, "Hastening an End to the Cold War."[9] NBC's Tim Russert said that Reagan's legacy was "winning the Cold War." Among the causes, Russert said that Reagan had dared the Soviets to an arms buildup they could not match and forced Mikhail Gorbachev to "reform" and the Kremlin to "cry uncle."[10]

Even Senator Ted Kennedy, who had once resorted to extraordinary lengths to defeat Ronald Reagan, now believed that the fortieth president "will be honored as the president who won the Cold War."[12] For Kennedy, it was a remarkable concession.

One group that Americans did not hear from that week was the hundreds of millions of individuals who once lived in the former Soviet bloc and USSR—those souls from Prague to Leningrad that Reagan sought to liberate. These were the "voiceless," as Reagan had described them. Now, they had voices, uncensored, and they stepped up to pay tribute to the man that many of them credit for their freedom.

In the Romanian newspaper, *Bucharest Ziua*, Dan Pavel wrote an op-ed calling Reagan, "The political leader who contributed the most to the fall of the totalitarian communist system." Pavel thanked Reagan for the "moral perceptiveness of defining the communist totalitarian regime as the 'Evil Empire,'" and for having "freed" him and his fellow members of the Soviet bloc from Communism. Pavel added: "A superficial analysis has conferred exaggerated merit on Soviet leader Gorbachev, but what the Soviets really wanted to do was to reform the system, not to overthrow it. They would not have gone as far as to achieve glasnost and perestroika if the Americans had not forced them to admit that they were outdated and defeated in the military, economic, political, and cultural competition."[12] Such a view was also advanced that week by Alexei Pankin in the *Moscow Times*: "My thesis is that thanks to Reagan, perestroika might never have happened."[13]

From Poland, Lech Walesa remembered his hero: "Our job was to overthrow this stupid and murderous Communist system." The now former Polish president told the Polish News Agency that Reagan was one of a few world leaders who contributed to doing just that. The news agency itself noted the obvious: "Reagan's contribution to the overthrow of Communism made him immensely popular in Poland."[14]

A writer for the *Budapest Business Journal*—its mere existence a testimony to Communism's defeat—recalled a 1996 breakfast he had with Marian

Krzaklewski, Walesa's successor as head of Solidarity, and Reagan defense secretary and longtime aide Caspar Weinberger. They sat on the top floor of the
forty-story Warsaw Marriott, which provided a sweeping vista of the rebuilt
Polish capital. Krzaklewski, whose face was already a bit flush, said to Weinberger with a tear in his eye: "You and Mr. Reagan saved my country. Without Reagan, we would have been finished!" Weinberger responded: "I was just
doing my job—the job the President asked me to do."[15]

From Russia itself, on June 7, two days after Reagan's death, the *Moscow
Times* ran a retrospective titled "Reagan Mourned in Former 'Evil Empire.'"
The article stated:

> Reagan is vividly remembered in Russia today as the force that precipitated
> the Soviet collapse. "Reagan bolstered the U.S. military might to ruin the
> Soviet economy, and he achieved his goal," said Gennady Gerasimov, who
> served as top spokesman for the Soviet Foreign Ministry during the 1980s.
>
> Even though Reagan's "Star Wars" never led to the deployment of an
> actual missile shield, it drew the Soviets into a costly effort to mount a re-
> sponse. Many analysts agree that the race drained Soviet coffers and trig-
> gered the economic difficulties that sped up the Soviet collapse in 1991.
>
> "Reagan's SDI was a very successful blackmail," Gerasimov said in an
> interview. "The Soviet Union tried to keep up pace with the U.S. military
> buildup, but the Soviet economy couldn't endure such competition."
>
> Retired General Vladimir Dvorkin, a senior Soviet arms control nego-
> tiator during the 1980s, said that trying to field a response to Reagan's Star
> Wars had "certainly contributed" to the Soviet economic demise but argued
> it didn't play the decisive role.[16]

Such sentiment was heard throughout the chain of command. "This is a
guy who changed the world," said Aleksandr Shakhnovich, fifty-seven, a former shipbuilder for the Soviet navy. "It wasn't only his speeches—it was his
actions. He cut down the economy of the USSR and it was one of the main
reasons the country just shut down. He did something that not only changed
my life, but changed the lives of everyone in the Soviet Union."[17]

Gratitude was also displayed in Afghanistan. President Hamid Karzai
stated: "The people of Afghanistan remember Mr. Ronald Reagan's assistance
to Afghanistan during the years of 'jihad' (holy war) against the Soviets."[18]
Under Reagan, wrote Pakistani columnist Mohammad Ashraf Azeem, the
United States "introduced the resistance strategy of the 'Afghan Jihad' for

checking the Soviet advance in Afghanistan. As a result of this [Afghan] war, the Soviet Union was disintegrated, and its dream of expanding its influence beyond Afghanistan was shattered once and for all."[19]

These voices, from Krakow to Kabul, were a glowing tribute to Reagan; they rose from precisely the freedom he sought for these once repressed peoples of the world. Collectively, they comprised his finest eulogy.

IN AUGUST 1976, RONALD REAGAN WROTE A LETTER FOR THAT time capsule he referenced in his speech that month at the Republican convention. In the letter, he spoke of the potential nuclear catastrophe looming over civilization. It occurred to him that those reading that letter a hundred years henceforth would know whether those missiles were fired, whether they had freedom, and how much that freedom had depended upon Reagan's generation.

We now know that that generation, and leaders like Reagan, met that challenge.[20] They averted nuclear catastrophe. Freedom came to a severely repressed part of the world. The challenge today, for professors and parents alike, is to adequately convey the degree to which the world had once peered into the darkness, feared nuclear annihilation, and, in much of the Soviet sphere, was deprived of the most basic rights.

With a confidence and can-do attitude that invigorated him like the waters of the Rock River, Ronald Reagan set out to right those wrongs. The extent to which eventual worldwide occurrences matched his extremely ambitious intentions is astonishing, and one of the great stories of the twentieth century and U.S. history. Those who do not see that reality need to; not simply because it is a quintessentially American story of doing the impossible, but also because, yes, the missiles were not fired and people are free.

"Wars end in victory or defeat," Ronald Reagan once said in 1961.[21] The Cold War ended in victory—or, to paraphrase Reagan from January 1977, "We won, they lost." It was a victory for which the world was thankful, especially given the tranquil way in which it ended—without the nuclear Armageddon that everyone so deeply feared and many expected. Within a decade of Reagan coming to power, the Cold War was over and the USSR ceased to exist, and both world-shaking developments occurred without a missile launched.

And it all started not at SAG or HUAC in 1947, not with *GE Theatre*, or at a Crusade for Freedom rally in the 1950s, not with a speech for Goldwater in 1964, and not even at that first trip to the Berlin Wall in 1978; oddly enough, it began at a park in Dixon, Illinois, the sight of murky, splashing water, where a young lifeguard named Dutch saved seventy-seven people over seven summers, and in the process went on to change the course of more than just a winding river.

EPILOGUE

WRITING IN 1986, COLD WAR HISTORIAN JOHN LEWIS GADDIS stated, "American officials at no point during the history of the Cold War seriously contemplated, as a deliberate political objective, the elimination of the Soviet Union as a major force in world affairs."[1] At that point in time, Gaddis could be excused for not knowing Ronald Reagan's cards; after all, the historian did not have copies of all of those classified NSDDs, nor a seat inside Bill Clark's National Security Council.

We now know, however, that such was precisely what Reagan had intended—and then some. What he pursued was truly revolutionary. He was not content to contain Soviet Communism. He wanted to kill it. He not only said so but committed himself and his administration to that very deliberate goal—a goal that stemmed from Reagan himself, not his advisers, long before 1981.

By the mid-1990s, with the presidency and Cold War over, Ronald Reagan might have spoken at length and repeatedly about his one-time intentions with the USSR. Unfortunately, his Alzheimer's quickly became the major obstacle that prevented a complete accounting of Reagan's goals with the Soviet Union. This meant that neither historians nor journalists could raise the issue with Reagan, though on more than one occasion everyday Americans—those Reagan heralded as America at its best—did. Bill Clark recalls a moment when he and Reagan were together again in the early 1990s. An admirer congratulated Reagan for "your success in ending the Cold War." Reagan smiled and replied deferentially, "No, not my success but a team effort by Divine Providence."[2] He saw God's hand in this "team effort" to "end" the Cold War.

On September 12, 1990, Reagan returned to the Berlin Wall as a private citizen. Just three years earlier he had called on Mikhail Gorbachev to tear down the structure, and was now armed only with a hammer to chisel off a chunk of the edifice as a memento of the Cold War that he had brought to a close. The ex-president that day kicked off a ten-day, four-country European trip, fittingly starting at the neutered wall. "It feels great," said the seventy-nine-year-old as he stepped carefully around the mangled steel rods that now protruded from the harmless, beat up barrier. "I don't think you can overstate the importance of it. I was trying to do everything I could for such things as this. . . . It happened earlier than I thought it would, but I'm an optimist."[3]

Only twenty months earlier, on Reagan's last day in the Oval Office, East German dictator Erich Honecker had defiantly proclaimed that the Berlin Wall would be standing 100 years henceforth, proudly "protecting our republic from robbers." But on that September 1990 day, Reagan was the robber, as he chopped hard at the divide with a blue-headed hammer, taking a piece home with him.[4]

Today, the largest chunk of the wall outside of Germany sits in Simi Valley, California, directly beyond the window at the welcome desk of the Reagan Library and Museum, a gift donated by the citizens of unified, free Berlin, again the capital of a united Germany. Inside that library sits a video cassette tucked in a box: it features footage of that May 1967 Reagan-Bobby Kennedy debate, when Ronald Reagan first publicly called for the dismantling of the Berlin Wall.

Alas, in a poignant moment later; indeed an instance of weakness born of a rapacious mental disease, in the summer of 1997 Ronald Reagan, by then fully in the throes of Alzheimer's, offered a private acknowledgment, a rare admission. It came at a time when Nancy was readying to close her husband off from the world, and when he would spend his final few years in a bewildered state. Apparently, some memories were momentarily retrievable; all was not lost—not yet:

That summer, Reagan strolled through Armand Hammer Park near his Bel Air home when he was approached by a tourist named Yakob Ravin and his twelve-year-old grandson, both Jewish Ukrainian émigrés living near Toledo, Ohio. They cheered Reagan as he got near and briefly spoke to the former president, who posed for a picture with the boy, which his grandfather proudly snapped. "Mr. President," said Ravin, "thank you for everything you did for the Jewish people, for Soviet people, to destroy the Communist empire." The slightly confused eighty-six-year-old Reagan paused and responded: "Yes, that is my job."[5]

That was his job—one he had assigned to himself long ago.

And then, after it all, after the task was complete, and after he was permitted, mercifully, a short window of time to comprehend and savor the accomplishment, it all quietly disappeared through the last ten years of his ninety-three years of life. And then, finally, Ronald Reagan's time on this earth terminated on June 5, 2004, as he ended that long, quiet drift into oblivion, and perhaps, again, drifted back to the Rock River.

Appendix

TEXT OF KGB LETTER
ON SENATOR TED KENNEDY

Special Importance
Committee on State Security of the USSR
14.05.1983 No. 1029 Ch/OV
Moscow

Regarding Senator Kennedy's request to the General Secretary of the Communist
Party Comrade Y.V. Andropov
Comrade Y.V. Andropov

On 9–10 May of this year, Senator Edward Kennedy's close friend and trusted
confidant J. Tunney was in Moscow. The senator charged Tunney to convey the following message, through confidential contacts, to the General Secretary of the Central Committee of the Communist Party of the Soviet Union, Y. Andropov:

Senator Kennedy, like other rational people, is very troubled by the current state
of Soviet-American relations. Events are developing such that this relationship coupled with the general state of global affairs will make the situation even more dangerous. The main reason for this is Reagan's belligerence, and his firm commitment to
deploy new American middle range nuclear weapons within Western Europe.

According to Kennedy, the current threat is due to the President's refusal to engage any modification on his politics. He feels that his domestic standing has been
strengthened because of the well publicized improvement of the economy: inflation
has been greatly reduced, production levels are increasing as is overall business

activity. For these reasons, interest rates will continue to decline. The White House has portrayed this in the media as the "success of Reaganomics."

Naturally, not everything in the province of economics has gone according to Reagan's plan. A few well known economists and members of financial circles, particularly from the north-eastern states, foresee certain hidden tendencies that may bring about a new economic crisis in the USA. This could bring about the fall of the presidential campaign of 1984, which would benefit the Democratic party. Nevertheless, there are no secure assurances this will indeed develop.

The only real potential threats to Reagan are problems of war and peace and Soviet-American relations. These issues, according to the senator, will without a doubt become the most important of the election campaign. The movement advocating a freeze on nuclear arsenals of both countries continues to gain strength in the United States. The movement is also willing to accept preparations, particularly from Kennedy, for its continued growth. In political and influential circles of the country, including within Congress, the resistance to growing military expenditures is gaining strength.

However, according to Kennedy, the opposition to Reagan is still very weak. Reagan's adversaries are divided and the presentations they make are not fully effective. Meanwhile, Reagan has the capabilities to effectively counter any propaganda. In order to neutralize criticism that the talks between the USA and the USSR are non-constructive, Reagan will grandiose, but subjectively propagandistic. At the same time, Soviet officials who speak about disarmament will be quoted out of context, silenced or groundlessly and whimsically discounted. Although arguments and statements by officials of the USSR do appear in the press, it is important to note the majority of Americans do not read serious newspapers or periodicals.

Kennedy believes that, given the current state of affairs, and in the interest of peace, it would be prudent and timely to undertake the following steps to counter the militaristic politics of Reagan and his campaign to psychologically burden the American people. In this regard, he offers the following proposals to the General Secretary of the Central Committee of the Communist Party of the Soviet Union Y.V. Andropov:

> 1. Kennedy asks Y.V. Andropov to consider inviting the senator to Moscow for a personal meeting in July of this year. The main purpose of the meeting, according to the senator, would be to arm Soviet officials with explanations regarding problems of nuclear disarmament so they may be better prepared and more convincing during appearances in the USA. He would also like to inform you that he has planned a trip through Western Europe, where he anticipates meeting England's Prime Minister Margaret Thatcher

and French President Mitterand in which he will exchange similar ideas regarding the same issues.

If his proposals would be accepted in principle, Kennedy would send his representative to Moscow to resolve questions regarding organizing such a visit.

Kennedy thinks the benefit of a meeting with Y.V. Andropov will be enhanced if he could also invite one of the well known Republican senators, for example, Mark Hatfield. Such a meeting will have a strong impact on American and political circles in the USA. (In March of 1982, Hatfield and Kennedy proposed a project resolution to freeze the nuclear arsenals of the USA and the USSR and published a book on this theme as well.)

2. Kennedy believes that in order to influence Americans it would be important to organize in August–September of this year, televised interviews with Y.V. Andropov in the USA. A direct appeal by the General Secretary of the Central Committee of the Communist Party of the Soviet Union to the American people will, without a doubt, attract a great deal of attention and interest in the country. The senator is convinced this would receive the maximum resonance in so far as television is the most effective method of mass media and information.

If the proposal is recognized as worthy, then Kennedy and his friends will bring about suitable steps to have representatives of the largest television companies in the USA contact Y.V. Andropov for an invitation to Moscow for the interview. Specifically, the president of the board of directors of ABC, Elton Raul and television columnists Walter Cronkite or Barbara Walters could visit Moscow. The senator underlined the importance that this initiative should be seen as coming from the American side.

Furthermore, with the same purpose in mind, a series of televised interviews in the USA with lower level Soviet officials, particularly from the military would be organized. They would also have an opportunity to appeal directly to the American people about the peaceful intentions of the USSR, with their own arguments about maintaining a true balance of power between the USSR and the USA in military terms. This issue is quickly being distorted by Reagan's administration.

Kennedy asked to convey that this appeal to the General Secretary of the Central Committee of the Communist Party of the Soviet Union is his effort to contribute a strong proposal that would root out the threat of nuclear war, and to improve Soviet-American relations, so that they define the safety of the world. Kennedy is very impressed with the activities of Y.V. Andropov and other Soviet leaders, who expressed their commitment to heal international affairs, and improve mutual understandings between peoples.

The senator underscored that he eagerly awaits a reply to his appeal, the answer to which may be delivered through Tunney.

Having conveyed Kennedy's appeal to the General Secretary of the Central Committee of the Communist Party of the Soviet Union, Tunney also explained that Senator Kennedy has in the last few years actively made appearances to reduce the threat of war. Because he formally refused to partake in the election campaign of 1984, his speeches would be taken without prejudice as they are not tied to any campaign promises. Tunney remarked that the senator wants to run for president in 1988. At that time, he will be 56 and his personal problems, which could hinder his standing, will be resolved (Kennedy has just completed a divorce and plans to remarry in the near future). Taken together, Kennedy does not discount that during the 1984 campaign, the Democratic party may officially turn to him to lead the fight against the Republicans and elect their candidate president. This would explain why he is convinced that none of the candidates today have a real chance at defeating Reagan.

We await instructions.

President of the committee
V. Chebrikov

NOTES

PREFACE

1. Vitaliy Korionov, "20th Century 'Crusaders,' " *Pravda*, July 14, 1982, 4, published as "Pravda's Korionov Denounces U.S. 'Crusades,' " in FBIS-26-JUL-82, July 26, 1982, A5–6; and Leonid Zamyatin, "The Washington Crusaders: The 'Ideological War' Declared by Reagan Against Communism and Socialism . . . ," *Literaturnaya Gazeta*, June 30, 1982, 14, published as "Zamyatin Scores Reagan 'Ideological War' Plan," in FBIS-14-JUL-82, July 14, 1982, A1–8. FBIS, which stands for Foreign Broadcast Information Service, is a U.S. government service that transcribed and translated all media from the USSR and around the world.

2. Genrikh Borovik speaking on Moscow Television Service, August 14, 1984, published as "Borovik Criticizes Statement," in FBIS-15-AUG-84, August 15, 1984, A7. For just one colorful example, see Gennady Vasilyev, "American Zero," *Pravda*, February 13, 1983, 4, in *The Current Digest of the Soviet Press*, 35, no. 7 (1983): 16.

3. Among available transcript copies, see Davis Houck and Amos Kiewe, *Actor, Ideologue, Politician: The Public Speeches of Ronald Reagan* (Westport, CT: Greenwood Press, 1993), 166.

4. Reagan, "Remarks at a Conservative Political Action Conference Dinner," February 26, 1982.

5. Reagan, "Remarks to Members of the Royal Institute of International Affairs in London," June 3, 1988.

6. Mikhail Gorbachev, *Memoirs* (New York: Doubleday, 1996), 328.

7. Vitaliy Korionov, "Production Line of Crimes and Hypocrisy," *Pravda*, January 10, 1984, 4, published as " 'Unprecedented Wave' of Lies Seen in U.S.," in FBIS-13-JAN-84, January 13, 1984, A1.

8. G. Arbatov, "The U.S.—Will There Be Changes?" *Pravda*, March 17, 1983, published as "Arbatov Assails US 'Propaganda Tricks,' " *The Current Digest of the Soviet Press* 35 no. 11, (1983): 4.

9. The author who has surpassed others in illuminating Reagan's personal role in the Cold War is Peter Schweizer. See Schweizer, *Victory* (Atlantic Monthly Press, 1994); and Schweizer, *Reagan's War* (New York: Doubleday, 2002).

10. Edmund Morris, *Dutch: A Memoir of Ronald Reagan* (New York: Random House, 1999), xiii.

11. Anthony Lewis interview on "Reagan," *The American Experience*, television documentary produced by PBS, WGBH-TV Boston, 1998. Hereafter cited as "Reagan," *The American Experience*, PBS.

12. Richard V. Allen, "The Man Who Changed the Game Plan," *National Interest*, Summer 1996, 60.

13. Clark in Peter Schweizer, ed., *The Fall of the Berlin Wall* (Stanford, CA: Hoover Institution Press, 2000), 76. Clark has personally told me this many times. I've also encountered the quote from him in a number of published statements he made in the 1980s.

14. Bob Schieffer and Gary Paul Gates, *The Acting President* (New York: E. P. Dutton, 1989), 374.

15. The point that Reagan knew only vaguely what his aides did is not necessarily inaccurate. Reagan believed that the best manner of management for an operation the size of the executive branch of the modern federal government was for the president to stay attuned to the big picture and to delegate the day-to-day details to those under him. In so doing, he delegated trust as well, and was burned more than once, likely leading to the Iran-Contra fiasco. This system of management brought both successes and failures. Sometimes it was a liability. Reagan, however, would argue that such was the nature of the beast.

16. The article attributes its authorship in this puzzling way: "By Richard Stengel. Reported by Laurence I. Barrett and Barrett Seaman/Washington." Full citation: Richard Stengel, "How Reagan Stays Out of Touch," *Time*, December 8, 1986, 34.

17. Don Oberdorfer speaking at 1993 Hofstra University conference on the Reagan presidency, in Eric J. Schmertz et al., eds., *President Reagan and the World* (Westport, CT: Greenwood Press, 1997), 129. Hereafter cited as "Hofstra conference (1993) proceedings."

18. Garry Wills, *Reagan's America: Innocents at Home* (New York: Doubleday, 1987), xv.

19. Among Wills' examples, see Wills, *Reagan's America: Innocents at Home*, xi.

20. On these sources, see Paul Kengor, "Reagan Among the Professors," *Policy Review*, December 1999/January 2000, no. 98, 15–27; and Paul Kengor, "The Legacy of Ronald Reagan: The Academic View," Conference on the Reagan Presidency, Saint Vincent College, Latrobe, Pennsylvania, October 10, 2001.

This is not to imply that all of these sources hail Reagan across the board, but only that they have treated him with fairness and have commended him and his presidency for various reasons, and often unexpectedly. For instance, James T. Patterson, a liberal, has been quite critical of certain Reagan domestic policies. On the other hand, John Sloan, also a liberal, has praised Reagan generously and personally rates him a "near-great" president. See John W. Sloan's book review of *The Reagan Presidency: Pragmatic Conservatism and Its Legacies*, published in *Presidential Studies Quarterly*, December 2004, 909–10.

21. Hugh Heclo, "Ronald Reagan and the American Public Philosophy," in W. Elliot Brownlee, ed., *The Reagan Presidency* (Lawrence, KS: University Press of Kansas, 2003).

22. In his seminal work, *The United States and the End of the Cold War*, in the chapter titled, "The Unexpected Ronald Reagan," Gaddis maintains that the president succeeded in "bringing about the most significant improvement in Soviet-American relations since the end of World War II." While he grants much of the credit to Mikhail Gorbachev's receptivity, Gaddis asserted, "it would be a mistake to credit him [Gorbachev] solely with the responsibility for what happened: Ronald Reagan deserves a great deal of the credit as well." Gaddis has urged

colleagues to put aside "preconceptions" in evaluating the Reagan record. See John Lewis Gaddis, *The United States and the End of the Cold War* (New York and Oxford: Oxford University Press, 1992), 130–31. Speaking of the Evil Empire, Gaddis wrote in *Foreign Affairs* in 1994: "Now that they are free to speak—and act—the people of the former Soviet Union appear to have associated themselves more closely with President Reagan's famous indictment of that state as an 'evil empire' than with more balanced academic assessments." This comes from his well-known and aptly titled piece, "The Tragedy of Cold War History." Gaddis, "The Tragedy of Cold War History," *Foreign Affairs*, 73, no. 1 (1994): 148.

In a separate work published in January 1989, Gaddis wrote similarly: "The time has come to acknowledge an astonishing development: . . . Ronald Reagan has presided over the most dramatic improvement in U.S.–Soviet relations—and the most solid progress in arms control—since the Cold War began." Gaddis pleaded: "it would be uncharitable—and historically irresponsible—to begrudge the strategic vision of an administration once thought by many of us to have had none at all." See Gaddis, "Hanging Tough Paid Off," *Bulletin of the Atomic Scientists*, January/February 1989, 11.

23. In a 1999 poll of presidential scholars by C-SPAN, Reagan ranked as the eleventh best president. He placed fourth in the category of "public persuasion," behind only FDR, Teddy Roosevelt, and Lincoln. In a 2000 survey of seventy-eight professors, by the *Wall Street Journal* and Federalist Society, Reagan rated eighth best president in history, reaching the "near great" category.

24. It has been argued that to the extent that Ronald Reagan should be granted any credit for winning the Cold War, credit ought to go not to him but to his advisers. At its most basic level, this suggestion is ridiculous. Even if Reagan never had a single idea or made not one decision affecting the Cold War, he would still deserve at least as much credit as his staff. Reagan, after all, was the one who got elected. It goes without saying that such is an incredibly difficult thing to accomplish. The United States has been filled with millions of brilliant, ambitious individuals traversing its centuries of existence, but Reagan was only the fortieth to make it to the Oval Office. With the exception of George H.W. Bush, who was the first vice president elected president in over 150 years, arguably as a stamp of approval on the Reagan presidency, no other talented person in the Reagan administration went on to capture the White House. Reagan had the charisma to get elected. His policy positions appealed to the right number of people at the right time. Richard V. Allen, Reagan's foreign-policy adviser in the latter 1970s, notes that Reagan's entire 1980 platform was rooted in Reagan's own thinking, not pollsters, aides, nor anyone else. See Allen in Schweizer, ed., *The Fall of the Berlin Wall*, 57. Allen's claim is well substantiated by the extensive volume of handwritten, policy-related Reagan material from the period—the lengthiest collection among modern presidents.

Reagan brought a Republican majority to the Senate on his coattails—a rarity at that point in American politics. Once elected, he had what it took to carry his agenda through Congress, including a partisan Democratic House. With the Oval Office secured, he attracted and assembled a team of capable advisers. With staff in place, he signed documents and made final decisions. Administration policy, up and down, side to side, was structured around his core beliefs, his conservatism, and his preferences.

Thus, even if Ronald Reagan never had an original idea to contribute to Cold War strategy, even if he sat motionless in meetings, the electoral triumph alone would be enough to make him the most important factor in any Reagan administration Cold War victory. The reality, however, is that he did contribute, and very much so.

CHAPTER 1

1. "News from Dixon YMCA," *Dixon Telegraph*, January 3, 1928; and "Water Carnival Monday Proved Great Success," *Dixon Telegraph*, September 4, 1928.

2. Ron Reagan, "My Father's Memories," *Esquire*, June 2003, 110.

3. July 23, 1932 *Dixon Telegraph*. Lou Cannon also lists the figure on 1,000 bathers at a time with no assistant. Cannon cites a July 3, 1931 *Dixon Telegraph* article. See Lou Cannon, *Governor Reagan* (New York: PublicAffairs, 2003), 22. The July 23, 1932 article updates Reagan's save tally to seventy-one lives.

4. Ron Reagan, "My Father's Memories," 111.

5. In the *Esquire* piece, his son Ron recalled one particularly interesting account: A large blind person, a towering hulk of a man, entered the area. Dutch spotted him right away, and worried how he would pull him from the water if he were fighting for his life. Sure enough, the man paddled away from the beach and within minutes was swept into the middle of the river. As Ron tells the story, bathers began yelling and pointing at the big arms slapping in vain at the surface. Dutch dove into the water and began chasing the bobbing head as it rushed downstream. He wondered if it might be his final rescue attempt. Bewildered and panicky, how would the terrified man react once someone grabbed him? "Dad imagined them in a grotesque embrace, rolling along the river bottom toward the next town downstream," said Ron. To Dutch's great relief, when he seized the man's shoulder, total compliance followed. Reagan attributed this response to the blind man's being accustomed to being led by others; he associated human touch with safety, and immediately relaxed.

6. This issue seven decades later evoked blushes from three elderly women who grew up with Reagan. With a wink, Marion Emmert Foster and sisters Olive and Savila Palmer explained that Dutch's physical appearance made him popular among the girls. "Oh, my!" exclaimed Savila. "You can imagine how that affected his confidence!" Interview with Olive and Savila Palmer and Marion Emmert Foster, Heritage Square nursing home, Dixon, Illinois, June 22, 2001.

7. Fran Swarbrick, ed., *Remembering Ronald Reagan* (Dixon, IL: Creative Printing, 2001), 8, 11–2. At a Ladies' Day event in Chicago in March 1965, the speaker introduced Reagan by teasing him about young girls nearly drowning themselves so he could rescue them in his lifeguard days. "Great Speech by Ronald Reagan Thrills Capacity Ladies' Day Meeting at the Hilton," *Executives' Club News*, 41, no. 20 (1965): 1–8, filed in "RWR-Speeches and Articles (1965–6)," folder, RRL, vertical files. "RRL" refers to the Ronald Reagan Library, which hereafter will be as "RRL." On women straying to the middle of the river so Reagan could save them, see Ron Reagan, "My Father's Memories," 111.

Throughout Reagan's life, many women were attracted to his looks, and this started at a young age. The letters and sources backing this up are striking. In coverage of his GE engagements, there were often comments from women about his looks. Also available are letters from women around the country who met him at events and take the time to recall how struck they were by his appearance. One especially amusing letter is from Audrey M. McAfee, February 4, 1999, filed in "RWR—Acting Career (1937–64)," vertical files, RRL.

8. For example, the June 18, 1931 *Telegraph* credited Reagan with fifty-one saves. Again, the July 23, 1932 edition updated the total to seventy-one.

9. Ruth Graybill quoted in Swarbrick, ed., *Remembering Ronald Reagan*, 7.

10. Kiron Skinner, Annelise Anderson, and Martin Anderson, eds., *Reagan: A Life in Letters* (New York: Free Press, 2003), 10.

11. Ronald Reagan with Richard Hubler, *Where's the Rest of Me?* (New York: Duell, Sloan & Pearce, 1965), 23; and Reagan letter to Mrs. John B. White, Peoria, Illinois, October 5, 1982, in Skinner, Anderson, and Anderson, eds., *Reagan: A Life in Letters*, 11.

12. Reagan and Hubler, *Where's the Rest of Me?*, 23.

13. Burrel J. Reynolds quoted in Swarbrick, ed., *Remembering Ronald Reagan*, 7.

14. Reagan and Hubler, *Where's the Rest of Me?*, 21. Ron Reagan, "My Father's Memories," 110.

15. Reagan speaking on "Ronald Reagan: A Legacy Remembered," History Channel productions, 2002.

16. This fact is not widely known. Reagan once shared it with Joseph A. Pecoraro, president of the U.S. Lifesaving Association, during a July 17, 1986 meeting in the Oval Office. Interview with Joseph Pecoraro, October 6, 2004; and letter from Pecoraro published in *American Lifeguard*, the official newsletter/publication of the U.S. Lifesaving Association, Autumn 1986.

17. In the early 1930s in Des Moines, Reagan reportedly saved one or two swimmers at the popular Camp Dodge pool, a spot where he cooled off frequently. Renda Lutz, "Ronald 'Dutch' Reagan got his big break while living in DM," *Des Moines Register*, June 15, 2004. On the latter: On July 4, 1967, Reagan hosted a staff garden party around the pool of the governor's residence in Sacramento in celebration of the completion of the legislative session. As the adults drank cocktails, chatted, and laughed, Dixon's former star lifeguard, now fifty-six years old and 2,000 miles west, could not help but keep an eye on the water. While he spoke to acquaintances, he privately tracked submersion times of kids dropping in and out of the water. A little girl quietly sank to the bottom. A fully dressed Reagan dashed along the concrete and plunged into the pool. The shocked guests tried to comprehend what was happening. Dutch emerged with her in his arms, gently sat her poolside, and, as a surviving picture attests, smilingly comforted her as his wet, wrinkled clothes stuck to his body. Ron Reagan, speaking on "Ronald Reagan: A Legacy Remembered," History Channel productions, 2002; and Morris, *Dutch*, 352–53.

18. Reagan and Hubler, *Where's the Rest of Me?*, 21; and Ron Reagan, "My Father's Memories," 111.

19. For a rescue that Reagan made on turf in Iowa, see Anne Edwards, *Early Reagan: The Rise to Power* (New York: Morrow, 1987), 138; and Lutz, "Ronald 'Dutch' Reagan got his big break while living in DM," June 15, 2004. For another, amusing example of Reagan's physical daring, see Reagan's tight-wire act, in William F. Buckley, Jr., "Reagan: A Relaxing View," *National Review*, November 28, 1967.

20. Interview with Bill Clark, July 17, 2003.

21. While all seventy-seven may not have died, many (perhaps even most) were in serious trouble, often at risk of death. On a separate matter, I've discussed Reagan's many rescues with a number of lifeguards and lifeguard organizations. Some lifeguards have admitted never saving a single life, whereas others relay actual incidents where a single lifeguard did a dozen or more rescues in a single day. Interviews with lifeguard Matt Scheff, June 5, 2003, and B. Chris Brewster, director, United States Lifesaving Association, October 5, 2004.

22. Paul Kengor, *God and Ronald Reagan: A Spiritual Life* (New York: HarperCollins ReganBooks, 2004).

23. Ronald Reagan, *An American Life* (New York: Simon and Schuster, 1990), 20–21. Reagan added on Nelle: "If something went wrong, she said, you didn't let it get you down: You stepped away from it, stepped over it, and moved on." Later on, "something good will happen and you'll find yourself thinking—'If I hadn't had that problem back then, then this better thing that *did* happen wouldn't have happened to me.'"

24. On the 'Teflon Presidency,'" see Peter B. Levy, *Encyclopedia of the Reagan-Bush Years* (Westport, CT: Greenwood Press, 1996), 359.

25. Melor Sturua, "Apologizing or Squirming? The US President Tries to Maneuver," *Izvestia*, August 24, 1984, 5, reprinted as "Sturua: Reagan Maneuvering Over Bombing Remark," in FBIS-27-AUG-84, August 27, 1984, A4.

26. U.S. Ambassador to the USSR Jack Matlock said of Reagan: "[H]is confidence did lead him to take chances because he felt the Soviet system could change." Published in William C. Wohlforth, ed., *Witnesses to the End of the Cold War* (Baltimore: Johns Hopkins University Press, 1996), 22. Transcript from the conference, in "Retrospective on the End of the Cold War," Woodrow Wilson School, Princeton University, February 25–27, 1993. For more from Matlock on Reagan's self security, see page 114.

CHAPTER 2

1. The strike was jurisdictional, meaning that it centered over which union would represent the workforce.

2. Ron and Allis Radosh, *Red Star Over Hollywood* (San Francisco: Encounter Books, 2005), 117–22.

3. "Threatened in '46 Strike, Ronald Reagan Testifies," *Los Angeles Times*, January 14, 1954, 3. Reagan subsequently wrote and talked about this a number of times.

4. Peggy Noonan, *When Character Was King: A Story of Ronald Reagan* (New York: Viking-Penguin, 2001), 56–57; and Schweizer, *Reagan's War*, 11–12.

5. Bill Clark has told me about the many threats during the gubernatorial years, which Clark said were too frequent to count. See Schweizer, *Reagan's War*, 51–53, 124, 178–79, 216.

6. John Meroney, "Rehearsals for a Lead Role," *Washington Post*, February 4, 2001, G8.

7. Joseph Shattan, *Architects of Victory: Six Heroes of the Cold War* (Washington, DC: Heritage Press, 1999), 236; Meroney, "Rehearsals for a Lead Role"; and Reagan and Hubler, *Where's the Rest of Me?*, 6–7.

8. Arthur F. McClure, C. David Rice, and William T. Stewart, eds., *Ronald Reagan: His First Career, A Bibliography of the Movie Years* (Lewiston, NY: Edwin Mellen, 1988), 12–17.

9. Doug McClelland, ed., *Hollywood on Ronald Reagan: Friends and Enemies Discuss Our President, the Actor* (Winchester, MA: Faber and Faber, 1983), 178.

10. John Meroney, an expert on Reagan's Hollywood years, correctly notes that today's conventional wisdom on Reagan's movie work was poisoned by politics. The "B movie actor" tag was not slapped on Reagan until the 1980s, long after he left the industry, when people who disliked his politics aimed to discredit him across-the-board. John Meroney, "Night Unto Reagan," *National Review Online*, August 4, 2005.

11. *Daily Variety*, February 9, 1950. Excerpted by McClelland, *Hollywood on Ronald Reagan*, 96. See Jack Gould, "Sweeping and Imaginative in Conception, 'Omnibus' of Ford Foundation Makes Video Debut," *New York Times*, November 10, 1952.

12. He joined SAG on June 30, 1937, was appointed to the board July 1941, was elected

president March 1947 (replacing Robert Montgomery), and resigned as president on June 6, 1960. See Anne Edwards, *Early Reagan*, 494. On the life membership card: This took place November 9, 1952. See "Radio and Television," *New York Times*, November 10, 1952, A32.

13. Reagan, "Interview with the President," December 27, 1981.

14. Meroney, "Rehearsals for a Lead Role."

15. Reagan, *An American Life*, 115.

16. This has been described in many accounts, including Reagan's own autobiographies, *An American Life* and *Where's the Rest of Me*. Among the better accounts are those by Anne Edwards in her *Early Reagan* and the excellent research of Stephen Vaughn in his Stephen Vaughn, *Ronald Reagan in Hollywood: Movies and Politics* (New York: Cambridge University Press, 1994), 124–26.

17. Ron and Allis Radosh, *Red Star Over Hollywood*, 121–22.

18. "The Refugees Still Wait," *The New York Times*, October 5, 1947, E8.

19. The figure was at least $100 million annually, in addition to the services of the U.S. soldiers. The camps were also located in Britain, Canada, Belgium, and Latin America. "Marshall Says DP Exit Would Ease U.S.–Russian Friction in Europe," *New York Times*, July 17, 1947, A6.

20. "U.S. Opposes Soviet on the Displaced," *New York Times*, November 4, 1947, A5.

21. "Marshall Says DP Exit Would Ease U.S.–Russian Friction in Europe," *New York Times*, July 17, 1947, A6.

22. "Rosenwald Urges U.S. to Take DP's," *New York Times*, May 13, 1947, A8.

23. "Bill on Displaced Faces Stiff Fight," *New York Times*, May 18, 1947, A29.

24. "Clark Urges U.S. to Take Refugees," *New York Times*, July 19, 1947, A5.

25. "Law Change Urged to Admit Refugees," *New York Times*, April 2, 1947, A11.

26. "Reagan Backs Bill for DP's," *New York Times*, May 8, 1947, A5.

27. John Howard Lawson, *Film in the Battle of Ideas* (Masses & Mainstream, 1953). See Marvin Olasky, "Reagan: A wonderful life," *World* magazine, February 7, 2004, 52.

28. This is not the place to debate the subsequent blacklist and the Hollywood Ten. That is a separate discussion.

29. "Star Witnesses," *Newsweek*, November 3, 1947, 23–25.

30. Reagan, "Testimony before the House Un-American Activities Committee," October 25, 1947.

31. Quoted by Vaughn, *Ronald Reagan in Hollywood*, 166.

32. "Film Stars' Lawyer Hits Kangaroo Court," *Daily Worker*, October 24, 1947, 3. (Note: the date may have been misprinted.)

33. Anne Edwards, *Early Reagan*, 350.

34. Schweizer, *Reagan's War*, 25–27, 33.

35. Gladwin Hill, "Reagan Weighing a New Role in Gubernatorial Race on Coast," *The New York Times*, January 23, 1965.

36. Meroney, "Rehearsals for a Lead Role."

37. Quoted in Vaughn, *Ronald Reagan in Hollywood*, 141.

38. Vaughn, *Ronald Reagan in Hollywood*, 144.

39. This line and anecdote has been widely quoted. Hayden's HUAC testimony, April 10, 1951; McCoogan, "How the Commies Were Licked," *New York Times*, April 11, 1951, 14; and Vaughn, *Ronald Reagan in Hollywood*, 212. Among more recently available sources, see Meroney, "Rehearsals for a Lead Role."

40. Earl B. Dunckel oral-history testimony, April 27, 1982, RRL, OHT, Vol. 14, Box 4, 16–17.

41. Quoted in "More Jobs in Films for Negroes Urged," *New York Times*, November 10, 1952, A32.

CHAPTER 3

1. Quoted in Joseph Lewis, *What Makes Reagan Run? A Political Profile* (New York: McGraw-Hill, 1968), 46; and Lou Cannon, *Reagan* (New York: Putnam, 1982), 141.

2. Reagan performed at the Last Frontier in February 1954. Later in 1954, he began the GE job. Information on Reagan at the Last Frontier was provided by the same hotel, which is now called the New Frontier; the name has been changed a number of times.

3. McClure et al., *Ronald Reagan: His First Career, A Bibliography of the Movie Years*, 188–93.

4. "General Electric Theater—1954–57," directory of show episodes on file at RRL.

5. Reagan said this in a 1980 campaign stop at a GE plant in Erie, Pennsylvania. Text located at RRL. The show began on September 12, 1954 and ran in thirty- to sixty-minute installments. In all, 200 episodes were made during the show's eight-year run. McClure et al., *Ronald Reagan: His First Career, A Bibliography of the Movie Years*, 188. See Morris, *Dutch*, 304.

6. Reagan, "Commencement Address at Eureka College," June 7, 1957.

7. Ibid.

8. This episode of *GE Theatre* was titled, "No Skin Off Me." It aired February 3, 1957. A copy of the video is located at the RRL.

9. McClure, Rice, and Stewart, eds., *Ronald Reagan: His First Career*, 188–93. See Schweizer, *Reagan's War*, 33.

10. Skinner, Anderson, and Anderson, *Reagan's Path to Victory* (New York: The Free Press, 2004), 228.

11. Matthews said it gave him his first "sense of Reagan the politician." Chris Matthews speaking at "The Reagan Legacy" conference, Ronald Reagan Library, Simi Valley, CA, May 20, 1996.

12. McClure, Rice, and Stewart, eds., *Ronald Reagan: His First Career*, 193–94; and Skinner, Anderson, and Anderson, eds., *Reagan: A Life in Letters*, 145n.

13. Reagan letter to Lorraine and Elwood Wagner, June 3, 1962, YAF collection.

14. Morris, *Dutch*, 314.

15. This is a UPI article that ran in the *New York Times*, May 9, 1961, titled, "Red Threat is Cited."

16. Among these, see, for example: "Reagan Spreads Warning About Reds in Hollywood," *The Independent* (Wilkes-Barre, PA), July 23, 1961. This was a UPI syndicated article.

17. Ronald Reagan, "Encroaching Government Controls," *Human Events*, July 21, 1961, 457.

18. Ibid.

19. "Reagan Warns U.S. Is In War," *Bartlesville Examiner Enterprise*, March 1, 1962.

20. "Reagan Says Free Word, Reds at War," *Dallas Times Herald*, February 27, 1962; and Editorial, "In Our Opinion—'Losing Our Freedom. . . . By Installments,'" *Angleton Times*, February 29, 1962. (The date on the paper seems inaccurate, since that year was not a leap year.)

21. "Reagan Warns U.S. Is In War," *Bartlesville Examiner Enterprise*, March 1, 1962.

22. Reagan, "A Foot in the Door," address to the Illinois Manufacturers' Costs Association, May 9, 1961. Text on file at RRL.

23. Reagan, "A Time for Choosing," October 27, 1964.

24. He said he told "Republican leaders" (his Kitchen Cabinet presumably) that if they would let him fulfill his speaking invitations in California, "I'd come back and tell them who should be running for Gov and I'd campaign for him. After a few months I discovered the candidate." Reagan letter to Lorraine and Elwood Wagner, March 9, 1992, Young Americans Foundation (YAF) collection.

25. The documentary was a presentation of a group called The National Education Program, the president of which was George S. Benson of the American Heritage Center in Searcy, Arkansas. I watched part two of the video at the Reagan Library. The library is not in possession of part one.

26. I've been told that this documentary is also about the Communist threat. While that is very likely the case, I have not been able to confirm that because I cannot locate a videotape copy.

27. Interview with Lew Uhler, July 1, 2005.

CHAPTER 4

1. Reagan and Hubler, *Where's the Rest of Me?*, 311–12.

2. Ibid.

3. In 1968, Reagan made this exact plea in a stump speech in California—still prior to the deals later brokered by détente. Quoted in Smith, *Who is Ronald Reagan?*, 90.

4. Lyn Nofziger, *Nofziger* (Washington, DC: Regnery Gateway, 1992), 43–44.

5. This quote was used vigorously in Reagan campaign ads at the time. See Gary G. Hamilton and Nicole Woolsey Biggart, *Governor Reagan, Governor Brown: A Sociology of Power* (New York: Columbia University Press, 1984), 165.

6. Quotes cited in Lee Edwards, *Ronald Reagan: A Political Biography* (Houston, TX: Nordland, 1980), 203, 209.

7. Debate between Ronald Reagan and Robert F. Kennedy, "The Image of America and the Youth of the World," CBS News, "Town Meeting of the World," internationally televised, May 15, 1967. A video of the debate is located at the Reagan Library. I have a transcript of the debate, which I obtained from Bill Clark, who has held a copy in his personal files for almost forty years.

8. "The Ronnie-Bobby Show," *Newsweek*, May 29, 1967, 26–27.

9. Steven F. Hayward, *The Age of Reagan* (Roseville, CA: Prima, 2001), 169; and Michael Knox Beran, *The Last Patrician: Bobby Kennedy and the End of American Aristocracy* (New York: St. Martin's Press, 1998), 150.

10. Lou Cannon, *Ronnie and Jesse: A Political Odyssey* (New York: Doubleday, 1969), 264.

11. Lewis, *What Makes Reagan Run?*, 196–97.

12. "The Ronnie-Bobby Show," *Newsweek*, May 29, 1967, 26–27.

13. Cannon, *Ronnie and Jesse*, 264.

14. Hayward, *The Age of Reagan*, 170; and Jules Witcover and Richard M. Cohen, "Where's the Rest of Ronald Reagan," *Esquire*, March 1976, 153.

15. Reagan, "Veterans Day Address at North Albany Junior High School," Albany, Oregon, November 11, 1967. On file at Reagan Library: "RWR—Speeches and Articles (1967)," folder, RRL, vertical files.

16. Ibid.

17. Reagan, "Speech to Republican State Central Committee Luncheon," Hilton Plaza, Miami, Florida, May 21, 1968. Speech filed at Reagan Library, "RWR—Speeches and Articles (1968)," vertical files.

18. "Speech to Republican State Central Committee Finance Dinner," Sheraton-Cleveland Hotel, Cleveland, May 22, 1968. Speech filed at Reagan Library, "RWR—Speeches and Articles (1968)," vertical files.

19. As Clark's biographer, I have had access to these materials.

20. Of all the work that has been done on this, the best source is a superb PBS documentary which interviewed all of the participants—Israeli, Egyptian, Soviet—at the highest levels. These individuals all spoke openly of the Soviet effort. "The 50 Years War: Israel and the Arabs," episode II, "The Six Day War," PBS, 1999.

21. The only transcript that I have been able to locate is held in Bill Clark's personal files.

CHAPTER 5

1. For more on this, see the excellent research of Schweizer in his *Reagan's War*, 108.

2. Allen in Schweizer, ed., *The Fall of the Berlin Wall*, 55–56; and Allen in Peter Hannaford, ed., *Recollections of Reagan* (New York: William Morrow, 1997), 6–8. Also, Allen spoke of the incident, and noted that it occurred specifically in the month of November, in an interview for the documentary, *In the Face of Evil: Reagan's War in Word and Deed* (American Vantage Films and Capital Films I, LLC, 2005).

3. Allen remarks, in Schweizer, ed., *The Fall of the Berlin Wall*, 55–56.

4. Ibid.

5. A superb source on Reagan's thinking toward the Soviets during the latter 1970s is Kiron Skinner, who has studied Reagan for years, first as a doctoral candidate in foreign policy at Harvard, then later as an assistant to George Shultz, a Hoover Institution fellow, and a fellow at the Council on Foreign Relations. She is the individual who discovered a box containing the 670 handwritten drafts of Reagan's radio broadcasts done between 1975–79, all researched and written entirely by Reagan. Some of these were published in the landmark volume, *Reagan, In His Own Hand*, coedited by Skinner and Martin and Annelise Anderson, as well as in subsequent volumes. Though hundreds of the broadcasts have been released and published, many remain unavailable. Skinner has carefully studied them all. Kiron F. Skinner, Annelise Anderson, and Martin Anderson, eds., *Reagan, In His Own Hand* (New York: The Free Press, 2001). A number of added follow-up volumes featuring more of these transcripts have been released by Skinner and the Andersons, or will be released in future planned volumes. The transcripts are also available at the Reagan Library among four boxes. My vetting of Reagan material initially included nearly all but the voluminous radio broadcasts from 1975–79. Once Skinner found and published the contents of those broadcasts, I found no inconsistency whatsoever with the Reagan record I studied. What she found in those hundreds of broadcasts conforms with what I found in some 10,000-plus pages of *Presidential Documents*, boxes of Reagan Library documents, and thousands of pages from other sources, from memoirs to secondary sources to interviews to oral histories and more. In sum, what happened in the 1980s—in terms of Reagan's intent to undermine the USSR and roll back Communism—matches Reagan's 1975–79 intentions, as clearly expressed in his radio broadcasts from the period.

6. Interview with Ed Meese, March 23, 1998.

7. Reagan, "Remarks on Soviet–U.S. Relations at the Town Hall of California Meeting," Los Angeles, August 26, 1987.

8. A.V. Mikhaylov, "Rejoinder: No-Good Subverters," *Pravda*, February 8, 1985, 5, published as "Reagan Attempt to 'Belittle' Yalta Accords Hit," in FBIS-SOV-11-FEB-85, February 11, 1985, A1.

9. Richard V. Allen, "The Man Who Changed the Game Plan," *National Interest*, Summer 1996, 61.

10. Interview with Caspar Weinberger, October 10, 2002.

11. Weinberger speaking at the conference, "Reagan's War and the War on Terrorism," hosted by the American Enterprise Institute, Washington, DC, November 13, 2002.

12. Weinberger speaking at a November 12, 1999 Ethics and Public Policy Center symposium titled, "Rebuilding American Power," held at the Ronald Reagan Building in Washington, DC. His remarks were published in the Ethics and Public Policy Center newsletter, Winter 2000, no. 69, 1.

13. Caspar W. Weinberger, "A Most Remarkable President," published in a July 2004 Ronald Reagan commemorative issue of *Libertas*, a publication by the Young America's Foundation.

14. Weinberger speaking during interview for documentary, *In the Face of Evil: Reagan's War in Word and Deed* (American Vantage Films and Capital Films I, LLC, 2005).

15. Yuri Zhukov, "Resurrection of a Dinosaur," *Pravda*, April 11, 1975, 4, published as "Zhukov Assails Reagan's London Speech," in FBIS, April 17, 1975, B7. Zhukov was quoting Reagan's own words from a speech in London that spring of 1975, the text of which is lost, with only two handwritten pages known to exist. Reagan gave the speech on April 7, 1975 to the Pilgrim Society in London. A full text of the speech does not exist. Two handwritten pages of the speech are in the possession of Kurt Ritter of Texas A&M University, who generously shared the two pages with me. They were given to Ritter by Reagan presidential speechwriter C. Landon Parvin. Ritter knows of no complete text of the speech, and neither does the Reagan Library or Hoover Institution (which has many Reagan speech texts). On April 8, 1975, the speech was reported by the *London Times* ("Warning that Europe is Main Soviet Prize") and *Daily Mirror* ("A Red Alert by Reagan").

16. Thatcher in Edwin Feulner, Jr., ed., *Leadership for America: The Principles of Conservatism* (Dallas, TX: Spence, 2000), 11.

17. Reagan said this in a series of interviews he did in the spring and summer of 1975 with author Charles Hobbs, publishing for a campaign book released in 1976. Charles D. Hobbs, *Ronald Reagan's Call to Action* (Nashville, TN and New York: Thomas Nelson, 1976), 50–51.

18. Interview with Ed Meese, December 5, 2001.

19. Of course, some would later dispute the morality of certain policies of President Reagan, which was understandable, but what they really disputed were means, not ends. Reagan thought that aiding the government of El Salvador, for example, would ultimately help much more than hurt. Domestically, he felt poverty was better addressed by the private sector than the public sector.

20. Allen in Schweizer, ed., *The Fall of the Berlin Wall*, 50.

21. Monica Crowley, *Nixon Off the Record* (New York: Random House, 1996), 26, 92–93.

22. Reagan letter to Lorraine and Elwood Wagner, August 3, 1971, YAF collection.

23. Letter quoted in Helene Von Damm, *Sincerely, Ronald Reagan* (Ottawa, IL: Green Hill Publishers, 1976), 75.

24. Ibid., 108. Reagan wrote to a friend saying that he supported a "bettering of relations with mainland China." However, he felt that "the dumping of a long time friend and ally, the Republic of China on Taiwan," was unforgivable. Reagan letter to Lorraine and Elwood Wagner, June 12, 1979, YAF collection.

During the 1980 campaign, Reagan thundered to an audience: "There will be no more Taiwans. . . . There will be no more betrayals of friends by the United States!" Quoted in Laurence I. Barrett, *Gambling with History: Ronald Reagan in the White House* (New York: Doubleday, 1983), 206–7.

25. Reagan, "Address to the Roundtable National Affairs Briefing," Dallas, Texas, August 22, 1980, located at Reagan Library, "Reagan 1980 Campaign Speeches, August 1980," vertical files.

26. Genrikh Aleksandrovich (Henry) Trofimenko in Hofstra conference (1993) proceedings, 136.

27. Peter Osnos, "Angola Stirs Questions on Détente Fine Print," *Washington Post*, January 16, 1976, A12.

28. Quoted by James Reston, "The Mood of the Capital," *New York Times*, February 27, 1976, 31.

29. Reagan commentary delivered on March 23, 1977. His source was a February 11, 1977 *Boston Globe* article by William Beecher, reprinted in *National Review* on March 4, 1977. According to the *Globe* article, in early 1973 British intelligence obtained a speech by Leonid Brezhnev given at a secret meeting of Eastern European Communist rulers in Prague. (An excerpt from the speech was quoted by R. Emmett Tyrrell, Jr. on the *New York Times* op-ed page on June 18, 1977. Tyrrell's piece also quoted the exact same words from Brezhnev's 1973 Prague speech.) The Brits rated the speech comparable in importance to Nikita Khrushchev's 1956 "Crimes of Stalin" speech. In the handwritten text of his radio commentary, Reagan complained that "the British informed our government of Brezhnev's speech, but apparently it didn't lessen our desire for 'détente.'" Brezhnev told his Communist bloc comrades: "We are achieving with détente what our predecessors have been unable to achieve using the mailed fist. We have been able to accomplish more in a short time with détente than was done for years pursuing a confrontation policy with NATO. . . . Trust us comrades, for by 1985, as a consequence of what we are now achieving with détente, we will have achieved most of our objectives in Western Europe. We will have consolidated our position. We will have improved our economy."

And then, Reagan continued to report, Brezhnev "added the bottom line which certainly should have guided our own policy for these intervening years. He said, '. . . a decisive shift in the correlation or forces will be such that come 1985, we will be able to extend our will wherever we need to.'" Brezhnev, said Reagan, "was optimistic about the future of Marxism in France" and said that "Finland was already in the Soviet pocket, trends in Norway were in the right direction, and Denmark was no longer a viable part of Western strength." Reagan expressed anger at Washington's nonresponse to the British intelligence report on the speech. According to the *Globe*, Secretary of State Kissinger had minimized the importance of the report. The only official reference to it came three years later (1976) in a CIA National Intelligence Estimate. Reagan wrote (in his unedited draft): "Maybe in 1973 there was some excuse for interpreting Brezhnev's remarks as a form of campaign rhetoric for in house consumption. But now we can look back over the four years since the speech was made and see how consistent with his words Soviet policy has been." Reagan then chronicled the Soviet advantage: He said that

Soviet forces on the "NATO front" had increased by fifty-four divisions, a 40 percent increase in tanks. He said the Soviets had developed six new strategic nuclear systems. He leveled more complaints and listed evidence of Soviet aggression. Located in "Ronald Reagan: Selected Radio Broadcasts, 1975–1979," January 1975 to March 1977, Box 1, RRL. For a full transcript, see Skinner, Anderson, and Anderson, eds., *Reagan, In His Own Hand*, 117–19.

A year later, in a September 1978 radio broadcast, Reagan probably referenced the same meeting when he spoke of an "intelligence report" of a secret meeting between Brezhnev and Communist Party leaders in which the general secretary allegedly said that (in Reagan's words) "détente was a stratagem to allow the Soviets time to build up their military so that by 1985 they could exert their will wherever they wished." Ronnie Dugger, *On Reagan: The Man & His Presidency* (New York: McGraw Hill, 1983), 536.

30. Reagan NSC member Constantine Menges says that between 1975 and 1980, eleven new "pro-Soviet regimes" were established. Tom Henriksen of the Hoover Institution says that from 1974–79 the Soviets "incorporated 10 countries into their orbit." Constantine C. Menges in Hofstra conference (1993) proceedings, 29–30; and Thomas Henriksen, "The lessons of Afghanistan," *Washington Times*, December 29, 1999.

31. For examples from the *New York Times* and the *Washington Post*, see *New York Times*, January 29, 1976, A21; February 26, 1976, A31; March 5, 1976, A1, A10; March 13, 1976, A10; March 14, 1976, A14; March 25, 1976, A1, A35; April 1, 1976, A31; April 8, 1976, A32; April 15, 1976, A1; May 14, 1976, A11; May 19, 1976, A46; May 28, 1976, A13; and June 10, 1976, A37. From *Washington Post*, see January 16, 1976, A12; February 11, 1976, A1; April 1, 1976, A10; and April 29, 1976, A6.

32. Reagan radio broadcast from October 31, 1975, titled "Détente." Dugger, *On Reagan: The Man & His Presidency*, 514.

33. Ronald Reagan, "Tactics for Détente," *Wall Street Journal*, February 13, 1976, A8.

34. Among others, see C. L. Sulzberger, "How The World Looks At Carter," *New York Times*, July 24, 1976, A23; William Safire, "Life of the Party," *New York Times*, March 18, 1976, A14; Cannon, *Reagan*, 219; and Laurence Barrett, *Gambling With History*, 288.

35. Anthony Lewis, "By His Own Petard," *New York Times*, April 19, 1976, A27.

36. Jon Nordheimer, "Reagan Attacks Ford's 'Timidity,'" *New York Times*, March 7, 1976, A40.

37. Editorial, "Mr. Reagan's Veto," *New York Times*, May 14, 1976, A26.

38. Editorial, "President Under Seige," *New York Times*, May 9, 1976, A14.

39. Quoted by Martin Anderson, *Revolution* (New York: Harcourt, Brace, Jovanovich, 1988), 43.

40. Reagan speaking on "Reagan," *The American Experience*, PBS.

41. "Where Reagan Stands, Interview on the Issues," *U.S. News & World Report*, May 31, 1976, 20.

42. Quoted in Jon Nordheimer, "Reagan, in Direct Attack, Assails Ford on Defense,'" *New York Times*, March 5, 1976, A10.

43. Lou Cannon noticed this as well. Cannon, *Reagan*, 219.

44. "Text of Platform Proposal," *New York Times*, August 17, 1976, A23.

45. Richard L. Madden, "Reagan's Plank Criticizes Ford-Kissinger Policies," *New York Times*, August 17, 1976, A1, A23.

46. Alexander Solzhenitsyn, *Alexander Solzhenitsyn Speaks to the West* (London: The Bodley Head, 1978), 73.

47. Solzhenitsyn once said to Reagan regarding his assassination attempt: "I unceasingly thank God that you were not killed by that villainous bullet." John O'Sullivan, "Friends at Court," *National Review*, May 27, 1991, 4.

48. This includes Reagan's attack on détente, Ford's ceasing to use the term and opting for peace through strength, and the morality plank. Soviet media archives from the time featured at least twenty separate newspaper articles, radio and TV transcripts, and press releases from the likes of TASS and the Moscow Domestic Service. They cover primarily the spring and summer 1976 period, but also touch late 1975 and early 1977.

49. Commentary by Valentin Zorin, Moscow Domestic Service, February 16, 1976, published as "Reagan Making Détente a 'Football Game,'" in FBIS-SOV-27-FEB-76, February 27, 1976, B5–6.

50. M. Sturua, "Reagan Applies the 'Corrective;' the Essence of the Amendments to the U.S. Republican Party Platform," *Izvestia*, August 25, 1976, 3, published as "Concessions to Right in Republican Platform Attacked," in FBIS-SOV-27-AUG-76, August 27, 1976, B1–3.

51. Morris, *Dutch*, 402.

52. Nancy Reagan recalled these words in an interview with "Reagan," *The American Experience*, PBS.

53. Reagan might have quickly uttered the word "to," though that is not clear.

54. A few weeks later, Reagan expanded on his remarks by turning them into a radio broadcast that he taped on September 1, 1976. For a handwritten copy of that broadcast, see Skinner, Anderson, and Anderson, eds., *Reagan, In His Own Hand*, 9–10 and inside cover of book.

55. Edmund Morris speaking on "Reagan," *The American Experience*, PBS.

56. Interview with Michael Reagan by telephone, May 9, 2005. Michael said his father "finally got that chance at Reykjavik, Iceland ten years later in October 1986 when he said ['nyet'] to Gorbachev over SDI."

57. Maureen Reagan, "A president and a father," *Washington Times*, June 16, 2000, A23.

58. Interview with Richard V. Allen, November 12, 2001.

59. Interview with Allen; and Richard Allen, "An Extraordinary Man in Extraordinary Times: Ronald Reagan's Leadership and the Decision to End the Cold War," Address to the Hoover Institution and the William J. Casey Institute of the Center for Security Policy, Washington, DC, February 22, 1999, in Schweizer, ed., *The Fall of the Berlin Wall*, 52.

60. Allen in Schweizer, ed., *The Fall of the Berlin Wall*, 58–59.

61. Here is a transcript of that exchange between myself and Allen:

Question to Allen: I'm trying to clarify that Reagan in fact had a specific intent to take on and defeat the USSR and the Soviet empire before his presidency even began. Was that his intent? *Allen*: Yes. *Q*: That's a big, big deal. Are you telling me that on that day in January 1977, Ronald Reagan told you that his goal was to take on and defeat the Soviet empire? That's what you're telling me? *Allen*: Yes. That's absolutely right. That's what I'm telling you. *Q*: So, you then, on that day, decided to join him for the purpose of taking on and defeating the Soviet empire? *Allen*: Yes. That's it exactly. Nothing longer and nothing shorter than that. *Q*: You joined Reagan because you were convinced that that was his intent. *Allen*: Yes. *Q*: And this was four years before his presidency began? *Allen*: That would be correct. January 1977. Four years. Interview with Richard V. Allen, November 12, 2001.

When Ed Meese was asked his response to Allen's statement about the January 1977 meeting, and asked if Reagan ever said such a thing to him prior to the presidency, Meese said, "Well, not in such stark terms . . . But what he said there is not surprising. He did believe that."

Bill Clark had the same reaction. Interviews with Ed Meese, December 5, 2001, and Bill Clark, July 14, 2005.

CHAPTER 6

1. Otto Kreisher, "Desert One," *Air Force Magazine*, January 1999, 82, no. 1.

2. Jimmy Carter, *Keeping Faith: Memoirs of a President* (University of Arkansas Press, 1995).

3. Jim Greeley, "Desert One," *Airman*, April 2001; and Kreisher, "Desert One."

4. Greeley, "Desert One."

5. Ibid.

6. "Transcript of President's Interview on Soviet Reply," *The New York Times*, January 1, 1980, 4.

7. Reagan wrote this January 1980 letter to a Professor Nikolaev, in Skinner, Anderson, and Anderson, eds., *Reagan: A Life in Letters*, 400.

8. Ibid., 433–34. Reagan wrote this January 1980 letter to a man named Edward Langley.

9. Carter responded with a series of actions: On January 3, 1980, he asked the U.S. Senate to suspend approval of the SALT II treaty he had signed with Brezhnev in Vienna the previous June. The next day he announced a sanctions package highlighted by an American boycott of the coming Olympic Games in Moscow as well as an embargo on U.S. grain exports to Russia. If the Soviets did not withdraw "within the next month," said Carter, the Olympic boycott would remain. The Soviets did not pull out.

Importantly, this sparked a major shift within the Carter administration, which embarked upon a tougher foreign policy and initiated notable increases in military spending. President Carter now sided with Zbigniew Brzezinski's more hawkish National Security Council, as opposed to Cy Vance's dovish State Department, which had previously won over the president.

10. Martin Schram, "Reagan Urges U.S. Mideast Presence," *Washington Post*, January 10, 1980, A3.

11. Morris in Wilson, ed., *Power and the Presidency* (New York: Public Affairs, 1999), 125–26.

12. For an extended analysis of this, see Kengor, *God and Ronald Reagan*.

13. Reagan, "Address to the Conservative Political Action Conference," Washington, DC, February 6, 1977, in James C. Roberts, ed., *A City Upon a Hill: Speeches by Ronald Reagan Before the Conservative Political Action Conference* (Washington, DC: The American Studies Center, 1989), 31–33.

14. Ibid., 34.

15. All of Reagan's first four CPAC speeches in the 1970s featured Shining City imagery. Reagan, "Address to the Conservative Political Action Conference," Washington, DC, February 6, 1977. Text appears in Roberts, ed., *A City Upon a Hill*, 31, 37.

16. Kissinger interviewed for CNN documentary, "The Reagan Years: The Great Communicator," Pt. II of series, CNN, February 2001.

17. Greenfield speaking on CNN documentary, "The Reagan Years."

18. Reagan, *An American Life*, 219.

19. For the text of these Reagan remarks, see Peter Hannaford, *The Reagans: A Political Portrait* (New York: Coward-McCann, 1983), 214–18.

20. "Ronnie's Romp," *Time*, March 10, 1980. See Reagan September 9, 1980 speech, "A

Strategy for Growth," to the International Business Council of Chicago. "RWR, Pres. Election-1980," folder, RRL.

21. Reagan, "Address at Liberty State Park," Jersey City, NJ, September 1, 1980. Speech text located at Reagan Library, "Reagan 1980 Campaign Speeches, September 1980," vertical files.

22. Credit again goes to the digging of Kiron Skinner and Annelise and Martin Anderson. For full text, see Skinner, Anderson, Anderson, eds., *Reagan, In His Own Hand*, 470–79. Prior to their discovery, one had to read about the speech in the form of quotes in newspapers and other documents. This is the first full copy made available.

23. The other speech was on August 18, 1980. "The greatest fallacy of the Lenin-Marxist philosophy is that it is the 'wave of the future,'" wrote Reagan in the speech draft. "Everything about it is as primitive as tribal rule." Speaking of boat people from Southeast Asia and Cuba, fleeing "the inhumanity of communism," he stated: "I believe it is our pre-ordained destiny to show all mankind that they, too, can be free without having to leave their native shore." Text available at Reagan Presidential Library. *Reagan-Bush 1980 Campaign Papers, 1979–80*, Box 949.

24. Skinner, Anderson, Anderson, eds., *Reagan, In His Own Hand*, 478.

25. Ibid., 479.

26. Ibid., 471.

27. Ibid., 472.

28. Ibid., 474, 478, 479.

29. Ibid.

30. Kiron Skinner found that nearly a third of the 670 Reagan radio transcripts addressed defense or foreign policy, of which "Reagan's main concern throughout . . . is the cold war." "According to Reagan," she deduced, "the main goal of the United States' cold war policy should be to hasten the end of communism. . . . Communism will not survive, he writes." She listed steps that Reagan felt would (in her words) "hasten the demise of communism." Skinner says Reagan wrote that (in her words) "a first step toward hastening the demise of Soviet communism was to distinguish the symptoms of the Cold War from its sources." She lays out the steps on pages 23–25 of *Reagan, In His Own Hand*.

31. As will be seen in the chapters ahead: The "strategic deterrent" Reagan insisted upon might be any of a number of nuclear missile programs in the 1980s, from the MX to Pershing IIs. The Naval superiority was seen in Reagan seeking his 600-ship Navy, restoring old, mothballed destroyers. Pay for military personnel was jacked up significantly once Reagan became president, sparking much higher morale and making the armed forces a destination rather than a last resort for young people. The science and technology thrust was embodied most saliently in SDI and other technological challenges to the Kremlin. Lastly, intelligence in the 1980s was refocused to search out and exploit Soviet economic vulnerabilities.

32. Reagan said this in Anaheim, California. "Reagan Proposes Arms Approach," *United Press International, Washington Post*, April 17, 1977, A5.

33. Quoted by Dugger, *On Reagan: The Man & His Presidency*, 395. To cite another example, in a May 1979 radio broadcast, Reagan complained: "Our President is telling us that SALT II holds out the promise of peace and an end to any costly arms race. But what does that do to us if we are the only ones racing?" In "Ronald Reagan: Pre-Presidential Papers: Selected Radio Broadcasts, 1975–1979," October 31, 1978 to October 1979, Box 4, RRL. This is taken from a Reagan radio broadcast titled simply "Miscellaneous I." For a full transcript, see Skinner, Anderson, and Anderson, eds., *Reagan, In His Own Hand*, 104–5.

34. Reagan was again on cue in a 1980 interview with the Associated Press: "They [the Soviets] know our industrial strength. They know our capacity. The one card that's been missing in these negotiations has been the possibility of an arms race. Now the Soviets have been racing, but with no competition. No one else is racing. And so I think we'd get a lot farther at the table if they know that as they continue, they're faced with our industrial capacity and all that we can do." In James S. Brady, ed., *Ronald Reagan: A Man True to His Word* (Washington: The National Federation of Republican Women, 1984), 38–39.

Also see speech written by Reagan and delivered in Chicago, August 18, 1980. Text available at Reagan Presidential Library. *Reagan-Bush 1980 Campaign Papers, 1979–80*, Box 949.

Analyzing the late 1970s record with access to more original documents than anyone else has had, Kiron Skinner (in a piece for *National Interest*) stated that Reagan believed "that Russia's inefficient economy and inferior technology ultimately could not survive competition with the United States over armaments. He discussed his hypothesis repeatedly, in his daily radio broadcasts and bi-weekly newspaper columns in the late 1970s."

35. "Reagan: 'It Isn't Only Washington . . . ,'" *National Journal*, March 8, 1980, 392.

36. Lou Cannon, "Arms Boost Seen as Strain on Soviets," *Washington Post*, June 19, 1980, A3. Among others, see Reagan, "Remarks and a Question-and-Answer Session via Satellite to Republican Campaign Events," October 14, 1982.

37. Cannon oral-history testimony at the University of Virginia's Miller Center. Kenneth W. Thompson, ed., *Leadership in the Reagan Presidency, Pt II: Eleven Intimate Perspectives* (Lanham, MD: United Press of America, 1993), 59, 65.

38. Lou Cannon in Hofstra conference (1993) proceedings, 468–69.

39. Letter in Skinner, Anderson, and Anderson, eds., *Reagan: A Life in Letters*, 374–75.

40. Reagan, "Acceptance Speech at Republican National Convention," July 17, 1980.

41. Thomas C. Reed, *At the Abyss: An Insider's History of the Cold War* (New York: Presidio, 2005), 234–35.

42. Schweizer, *Reagan's War*, 215–16.

43. Michael Deaver, *A Different Drummer: My Thirty Years with Ronald Reagan* (New York: HarperCollins, 2001), 76–77.

44. Earl Dunckel in oral-history testimony, April 27, 1982. RRL, Oral History Testimony (OHT), Volume 31, Box 7, 19.

45. John Sears, "A Man Who Knows Himself," *Washington Post*, July 13, 1980, E7.

46. George F. Will, "The best do not linger," op-ed, *Washington Post*, August 20, 1995.

47. Quoted by Larry Berman in Berman, ed., *Looking Back on the Reagan Presidency* (Baltimore: The Johns Hopkins University Press, 1990), 7. This was not the first time he used such a line. At the December 1987 Washington Summit, a reporter noticed a gaggle of media fawning over Soviet leader Mikhail Gorbachev, while Reagan strangely appeared off alone with no one interested in him. Asked if he felt upstaged by Gorbachev, Reagan replied: "Good Lord, no. I've been on the same stage with Errol Flynn." Reagan, "Remarks and a Question-and-Answer Session With Area High School Seniors," Jacksonville, Florida, December 1, 1987.

48. Interview with Ben Elliott, September 20, 2001.

49. Peter W. Rodman, *More Precious Than Peace* (New York: Charles Scribner's Sons), 234.

50. Commentator Fred Barnes adds that, "One of the amazing things about Reagan, and one of the traits that is the least commented on, was his amazing ability to just block out the buzz in Washington and in the rest of the world for that matter." Kenneth W. Thompson, ed.,

Leadership in the Reagan Presidency: Seven Intimate Perspectives (Lanham, MD: Madison Books, 1992), 96.

51. Dinesh D'Souza, *Ronald Reagan: How an Ordinary Man Became an Extraordinary Leader* (New York: Free Press, 1997), 236. D'Souza's work is a solid source on the confidence issue.

52. By Delchamps' description, "I spent a lot of time in the campaign, with the Reagan people and the people running the campaign." Interview with Ollie Delchamps, May 7, 2004.

During the Reagan presidency, Delchamps became chairman of the U.S. Chamber of Commerce. Delchamps was not chairman for the entire Reagan presidency. The "U.S. Chamber," as it is called, is a private entity that is not a part of the U.S. Department of Commerce. Delchamps said he worked "very closely" with the White House. He frequently had lunch with budget director Jim Miller and had "a lot of connections" to the White House, connections he describes in detail. He recalls that the economy was the overriding issue at the start of the Reagan administration. "The economy had to be improved first so we could rebuild the military that Carter had gutted so bad," said Delchamps, expressing the conventional thinking. "Then that would put us back into the race against the Soviets. That was the thinking. That was the plan."

53. Callahan chose to share his recollection of the incident with me for a number of reasons: He read and enjoyed my book, *God and Ronald Reagan*, and also has tremendous respect for Grove City College, the college where I teach and where Callahan sent one of his children. He also learned from *God and Ronald Reagan* that I was continuing to write a second book on Reagan and the Cold War (this book), and thought his account of his meeting with Delchamps would be of interest to my research. As a result, he called me one day (May 4, 2004) and eagerly shared this story.

54. Interview with Bob Callahan, Sr., May 4, 2004.

55. For the record, Delchamps, now saddled by the enemy of old age, said he could not recall the incident with Callahan. "I'm sorry," Delchamps told me, "but there are only so many things I can remember." However, he emphasized, if Callahan "said I said that, then I'm sure I did." Interview with Ollie Delchamps, May 7, 2004. As noted, Callahan never forgot the incident, and longed to share it. Callahan, a highly respected individual who counts Margaret Thatcher among his personal friends, is certainly credible.

CHAPTER 7

1. A complete handwritten text of the inaugural address is on file at the Reagan Library.

2. On this meeting, see Schweizer, *Victory*, 5–8.

3. Part of this plan was to "intentionally," "deliberately," "initially take us into a confrontation with the Soviet Union." Allen pointed to Reagan's confidence and perseverance in pursuing such a strategy in the face of criticism all around him from observers like George F. Kennan and Strobe Talbott: "Reagan was confident his strategy would work." Allen added: "Eventually, though, he put no timeline on it and certainly did not see it as something to be exploited politically. . . . Rather, Reagan thought of the eventual demise of the Soviet Union as a good to be pursued in its own right." Interview with Richard V. Allen, December 7, 2001. See: Richard V. Allen, "The Man Who Changed the Game Plan," *National Interest*, Summer 1996, 60, 62, 65.

4. Interview with NSC staff member Norman A. Bailey, May 24, 2005. Bailey recalled these exact words from Allen to the NSC staff.

5. I have spoken to Schweizer about the document, which indeed appears authentic.

6. Schweizer, *Reagan's War*, 141.

7. Interview with Louis H. Evans, February 22, 2006.

8. Evans remembers the gathering place as a kind of "reception room" with a couch and some chairs. He is not sure of the name of the room, but believes it was not part of the White House living quarters. He says that they tracked down Moomaw at a conference that Moomaw had been attending in the Caribbean.

9. For an extended discussion, see Kengor, *God and Ronald Reagan*.

10. Interview with Louis H. Evans, February 22, 2006.

11. On this, see *God and Ronald Reagan*. Bill Clark especially stresses the humility element.

12. For a lengthy analysis, see *God and Ronald Reagan*. The speech was written by Reagan speechwriter Tony Dolan, with few edits from the president. Draft is located in PHF, PS, RRL, Box 1, Folder 7. Reagan, "Address at Commencement at the University of Notre Dame," South Bend, Indiana, May 17, 1981.

13. For a very early example of this theme, see Reagan, "America the Beautiful," commencement address, William Woods College, June 1952.

14. This notion of divine challenge is also discussed at length in *God and Ronald Reagan*.

15. Reagan, "The President's News Conference," June 16, 1981.

16. Reagan, "Speech to Members of Platform Committee," Republican National Convention, July 31, 1968. Speech filed at Reagan Library, "RWR—Speeches and Articles (1968)," vertical files.

17. This is from an October 12, 1972 Reagan speech, titled "The Obligations of Liberty," in Alfred Balitzer, ed., *A Time for Choosing: The Speeches of Ronald Reagan* (Chicago: Regnery Gateway, 1983), 98–106.

18. Full text in Skinner, Anderson, and Anderson, eds., *Reagan, In His Own Hand*, 4–9.

19. Reagan, *An American Life*, 333.

20. Reagan, "Address at Commencement Exercises at the United States Military Academy," May 27, 1981.

21. For an excellent analysis, see Andrew E. Busch, "Ronald Reagan and Economic Policy," in Kengor and Schweizer, eds., *The Reagan Presidency: Assessing the Man and His Legacy* (Lanham, MD: Rowman-Littlefield, 2005).

22. Baker, Deaver, Duberstein, and Meese interviewed for CNN documentary, "The Reagan Years: Inside the White House," Pt. II of series, CNN, February 2001.

23. "How Reagan Decides," *Time*, December 13, 1982, 12.

24. Reagan, *An American Life*, 234–35.

25. Ibid., 333.

CHAPTER 8

1. Steven Strasser, Theodore Stanger, and Douglas Stanglin, "Crackdown on Solidarity," *Newsweek*, December 21, 1981, 28.

2. David Cross, "Shooting reported in Poland as troops break wave of strikes," *London Times*, December 16, 1981.

3. Ibid.

4. Interview with Joseph Dudek, conducted by Margie Dudek, November 2004.

5. Strasser et al., "Crackdown on Solidarity," December 21, 1981; and Cross, "Shooting reported in Poland," December 16, 1981.

6. Richard Owen, "How Army has filled vacuum left by party," *London Times*, December 14, 1981.

The last time martial law was introduced in Eastern Europe was during the uprising in Hungary in 1956. The difference in this case twenty-five years later was that the armed forces took over the reins of government. In any other context, wrote Richard Owen in the *London Times* the next day, these actions would be classified as a military coup. Instead, Poland implemented a sort of hybrid government between the military and Communist politicians.

7. Transcript was published in major newspapers around the world on December 14 and 15, 1981.

8. Cross, "Shooting reported in Poland as troops break wave of strikes," December 16, 1981.

9. "Ex-Prime Minister among those held," *London Times*, December 14, 1981; and Cross, "Shooting reported in Poland as troops break wave of strikes," December 16, 1981.

10. It was the kind of prudent religious sensitivity by Polish Communists that the brute Lenin and Soviet Communists never countenanced.

11. Official Soviet TASS statement, published in the *London Times*, December 14, 1981, 6.

12. Cross, "Shooting reported in Poland as troops break wave of strikes," December 16, 1981.

13. Schweizer, *Victory*, 29, 31.

14. Agostino Bono, "Officials say pope, Reagan shared Cold War data, but lacked alliance," *Catholic News Service*, November 17, 2004.

15. Bill Clark, "President Reagan and the Wall," Address to the Council of National Policy, San Francisco, California, March 2000, 7–8.

16. This is fact easily and immediately confirmable by speaking to Poles from the era. One scholar who demonstrates it nicely is Timothy Garton Ash in *The Polish Revolution: Solidarity*, 2nd ed. (New Haven and London: Yale University Press, 2002), 3, 7.

17. Interview with Bill Clark, August 24, 2001.

18. Among them was a compelling November 2, 1976 piece about the Katyn Forest massacre.

19. Gerhard Simon, "The Catholic Church and the Communist State in the Soviet Union and Eastern Europe," in Bociurkiw and Strong, eds., *Religion and Atheism in the USSR and Eastern Europe* (London: Macmillan, 1975), 212–13. Simon states that Catholic Church organization in the USSR was completely destroyed by the 1930s, and Catholicism was not permitted to reestablish a central apparatus after World War II, unlike some other churches. Writing in the mid-1970s, Simon reported that there was not a single Catholic monastery, convent, school, or welfare institution in the entire Soviet Union. Poland, however, was another story.

20. Ibid., 212–15, 242–51.

21. Fr. Robert A. Sirico, "The Cold War's Magnificent Seven, Pope John Paul II: Awakener of the East," *Policy Review*, no. 59 (Winter 1992): 52.

22. Malachai Martin, *The Keys of This Blood* (New York: Simon and Schuster, 1990), 102.

23. Ibid., 103.

24. This particular broadcast was titled simply, "The Pope in Poland." Located in "Ronald Reagan: Pre-Presidential Papers: Selected Radio Broadcasts, 1975–1979," October 31, 1978 to October 1979, Box 4, RRL. For a full transcript, see Skinner, Anderson, and Anderson, eds., *Reagan, In His Own Hand*, 174–75.

25. Reagan, "Address to the Roundtable National Affairs Briefing," Dallas, Texas, August

22, 1980. Speech text located at Reagan Library, "Reagan 1980 Campaign Speeches, August 1980," vertical files.

26. Located in "Ronald Reagan: Pre-Presidential Papers: Selected Radio Broadcasts, 1975–1979," October 31, 1978 to October 1979, Box 4, RRL. For a full transcript, see Skinner, Anderson, and Anderson, eds., *Reagan, In His Own Hand,* 176–77.

27. Quoted in Schweizer, *Victory,* 35–36, 59, 69, 159–61. Allen said that, "Reagan had a deep and steadfast conviction that this pope would help change the world." Quoted in Carl Bernstein and Marco Politi, *His Holiness: John Paul II and the Hidden History of Our Time* (New York: Doubleday, 1996), 270. Ed Rowny, a Polish-American and one of Reagan's chief advisers on nuclear arms, who would brief John Paul II four times on Reagan's behalf, also confirms that the president believed that the pope would be an important factor in the eventual liberation of Poland. Agostino Bono, "Officials say pope, Reagan shared Cold War data, but lacked alliance," *Catholic News Service,* November 17, 2004.

28. Interview with Bill Clark, August 24, 2001.

29. Editorial, "The Polish Pope in Poland," *New York Times,* June 5, 1979, A20. Reagan was not entirely alone. George Weigel, biographer of John Paul II, states that the import of those nine days in Poland was not lost among two key Slavic observers: In Moscow, KGB head and future Soviet general secretary Yuri Andropov was gravely concerned; prior to the Poland visit, within just six weeks of Karol Wojtyla's election as Pope, Andropov ordered up a "massive" (Weigel's word) KGB analysis on the potential impact of the new chief at the Holy See. (There have been reports, not to mention statements from John Paul II himself, that the USSR was so worried about the Polish pope that the KGB began considering assassination options.) Another Russian of very different ideological bent, Alexander Solzhenitsyn, witnessed the Pope's reception in Poland from his exiled home in Vermont. He was elated: "This is the greatest thing to happen to the world since World War I," he declared. "It's the first real sign of hope since the Bolshevik revolution." Quoted by George Weigel, "And the Wall Came Tumbling Down: John Paul II and the Communist Crack Up," Address at Grove City College, February 15, 2001.

30. Information provided by Tomasz Pompowski, senior editor and reporter at *Fakt* ("Fact"), Poland's largest daily newspaper, via e-mail correspondence September 5, 23, and 29, 2005.

31. Quoted in Rowland Evans and Robert Novak, *The Reagan Revolution* (New York: Dutton, 1981), 11–12.

32. Reagan, "The President's News Conference," June 16, 1981.

33. Acknowledging this spiritual link, Russia expert James Billington notes that as a "bottom-up mass movement rooted in religion," Solidarity was not the typical movement that apparatchiks could domesticate by decapitation or by offering carrots and sticks to its members. James H. Billington, "The Foreign Policy of President Ronald Reagan," Address to the International Republican Institute Freedom Dinner, Washington, DC, September 25, 1997, 2.

34. Arthur R. Rachwald, *In Search of Poland* (Stanford, CA: Hoover Institution Press, 1990), 3.

35. Cited by Brian Crozier, *The Rise and Fall of the Soviet Empire* (Rocklin, CA: Forum, 1999), 359.

36. Among others, see Reagan, "Proclamation 4891—Solidarity Day," January 20, 1982.

37. Reagan, "Address to the International Brotherhood of Teamsters," Columbus, Ohio,

August 27, 1980. Speech text located at Reagan Library, "Reagan 1980 Campaign Speeches, August 1980," vertical files.

38. Schweizer, *Victory*, 29, 31.

39. Benjamin Weiser, *A Secret Life: The Polish Officer, His Covert Mission, and the Price He Paid to Save His Country* (New York: PublicAffairs Books, 2004), 3. I also learned of Kuklinski's importance from Gus Weiss, who told me only vaguely that Kuklinski "earned his salary during the crisis." Interview with Weiss, November 26, 2002.

40. For instance, Polish state television as early as November 1980 had broadcast ominous film of joint Polish-Soviet military maneuvers, using what was likely old video footage (the trees in the video had leaves suggestive of summertime). The point of these broadcasts was to directly suggest Soviet military intervention, even if many Poles were uncertain about the seriousness of the suggestion. Timothy Garton Ash, *The Polish Revolution: Solidarity*, 88.

41. Jerrold Schecter and Leona Schecter, *Sacred Secrets: How Soviet Intelligence Operations Changed American History* (Washington, DC: Brassey's, 2002), 305.

42. December 16, 1980 letter in George Weigel, *Witness to Hope: The Biography of Pope John Paul II* (New York: Harper Perennial, 2001), 406–7.

43. Richard Pipes, *Vixi: Memoirs of a Non-Belonger* (New Haven: Yale University Press, 2003), 168–69.

44. This information was reported and presented by Mark Kramer, director of the Harvard Project on Cold War Studies, in a number of his papers and bulletins, including the *Cold War International History Project Bulletin*, specifically issue no. 5 (Spring 1995) and issue no. 11 (Winter 1999).

Some Soviet officials insisted on an invasion while others, like Mikhail A. Suslov, asserted: "There is no way that we are going to use force in Poland." Edward Shevardnadze, *The Future Belongs to Freedom* (New York: The Free Press, 1991), 121.

45. Shevardnaze, *The Future Belongs to Freedom*. Edward Shevardnadze said Jaruzelski probably spared his country an invasion. See Pipes, *Vixi*, 169. That said, the latest research on this is quite interesting and much more complicated: Mark Kramer reports that as the decisive moment approached in December 1981, Jaruzelski "lost his nerve" and began urging Moscow to send Soviet troops to Poland to help him introduce martial law. By that point, Moscow was firmly against sending troops and, writes Kramer, "tersely brushed aside [Jaruzelski's] repeated pleas." Kramer's findings completely alter the accepted view on Jaruzelski.

46. In illuminating this extraordinary consideration, a clarification is essential: Aside from NSDD-32, there were other formal Reagan directives that addressed the Poland issue, such as NSDD-75, released January 17, 1983, which (as noted) explicitly aimed to undermine the Soviet empire. Among the most dramatic wording in all of the NSDDs was this line on the middle of page three of NSDD-75: "In the longer term, should Soviet behavior worsen, for example, an invasion of Poland, we would need to consider extreme measures." (NSDD is on file at Reagan Library.) NSDD-75 was safely declassified on July 16, 1994, five years after the end of the Soviet grip on Poland and three years after the end of the USSR itself. The directive does not specify what those "extreme measures" might be. Could such measures include U.S. military force in Poland? Those interviewed on the "extreme measures" phrase offered mixed, puzzled reactions, with most claiming ignorance. Some did not recall that seemingly provocative line, including no less than Richard Pipes, regarded as the primary author. This is confusing, until one reads the little known, thirty-five page "support paper" to NSDD-75, which was released on December 6, 1982 with the simple title, "U.S. Relations with the USSR," declassified in 1997. On page thirty

is the reference to "extreme measures." Unlike NSDD-75, however, the support paper explains what is meant by "extreme measures." It states: "the West would need to consider extreme measures such as a total trade boycott, including grain." In other words, "extreme measures" refers to trade, not war.

47. Interview with Richard Pipes, September 9, 2002.

48. Interview with Richard V. Allen, September 18, 2002.

49. Interview with Caspar Weinberger, October 10, 2002.

50. In his memoirs, Weinberger used almost verbatim language, saying that at this particular meeting, Reagan told him: "Cap, I know that, and we must never be in this position again. We must regain our military strength quickly if we want to secure any kind of peace." Caspar Weinberger, *In the Arena* (Washington, DC: Regnery, 2001), 280.

51. Interview with Caspar Weinberger, October 10, 2002. Weinberger used some of this same language in very vaguely recalling this incident a month later at a conference in Washington. His statement was quite unclear, to the point that I was probably the only one in the room to fully comprehend what he was alluding to. Weinberger speaking at the conference, "Reagan's War and the War on Terrorism," hosted by the American Enterprise Institute, Washington, DC, November 13, 2002.

52. Interview with Caspar Weinberger, October 10, 2002.

53. Interview with Caspar Weinberger, October 10, 2002.

54. "The Pope and the President: A key adviser reflects on the Reagan Administration," interview with Bill Clark, *Catholic World Reporter*, November 1999.

55. Bill Clark, "President Reagan and the Wall," Address to the Council of National Policy, San Francisco, California, March 2000, 7–8.

56. Interview with Bill Clark, August 24, 2001.

57. Interview with Bill Clark, July 17, 2003.

58. Alas, in April 2005, immediately after the death of Pope John Paul II, Clark talked to one media source. He gave an exclusive interview to *Newsmax*, which was totally ignored by the mainstream press. In promoting the exclusive, *Newsmax* said that Clark had revealed for the first time that "Reagan told the pope he would use military force and go to war if Russia attempted to invade Poland." I contacted Clark to discuss the article; he offered what he noted was a key clarification: "We hadn't actually prepared to go to war. That stretches it. More accurately, we were prepared to recommend the use of force. There's an important difference." Interview with Bill Clark, April 6, 2005. *Newsmax* reporter Phil Brennan quoted Clark accurately in the article; it was the promotional material that went a bit beyond what Clark said he intended. The promotional material appeared at Newsmax.com on April 5, 2005. It ran with an accompanying article by Brennan, titled, "Adviser: Reagan Threatened War Over Poland," April 5, 2005, Newsmax.com.

59. Interview with Ed Meese, September 9, 2002.

60. Reagan, "Excerpts From a Telephone Conversation With Pope John Paul II About the Situation in Poland," December 14, 1981.

61. Documents located in ES, NSC, HSF: Records, Vatican: Pope John Paul II, RRL, Box 41, Folders "Cables 1 of 2" and 8107378–8200051.

62. December 29, 1981 letter from Ronald Reagan to Pope John Paul II. Document is located at Reagan Library, ES, NSC, HSF: Records, Vatican: Pope John Paul II, RRL, Box 41, Folder 8107378–820051. Document was declassified on July 18, 2000.

63. Reagan, *An American Life*, 303.

64. Ibid., 301.

65. Pipes, *Vixi*, 170–72.

66. Ibid.

67. Ibid.

68. Genrikh Borovik, "Plot Against Poland," *Literaturnaya Gazeta*, December 23, 1981, 14, published as "U.S. 'Lies, Hypocrisy,'" in *FBIS*, FBIS-SOV-7-JAN-82, January 7, 1982, F11–12.

69. Pipes, *Vixi*, 172–73.

70. Reagan, "Interview With the President," December 23, 1981.

71. Among other examples, see Reagan, "Proclamation 4891—Solidarity Day," January 20, 1982.

72. Reagan, "Address to the Nation About Christmas and the Situation in Poland," December 23, 1981.

73. Michael Deaver, *Behind the Scenes* (New York: William Morrow, 1988), 142–43.

74. Y. Nilov, "No Scruples . . . ," *Novoye Vremya*, January 1, 1982, 8–9, published as "Weinberger's Remarks on Poland, Church Assailed," in *FBIS*, FBIS-SOV-19-JAN-82, January 19, 1982, F1–3.

75. Valentin Zorin, "Moscow Viewpoint," December 27, 1981, published as "Zorin Commentary," in *FBIS*, FBIS-SOV-30-DEC-81, December 30, 1981, F3–5.

76. Moscow TASS, "Imperialist Interference Rebuffed," December 29, 1981, published as "Zolnierz Wolnosci on Reagan Address," in *FBIS*, FBIS-SOV-30-DEC-81, December 30, 1981, F2–3.

77. Ibid.

78. Interview with Barbara Dudek, conducted by Margie Dudek, November 2004.

79. Interview with Jan Pompowski, October 31, 2005, translated by Tomasz Pompowski.

80. Interview with Radek Sikorski, March 3, 2003.

81. Dispatch by special correspondent "UL," "Impudent Ultimatum," *Rude Pravo*, December 28, 1981, 7, published as "Events in Knurow Described," in *FBIS*, FBIS-EE-31-DEC-81, December 31, 1981, D6.

82. Reagan, "Address to the Nation About Christmas and the Situation in Poland," December 23, 1981.

83. Reagan, *An American Life*, 304.

84. I believe that the letter was a combination of two drafts, one drafted by Richard Pipes (which formed the opening paragraphs) and one drafted by the State Department. See Pipes, *Vixi*, 172–73.

85. December 23, 1981 letter from Ronald Reagan to Leonid Brezhnev, ES, NSC, HSF: Records, USSR: GSB (8190210), Box 38, RRL. Document was declassified on October 22, 1999.

86. December 23, 1981 letter, 1.

87. December 23, 1981 letter, 2.

88. December 23, 1981 letter, 2–3.

89. December 23, 1981 letter, 3.

90. Quite the contrary, the founder of the Soviet state had different means for marking the Christmas holiday. On December 25, 1919, Bolshevik godfather Vladimir Lenin had ordered: "To put up with 'Nikola' [the religious holiday commemorating the relics of St. Nikolai] would

be stupid—the entire Cheka must be on the alert to see to it that those who do not show up for work because of 'Nikola' are shot." Cited by Alexander N. Yakovlev, *A Century of Violence in Soviet Russia* (New Haven and London: Yale University Press, 2002), 157.

91. Reagan, *An American Life*, 304.

92. Ibid., 305.

93. The internal response is located at the Reagan Library in ES, NSC, HSF: Records, USSR, folder "USSR: General Secretary Brezhnev," RRL, NSC case file, Folders 8190211-8190212. See Reagan, *An American Life*, 305.

94. Ibid., 305.

95. Reagan, "Statement on U.S. Measures Taken Against the Soviet Union Concerning its Involvement in Poland," December 29, 1981.

96. These are the words of reporter Aleksandr Mozgovoy in *Sovetskaya Rossiya*, quoted by TASS, January 12, 1982, published as "U.S. Slander," in *FBIS*, FBIS-SOV-12-JAN-82, January 12, 1982, F5. Throughout subsequent years, TASS itself would dub the Reagan administration's efforts in Poland a "subversive policy." Among others, see Aleksandr Bovin, "A Face Not a Policy," *Izvestia*, January 10, 1982, 5, published as "Bovin on U.S. Poland Policy," in *FBIS*, FBIS-SOV-10-JAN-82, January 10, 1982, F5; and TASS statement from January 23, 1984, published as "Polish Press on Reagan 'Softening' Sanctions" in *FBIS*, FBIS-SOV-23-JAN-84, January 23, 1984, F7.

For Communist bloc sources on Reagan exploiting martial law as a "pretext" to undermine Communism, see Sofia BTA, by BTA observer Krasimir Drumev, "Illusory and Dangerous Course," December 30, 1981, published as "Reagan's Course: 'Illusory and Dangerous,'" in *FBIS*, FBIS-EE-31-DEC-81, December 30, 1981, C1; Prague Domestic Service, commentary by Editor Antonin Kostka, December 30, 1981, published as "U.S. Anti-Soviet Sanctions Further Denounced," in *FBIS*, FBIS-EE-31-DEC-81, December 30, 1981, D1; Prague Domestic Service, commentary by Correspondent Michal Stasz, December 30, 1981, published as "Only America Will Suffer," in *FBIS*, FBIS-EE-31-DEC-81, December 30, 1981, D2–3; Bratislava *Pravda* in Slovak, by Milan Rusko, "American Hegemonism, Reagan Version," December 28, 1981, published as "Reagan Message Criticized," in *FBIS*, FBIS-EE-31-DEC-81, December 31, 1981, D3; Moscow TASS statement, December 30, 1981, published as "Reagan's 'Discriminatory Measures' Condemned," in *FBIS*, FBIS-SOV-30-DEC-81, December 30, 1981, F2; and Zdenek Porybny, "Pressure Will Not Succeed," *Rude Pravo*, December 27, 1981, 7, published as "U.S. Declares Economic War on Poland," in *FBIS*, FBIS-EE-31-DEC-81, December 31, 1981, D5.

97. Statement by TASS, January 7, 1982 (0921 GMT), published as "U.S. 'Interference' in Poland Criticized," in *FBIS*, FBIS-SOV-7-JAN-82, January 7, 1982, F8–9.

98. Moscow Domestic Television Service, "Studio 9," January 30, 1982, published as "'Studio 9:' Reagan, Roosevelt Compared," in *FBIS*, FBIS-SOV-1-FEB-82, February 1, 1982, CC4.

99. Reagan, "Statement on U.S. Measures Taken Against the Soviet Union Concerning its Involvement in Poland," December 29, 1981.

100. Reagan, "Excerpt From an Exchange With Reporters on the Situation in Poland," December 29, 1981.

101. Barrett, *Gambling With History*, 298.

102. According to Richard Pipes, at a conference years later, Jaruzelski revealed that these

and subsequent sanctions had cost Poland $12 billion, which was an enormous sum to a country of that size. Pipes, *Vixi*, 173.

103. The December 31, 1981 telegram is now filed in the Reagan Library in Folder 22, Box 2, PHF: Presidential Records.

104. See: Memorandum of Conversation, "Meeting with President Mitterand of France," October 27, 1982, prepared by Ambassador Evan Galbraith, declassified July 26, 2000, on file at the Reagan Library.

105. Schweizer, *Victory*, 29, 31.

106. Bernstein and Politi, *His Holiness*, 262.

107. Moscow TASS statement, December 30, 1981, published as "Reagan's 'Discriminatory Measures' Condemned," in *FBIS*, FBIS-SOV-30-DEC-81, December 30, 1981, F1–2.

108. Ibid.

109. This Genscher statement was quoted in the Bulgarian press. Sofia BTA, by BTA observer Krasimir Drumev, "Illusory and Dangerous Course," December 30, 1981, published as "Reagan's Course: 'Illusory and Dangerous,'" in *FBIS*, FBIS-EE-31-DEC-81, December 30, 1981, C2.

110. Zdenek Porybny, "Pressure Will Not Succeed," *Rude Pravo*, December 27, 1981, 7, published as "U.S. Declares Economic War on Poland," in *FBIS*, FBIS-EE-31-DEC-81, December 31, 1981, D5.

111. Article by Aleksey Petrov, "Dullness and Not Adamance," *Pravda*, January 11, 1982, circulated by Warsaw PAP, January 12, 1982, published as "U.S. Desire for Socialist 'Split' Scored," in *FBIS*, FBIS-SOV-13-JAN-82, January 13, 1982, G14.

112. Kornilov statement circulated by TASS, May 11, 1982, published as "Reagan Fabrications on Soviet-Polish Ties," in *FBIS*, FBIS-SOV-12-MAY-82, May 12, 1982, A1.

113. For these reasons, said Rostowski, writing in caps for emphasis, the Reagan administration was imposing sanctions: "THESE ARE PROBABLY THE ACTUAL SOURCES AND CAUSES OF THE ECONOMIC RESTRICTIONS AGAINST POLAND AND THE ENTIRE POLISH NATION. THEY ARE ALSO THE CAUSES OF ALL THOSE POLITICALLY IRRESPONSIBLE AND NERVOUS DECISIONS ADOPTED BY THE U.S. ADMINISTRATION TO EXPAND ECONOMIC AND POLITICAL SANCTIONS ON THE SOVIET UNION." Rostowski said that the U.S. goal was to weaken the economic and political links between Eastern Europe and the Soviet Union. Adam Rostowski, "The Camouflaged U.S. Political Plans," *Zolnierz Wolnosci*, January 9–10, 1982, published as "U.S. Designs, Intentions," in *FBIS*, FBIS-SOV-13-JAN-82, January 13, 1982, G16–17.

CHAPTER 9

1. These three articles were: "Solidarity Activist Appeals to Troops To Defy Authority;" "Stunned Militant: 'It's Our Own Army. We Are Confused;' " and "Allen's Job Expected to Go to Clark," *Washington Post*, January 1, 1982, A1.

2. Reagan, "Remarks to the People of Foreign Nations on New Year's Day," January 1, 1982.

3. Ibid.

4. Clark in Schweizer, ed., *Fall of the Berlin Wall*, 69.

5. See: Edmund Morris interviewed by *American Enterprise*, November/December 1999; Maureen Dowd, "The Man with the President's Ear," *Time*, August 8, 1983; and Steven R. Weisman, "The Influence of William Clark," *New York Times Magazine*, August 14, 1983.

6. Of these two Clark memos to Reagan, one is dated January 11, 1982 and the other is not dated. It is unquestionably a February 1982 document, probably February 23. The documents are located at the Reagan Library, ES, NSC, HSF: Records, Vatican: Pope John Paul II, RRL, Box 41, Folders 8107378-820051 and 8200555-8204184. The two documents were declassified on July 18, 2000.

7. February 1982 letter from Ronald Reagan to Pope John Paul II. The exact date was not featured on the draft on file. The date was probably February 23. Document is located at Reagan Library, ES, NSC, HSF: Records, Vatican: Pope John Paul II, RRL, Box 41, Folder 8200555-8204184. Document was declassified on July 18, 2000.

8. Bernstein and Politi, *His Holiness*, 260.

9. Interview with Bill Clark, July 17, 2003; and Bill Clark, "President Reagan and the Wall," Address to the Council of National Policy, San Francisco, California, March 2000, 7–8.

10. Interview with Bill Clark, February 14, 2005.

11. Bill Clark, "President Reagan and the Wall," Address to the Council of National Policy, San Francisco, California, March 2000, 2.

12. Ibid., 3–5.

13. Norman A. Bailey, *The Strategic Plan That Won the Cold War: National Security Decision Directive 75* (McLean, VA: The Potomac Foundation, 1999), ii.

14. Another Californian from the gubernatorial days, speaks to Clark's role. Noting that Reagan was ready to make "a concerted effort, as part of a coherent overall strategy, to win the Cold War and consign the Soviet system to 'the ash heap of history,'" Weinberger says that this goal now proceeded full throttle under Bill Clark. Weinberger, *In the Arena*, 285–87.

15. Pipes, *Vixi*, 204.

16. Aleksey Petrov, "Not Firmness But Thick-Headedness," *Pravda*, January 9, 1982, circulated by TASS, January 10, 1982, printed as "Petrov Article Assails U.S. Course on Poland," in FBIS-SOV-11-JAN-82, January 11, 1982, F2.

17. Moscow Domestic Service, March 17, 1983. Transcript published as "Directive 75 'Subversive' Anti-Soviet Plan" in FBIS-SOV-18-MAR-83, March 18, 1983, A8.

18. Richard Pipes, "Misinterpreting the Cold War," *Foreign Affairs*, January/February 1995, 157.

19. Pipes, *Vixi*, 165, 168.

20. Reagan, "Proclamation 4891—Solidarity Day," January 20, 1982.

21. See, among others: Reagan, "Remarks at the New York City Partnership Luncheon," New York, NY, January 14, 1982; "State of the Union," 1982; and Reagan, "Remarks at the Centennial Meeting of the Supreme Council of the Knights of Columbus," Hartford, Connecticut, August 3, 1982.

22. For another example, see Reagan, "Address to the Conservative Political Action Conference," Washington, DC, February 26, 1982, in Roberts, ed., *A City Upon a Hill*, 70.

23. Ibid., 70.

24. Reed, *At the Abyss*, 226–28, 239.

25. The speech is dated circa 1963. It is called "Are Liberals Really Liberal?" A copy is located in the Hoover Institution Archives in Ronald Reagan Subject Collection, Box 1, which contains pre-1966 speeches and writings. A full transcript in Skinner, Anderson, and Anderson, eds., *Reagan, In His Own Hand*, 438–42.

26. Schweizer, *Victory*, 5.

27. Reagan, *An American Life*, 237.

28. This Reagan's diary entry from March 26, 1982. Ibid, 316.

29. Ibid, 237–38.

30. Ibid, 237.

31. Ibid, 552.

32. Owen Smith, Casey's son-in-law, observes that Casey was an actual "DCI"—the formal title of the head of the CIA—more than a "DCIA," in the sense that he was more a Director of Central Intelligence (DCI) than a Director of the Central Intelligence Agency (DCIA). Conversation with Owen Smith, Washington, DC, May 10, 2000.

33. This was a natural extension of Casey's early career training, having practiced economic warfare against the Nazis when he was at the Office of Strategic Services, the CIA's World War II predecessor organization. He saw the Bolsheviks as replacing one totalitarian system (Hitler's) with their own. Pointing to Communist countries like Cambodia, where 2 to 3 million out of a population of 5 to 7 million was killed or starved to death in just four years in the latter 1970s, Casey said that Marxist regimes were responsible for a "holocaust comparable" to that of Nazi Germany. Casey stated this during an October 27, 1986 speech to the John Ashbrook Center for Public Affairs at Ashland College in Ashland, OH. In Mark B. Liedl, ed., *Scouting the Future: The Public Speeches of William J. Casey* (Washington, DC: Regnery, 1989), 31–39.

34. In a May 1981 speech (Casey wrote his own speeches)—very early in the life of the Reagan administration—Casey said, "The Soviet economy is gasping under its inherent inefficiencies." While informing his audience that his CIA would devote "a large slice" of funding to conventional activities, he added: "But they will have to be supplemented by increased efforts to assess economic vulnerabilities." See Casey address to the Annual Meeting of the Business Council, HS, VA, May 9, 1981, in Liedl, *Scouting the Future*, 8, 16–24.

35. Clark has spoken to this a number of times. Among others, see Schweizer, *Victory*, xvi, xix. Reagan gave Casey the green light to do countless things that a DCI is not supposed to do. Some will protest that that was a big problem. Others will say it was critical in damaging the USSR. None can deny the influence Casey was granted.

36. "Untouchable Crook: Political Profile of CIA Director W. Casey," *Sovetskaya Rossiya*, December 17, 1986, published as "Sovetskaya Rossiya Profiles CIA Director Casey," in *FBIS*, FBIS-SOV-31-Dec-86, December 31, 1986, A2–A5.

37. Casey said this in his October 1986 speech at Ashland College. A *Pravda* article (which follows) said he made the claim in San Antonio, Texas as well.

38. V. Korionov, "Rejoinder: Just You Try!," *Pravda*, May 22, 1985, 5, published as "Casey's Speech in San Antonio Attacked in Pravda," in *FBIS*, FBIS-24-MAY-85, May 24, 1985, A2.

39. Casey delivered this speech in October 1986. See Liedl, *Scouting the Future*, 36.

40. Derek Leebaert, *The Fifty-Year Wound* (Boston: Little, Brown, 2002), 503, 694n. Leebaert's sources are Norm Bailey and Richard Perle.

41. Ibid., 524.

42. Quoted in Schweizer, *Victory*, xv.

43. Clark notes that Reagan "clearly saw that security issues and the economy were inextricably linked." This was certainly true at home, but he also understood the link as it applied to the USSR. Reagan, said Clark, "pronounced this many times, both as it related to the Soviet Union and to ourselves." Bill Clark, "President Reagan and the Wall," Address to the Council of National Policy, San Francisco, California, March 2000, 3–5.

44. As Clark put it, Reagan "worked closely" with these men "to configure a security-minded economic strategy that would constrict financial and other forms of Western life-

support being tapped by the Kremlin." See Bailey, *The Strategic Plan That Won the Cold War*, ii.

45. Interviews with Roger Robinson, June 6 and 8, 2005.

46. On Reagan addressing this, see Reagan, "Interview With Julius Hunter of KMOX-TV," St. Louis, Missouri, July 22, 1982.

47. Interview with Cap Weinberger, October 10, 2002.

48. Interview with Gus W. Weiss, November 13, 2002.

49. The information in this section is taken from material provided to me by Weiss as well as my interviews with Weiss. Among the materials, I've relied primarily on an article titled, "Duping the Soviets: The Farewell Dossier," published by Weiss in 1996 in a CIA journal called *Studies in Intelligence*. Weiss also provided me with a strange, poorly written and poorly organized forty-eight-page bound document titled, "The Farewell Dossier: Strategic Deception and Economic Warfare in the Cold War," published by Weiss through the American Tradecraft Society, which Weiss told me was a kind of memoir, and which I did not rely upon. I did interviews with Weiss on November 13, 14, 17, 26, and 30, 2002 and on February 18, 2003 (among others), mainly through e-mail, and spoke to him in person at a function at the American Enterprise Institute in Washington, DC in early November 2002.

Information on the *Farewell Dossier* has since been published in a *Washington Post* article by David E. Hoffman and in books by Thomas Reed and Derek Leebaert. Like Weiss, Reed is a primary source. Reed was granted permission from the CIA to share his information. All non-footnoted material in this section comes from the materials that Weiss shared with me and from my interviews with him.

50. Weiss, "Duping the Soviets: The Farewell Dossier," *Studies in Intelligence*, 124; and Reed, *At the Abyss*, 266–67.

51. Decades earlier, the Soviets had great success in penetrating American atomic research. Now, they were interested in all technologies, not just for military reasons but mainly for economic ones.

52. Weiss, "Duping the Soviets: The Farewell Dossier," *Studies in Intelligence*, 124.

53. Reed, *At the Abyss*, 267.

54. Interviews with Weiss and Reed, *At the Abyss*, 267–68, 270.

55. On this, see Weiss, "Duping the Soviets: The Farewell Dossier," *Studies in Intelligence*, 125; David E. Hoffman, "Reagan Approved Plan to Sabotage Soviets," *Washington Post*, February 27, 2004, A1; and Reed, *At the Abyss*, 266–70.

56. Interviews with Weiss and Reed, *At the Abyss*, 267–68, 270.

57. Information on the directives has been published in a variety of sources. Perhaps the best is Christopher Simpson, *National Security Directives of the Reagan and Bush Administrations: The Declassified History of U.S. Political and Military Policy, 1981–1991* (Boulder, CO: Westview Press, 1995). The directives are also on file at the Reagan Library, which is where I read most of them.

58. Bill Clark, "President Reagan and the Wall," Address to the Council of National Policy, San Francisco, California, March 2000.

59. Interview with Bill Clark, August 24, 2001.

60. Pipes, "Misinterpreting the Cold War," 157.

61. Shultz interviewed on "The Point with Greta Van Susteren," CNN, February 6, 2001. Shultz said the same to me in a July 15, 2003 interview in Palo Alto, California.

62. NSDD-24, February 9, 1982. NSDD is on file at the Reagan Library. See Simpson, *NSDDs of Reagan and Bush*, 58; and Richard Perle, "Department of Defense Position on the Soviet-West European Natural Gas Pipeline," November 21, 1981, *AFPCD 1981*, 431–34.

63. NSDD-24 was a follow-up to the White House's December 30, 1981 ban on U.S. exports that might assist the pipeline project, an embargo implemented in response to the Communist imposition of martial law in Poland.

64. It is not clear who, if anyone, Reagan was quoting here, though he appeared to be quoting someone. Reagan, "Remarks at the AFL-CIO," April 5, 1982.

65. Reagan, "Remarks to Citizens in Hambach, Federal Republic of Germany," May 6, 1985.

66. Reagan, "Address at Commencement Exercises at Eureka College," Eureka, Illinois, May 9, 1982.

67. NSDD-32 was a summary of a report released in April 1982. The NSDD and the report are both on file at the Reagan Library. The NSDD was declassified in February 1996.

68. Carl Bernstein, "The Holy Alliance," *Time*, February 24, 1992, 31.

69. Keith Schneider, "Reagan-Pope Plan to Topple Warsaw is Reported," *New York Times*, February 18, 1992; Bernstein, "The Holy Alliance," 28–35; Michael Ledeen, "This Political Pope," *The American Enterprise*, 4, no. 4 (July 1993): 40–43; Schweizer, *Victory*, xviii, 68–69; and Raymond L. Garthoff, *The Great Transition: American-Soviet Relations and the End of the Cold War* (Washington, DC: Brookings Institution, 1994), 31–32.

70. Bernstein, "The Holy Alliance," 31.

71. Joseph E. Persico, *Casey: From the OSS to the CIA* (New York: Viking, 1990), 236; Bernstein, "The Holy Alliance," 28–35; and Schweizer, *Victory*, xviii, 68–69.

72. This aim of NSDD-32, which related to Poland and the Communist bloc generally, was evident in the larger April 1982 study from which NSDD-32 was derived. The study staked out this U.S. objective: "to increase the costs of Soviet repression of popular movements and institutions in Poland and other East European countries; and to maximize prospects for their independent evolution." This, too, was striking language. Report, "U.S. National Security Strategy," April 1982, 6.

73. Yuri Zhukov, "A Doctrine of Interference," *Pravda*, June 25, 1982, 4, published as "Reagan Policy Called 'Doctrine of Interference,' " in FBIS-6-JUL-82, July 6, 1982, A1–3.

74. Ye. Rusakov, "War Games and 'Peace' Games: The Path Chosen by the Reagan Administration," *Pravda*, June 1, 1984, 4, published as "Rusakov: Reagan Playing Games With War, Peace," in FBIS-5-JUN-84, June 5, 1984, A3.

75. Richard Halloran, "Reagan Aide Tells of New Strategy on Soviet Threat," *New York Times*, May 22, 1982, A1.

76. Reed delivered his speech on June 16, 1982. Reed, *At the Abyss*, 237.

77. Richard Halloran, "Pentagon Draws Up First Strategy for Fighting a Long Nuclear War," *New York Times*, May 30, 1982, A1.

78. Ibid.

79. Ibid.

80. Ibid.

81. Ibid.

CHAPTER 10

1. On this, see Kengor, *God and Ronald Reagan*.

2. Reagan expressed his shock and prayers to the pope in a May 13, 1981 cable on the

assassination attempt, in ES, NSC, HSF: Records, Vatican: Pope John Paul II, RRL, Box 41, Folder "Cables 1 of 2."

3. Interview with Bill Clark, August 24, 2001.

4. Bernstein, "The Holy Alliance," 31

5. Ibid., 28, 30.

6. See "The Pope and the President: A key adviser reflects on the Reagan Administration," interview with Bill Clark, *Catholic World Reporter*, November 1999; and Bernstein, "Holy Alliance," 30.

7. Bernstein, "The Holy Alliance," 29, 35.

8. Ibid.

9. Quoted in Schweizer, *Reagan's War*, 213.

10. Bernstein, "The Holy Alliance," 28–35.

11. Reagan, "Remarks Following a Meeting With Pope John Paul II in Vatican City," June 7, 1982.

12. Ibid.

13. Ibid.

14. Bernstein, "The Holy Alliance," 28–35.

15. Ibid.

16. Bernstein and Politi, *His Holiness*, 260.

17. Ibid., 12.

18. This probably began as early as the spring of 1981, well before the June 1982 meeting.

19. Bernstein and Politi, *His Holiness*, 268–9.

20. Ibid.

21. Weigel, *Witness to Hope*, 622.

22. The cable is dated only "FEB 84." The exact date is unclear, though it is before Reagan's February 22 response letter to the pope. The cable is located in ES, NSC, HSF: Records, Vatican: Pope John Paul II, RRL, Box 41, Folder 8490136-8490538. The cable was declassified on July 18, 2000.

23. ES, NSC, HSF: Records, Vatican: Pope John Paul II, RRL, Box 41, Folder 8490136-8490538.

24. Ibid.

25. This February 22, 1984 letter is located in ES, NSC, HSF: Records, Vatican: Pope John Paul II, RRL, Box 41, Folder 8490136-8490538. The letter was declassified on July 18, 2000.

26. Though Reagan did not lift all of the sanctions, he modified them.

27. Ibid.

28. Calculating the exact number is difficult. Nancy Reagan says that she met with the pope seven times. In all likelihood, her husband equaled or exceeded that number. Nancy Reagan interview with CNN's Wolf Blitzer, "Nancy Reagan: Pope 'was very special,'" CNN, April 3, 2005.

29. Ibid., Nancy said that the pope told her this in a letter that he sent upon Reagan's death in June 2004.

30. The individual said this to me personally. Obviously, he preferred to remain anonymous. I include the quote as a reflection of the harsh feelings. This person was one of Reagan's very top aides.

31. This person also preferred anonymity. I include the words for the same reason cited in the previous note.

32. Agostino Bono, "Officials say pope, Reagan shared Cold War data, but lacked alliance," Catholic News Service, November 17, 2004.

33. George Weigel, "The President and the Pope," *National Review*, June 28, 2004.

34. Interview with Bill Clark, August 24, 2001.

35. Asked about Dolan's and Reagan's tough words, Ben Elliott, director of White House speechwriting, speaks to their larger motivation: "There is no doubting that it was implicitly understood by us that, unlike the '70s under Nixon, Ford, and Carter, there was a new policy and it was: we win, they lose." Interviews with Ben Elliott, October 13, 2004 and January 11, 2005.

36. The May 19 draft is located at the Reagan Library in PHF, PS, RRL, Box 5, Folder 83.

37. The phrase "ash heap of history" was a clever twist of Trotsky-speak. The phrase has also been attributed to Lenin, though it is more commonly credited to Trotsky.

38. The May 24 draft is located at the Reagan Library in PHF, PS, RRL, Box 5, Folder 85.

39. Cannon, "Reagan Radiated Happiness and Hope," *George* magazine, August 2000, 58.

40. Quoted by Steven Rottner, "Britons Reassured by Reagan's Visit," *New York Times*, June 10, 1982, A17.

41. Richard Pipes used these words. Pipes chronicled Soviet reactions to the speech for the NSC. When he reported the reactions to Reagan, the president responded; "So we touched a nerve." Pipes, *Vixi*, 200.

42. S. Volovets, "Megaphone Diplomacy," *Sovetskaya Rossiya*, June 17, 1987, 5, published as "Reagan Practices 'Megaphone Diplomacy' in W. Berlin," in *FBIS*, FBIS-SOV-18-Jun-87, June 18, 1987, A1.

43. Statement by the Moscow Domestic Service, February 7, 1984, published as "U.S. 'Democracy Campaign' Increasing World Tension," in *FBIS*, FBIS-8-FEB-84, February 8, 1984, A5. I've included here a small sample of similar reactions by Soviets in regard to the Westminster speech.

44. Vitaliy Korionov, "20th Century 'Crusaders,'" *Pravda*, July 14, 1982, 4, published as "Pravda's Korionov Denounces U.S. 'Crusades,'" in *FBIS*, FBIS-26-JUL-82, July 26, 1982, A5–6.

45. Report language is quoted by Leonid Zamyatin in his article, "The Washington Crusaders: The 'Ideological War' Declared by Reagan Against Communism and Socialism . . . ," *Literaturnaya Gazeta*, June 30, 1982, 14, published as "Zamyatin Scores Reagan 'Ideological War' Plan," in *FBIS*, FBIS-14-JUL-82, July 14, 1982, A1–8.

46. "Reagan Makes 'Anti-Soviet' Speech to UK MPs," Warsaw Domestic Service, June 8, 1982, printed in FBIS, Eastern Europe, June 9, 1982, G1.

47. Editorial, "Another Anticommunist Fit," *Pravda*, July 22, 1982, 4, published as "Reagan's 'Captive Nations' Speech Slammed," in *The Current Digest of the Soviet Press*, 34, no. 29 (1982): 10–11.

48. A review of 1982 by *Pravda*, published in the January 3, 1983 edition, sweepingly said that, "The year 1982 [was] . . . a year when the U.S. President publicly called for a 'crusade' against the USSR, that is, for a political, economic, and ideological offensive against real socialism." Vsevolod Ovchinnikov, "International Review," *Pravda*, January 3, 1983, 4. Text is published as "Pravda Views International Events of 1982,'" in FBIS-SOV-5-JAN-83, January 5, 1983, CC1.

CHAPTER 11

1. In May, General Secretary Brezhnev said at a major KGB conference that Reagan was committed to "a further expansion of the arms race and. . . . working to undermine the Soviet economy." He said that Reagan wanted to "erase the gains of international socialism through provocations" and "economic warfare." Minutes after Brezhnev's statement, KGB head Yuri Andropov stepped to the podium to call the advent of the Reagan administration a sign of "dangerous times." Schweizer, *Victory*, xi, 40–41.

2. The Clark quote from Dobrynin comes from: Bill Clark, "President Reagan and the Wall," Address to the Council of National Policy, San Francisco, California, March 2000, 3–5. Clark does not recall the exact date.

3. Haig's reaction is recounted in Pipes, *Vixi*, 179.

4. This is one of a number of examples offered by Cannon himself which undermine Cannon's portrayal of Reagan in *Role of a Lifetime* as an easily manipulated leader led around by his advisers. The president with almost no ideas of his own, according to Cannon's first edition of the book, who was controlled by his advisers, was yet again, by Cannon's own account, in the same book where he alleged the lack of ideas and control by Reagan, was in fact himself the main obstacle. In other words, Reagan himself was in control of the decision and direction. Lou Cannon, *President Reagan: The Role of a Lifetime* (New York: PublicAffairs Books, second edition, 2000), 167.

5. Reagan, "Interview With Julius Hunter of KMOX-TV," St. Louis, Missouri, July 22, 1982.

6. Reagan, "Responses to Questions Submitted by Bunte Magazine," April 25, 1983.

7. Reagan, "Interview With Julius Hunter of KMOX-TV," St. Louis, Missouri, July 22, 1982.

8. The 70 percent estimate was provided by Roger Robinson. Interview with Roger Robinson, June 6, 2005. The 50 percent figure was provided by Richard Pipes, who says that it was Bill Casey's estimate. Pipes, *Vixi*, 180.

9. Data provided by Roger Robinson. Interview with Roger Robinson, June 6, 2005.

10. Interviews with Roger Robinson, June 6 and 8, 2005.

11. Located in "Ronald Reagan: Pre-Presidential Papers: Selected Radio Broadcasts, 1975–1979," October 31, 1978 to October 1979, Box 4, RRL. For transcript, see Skinner, Anderson, and Anderson, eds., *Reagan, In His Own Hand*, 73–74. Alexander Solzhenitsyn was fond of making this same point. See Solzhenitsyn, *Alexander Solzhenitsyn Speaks to the West*, 69.

12. In a July 9, 1979 radio broadcast, Reagan said: "Several weeks ago a Commerce Department official whose job is to monitor the sale of advanced technology to the Soviet Union so as to guard against giving them something that would be used militarily, blew the whistle on his own department." He continued: "He said our system of export controls is a 'total shambles.'" Reagan spoke of the debate in the Carter administration over selling "advanced American products to the Soviets." He took aim at the dovish Carter State Department and Secretary of State, who favored increased trade. He sided with the more hawkish Carter DOD and Zbigniew Brzezinski's NSC: "Understandably aides in the Defense Department and the National Security Council are opposed." He concluded with this dire warning: "Maybe we should remember World War II when a former trading partner returned tons of our scrap iron in the form of shrapnel that killed our young men. What we are talking about today isn't scrap iron.

Mr. [Lawrence] Brady [director of the Office of Export Administration] says that last year only a few hundred of the 7000 requests to sell our products to the Soviet-bloc nations were turned down. We are supposed to insure that nothing we sell can be diverted to military use but that is virtually impossible to do. Truck motors turning up in assault vehicles is proof of that." Skinner, Anderson, and Anderson, eds., *Reagan: A Life in Letters*, 73–74. This was the polar opposite of the sort of economic warfare Reagan wanted to pursue.

13. Reagan, *An American Life*, 558.

14. Ibid., 316.

15. Reagan, "Interview in New York City With Members of the Editorial Board of the *New York Post*," March 23, 1982.

16. Quoted by Schweizer, *Victory*, xiv.

17. Pipes, *Vixi*, 179–80.

18. Reagan, *An American Life*, 320.

19. Jim Hoagland, "France Refuses to Wage Economic War on Soviets," *Washington Post*, June 15, 1982, A1.

20. Reagan stood and left the room, as expressions of momentary pleasure quickly transformed into looks of horror. "They were crestfallen," Roger Robinson remembers. "Those were [Reagan's] last words at the meeting. And I will never forget them." Interviews with Roger Robinson, June 6 and 8, 2005.

21. "A. A. Gromyko's Press Conference in New York," TASS, printed in both *Pravda* and *Izvestia*, June 22, 1982, 5 and 4, respectively. Published as "Gromyko: US Is 'Blowing Up Bridges," in *The Current Digest of the Soviet Press*, 34, no. 25 (1982): 6.

22. "Lies as Policy, or the Policy of Lies," *Za Rubezhom*, August 5, 1982, 5, published as "U.S. Resorts to Foreign 'Policy of Lies,'" in *FBIS*, FBIS-13-AUG-82, August 13, 1982, A1–2. The author of this article was listed only as "D. K."

23. The editorial continued: "As yet it is hard to say whether or not such a solution will be found. One thing is clear: The present American leadership is refusing to abandon the shameful role of destroyer of everything positive accumulated over the preceding period in relations between states with different social systems." Editorial, "Normal International Contact Versus Adventurism of 'Sanctions,'" *Za Rubezhom*, September 9, 1982, 1, published as "Za Rubezhom Scores Reagan Policy on Pipeline," in *FBIS*, FBIS-15-SEP-82, September 15, 1982, CC1–2.

24. Reagan, "The President's News Conference," June 30, 1982.

25. Reagan, "Interview With Julius Hunter of KMOX-TV," St. Louis, Missouri, July 22, 1982.

26. The Soviets were "very hard pressed economically," he said openly on October 18, 1982. The pipeline's construction would help bail them out. Reagan, "Remarks and a Question-and-Answer Session via Satellite to Republican Campaign Events," October 18, 1982.

27. Reagan, "Remarks at an Illinois Republican Party Rally," Peoria, Illinois, October 20, 1982.

28. Reagan said this at the June 18, 1982 meeting. Quoted by Laurence I. Barrett, *Gambling with History: Ronald Reagan in the White House* (New York: Doubleday, 1983), 300.

29. V. Shmyganovsky, "On the Basis of Mutuality," *Izvestia*, October 19, 1983, 2. Reprinted in *The Current Digest of the Soviet Press*, 35, no. 42 (1983): 16.

30. I will be detailing this very candid meeting at length in my upcoming biography of Bill Clark. Also on this, see Barrett, *Gambling with History*, 299–302.

31. Among others, see Pipes, *Vixi*, 181, 208; Reed, *At the Abyss*; and Milt Bearden, *The Main Enemy* (New York: Random House, 2003).

32. Interview with Bill Clark, February 14, 2005.

33. One can find this exact objection particularly in TASS press releases at the time and in the pages of *Pravda* and *Izvestia*.

34. Memorandum of Conversation, "Meeting with President Mitterand of France," October 27, 1982, prepared by Ambassador Evan Galbraith, declassified July 26, 2000, on file at the Reagan Library.

35. Barrett, *Gambling with History*, 301.

36. Ibid., 302.

37. Reagan radio address, "East-West Trade Relations and the Soviet Pipeline Sanctions," November 13, 1982, in Fred L. Israel, ed., *Ronald Reagan's Weekly Radio Addresses, Vol. 1* (Wilmington, DE: Scholarly Resources, 1987), 45.

38. Leebaert, *The Fifty-Year Wound*, 524.

39. Weiss, "Duping the Soviets: The Farewell Dossier," *Studies in Intelligence*, 125.

40. Reed, *At the Abyss*, 268–69.

41. See David E. Hoffman, "Reagan Approved Plan to Sabotage Soviets," *Washington Post*, February 27, 2004, A1; and Reed, *At the Abyss*, 266–70.

42. Weiss, "Duping the Soviets: The Farewell Dossier," *Studies in Intelligence*, 125.

43. Referring to Reagan's Captive Nations speech, a July 1982 editorial in *Pravda* stated, "Without a trace of embarrassment, the speech predicts the impending collapse of this ideology," referring to Marxism-Leninism. Editorial, "Another Anticommunist Fit," *Pravda*, July 22, 1982, 4. Published as "Reagan's 'Captive Nations' Speech Slammed," in *The Current Digest of the Soviet Press*, 34, no. 29 (1982): 10.

44. Schweizer, *Reagan's War*, 184–86.

45. NSDD-41, June 22, 1982. NSDD is on file at the Reagan Library. See Simpson, *NSDDs of the Reagan and Bush Administrations*, 68–69; and George Shultz, *Turmoil and Triumph* (New York: Scribner, 1993), 136.

46. NSDD-48, July 23, 1982. NSDD is on file at the Reagan Library.

47. Simpson, *NSDDs of the Reagan and Bush Administrations*, 71–72.

48. Interviews with Roger W. Robinson, June 6 and 8, 2005.

49. Ibid.

50. Leebaert, *The Fifty-Year Wound*, 523.

51. Interviews with Roger W. Robinson, June 6 and 8, 2005.

52. Bailey, *The Strategic Plan That Won the Cold War*, 8.

53. Ibid., 5.

54. Reagan, *An American Life*, 559.

55. NSDD-54, September 2, 1982. NSDD is on file at the Reagan Library.

56. NSDD-54, 2–4.

57. "After Détente, the Goal is to Prevail," *New York Times*, September 23, 1982, B16.

58. Ibid.

59. Reagan, *An American Life*, 559.

60. Bill Clark frequently emphasizes that almost all of the products that came out of his NSC, from drafts of directives to speeches and more, were a "team effort." He is reluctant to credit single individuals as sole authors, including himself. Various interviews with Bill Clark.

61. Interviews with Roger W. Robinson, June 6 and 8, 2005.

62. Peter Schweizer, *Victory*, 125–27.

63. Roger Robinson is the source for this item. He says that the group of individuals was roughly a dozen. See Robinson interviewed on the documentary *In the Face of Evil: Reagan's War in Word and Deed* (American Vantage Films and Capital Films I, LLC, 2005).

64. Ibid. Robinson makes this point. These few individuals are quoted throughout this book, and obviously include the likes of Bill Clark and Bill Casey.

65. The latter was the title used by Christopher Simpson on page 80 of his book on Reagan's NSDDs.

66. NSDD-66, November 29, 1982, 1. NSDD is on file at the Reagan Library.

67. Ibid., 1–2.

68. "Summary of Conclusions" addendum to NSDD-66, 1. On file at Reagan Library with NSDD-66.

69. NSDD-66, 2.

70. Schweizer, ed., *Fall of the Berlin Wall*, 34–38.

71. U.S. Department of Commerce, International Trade Administration, "Quantifications of Western Exports of High-Technology Products to Communist Countries Through 1983," 1985, 12, 28–29; and "Geheimclub COCOM," *Die Zeit*, October 10, 1983, 34. Credit goes to Peter Schweizer for locating these sources.

72. For more, see Schweizer's various works.

73. Quoted in Don Oberdorfer, *The Turn* (New York: Poseidon, 1991), 76.

74. NSDD-70, November 30, 1982, 1–4. NSDD is on file at the Reagan Library.

CHAPTER 12

1. Vladimir Serov, "The Plans of Nuclear 'Crusaders,' " distributed by TASS, January 17, 1983. Text is published as "Serov: Directive 'Crusade Against Communism,' " in FBIS-SOV-18-JAN-83, January 18, 1983, A2.

2. Though formally issued January 17, 1983, the document was approved by Reagan at an NSC meeting on December 17, 1982. See Bailey, *The Strategic Plan That Won the Cold War*, 13.

3. Those words are actually the full name of a monograph on NSDD-75 by Bailey.

4. Reed, *At the Abyss*, 240.

5. Interview with Richard Pipes, September 9, 2002. Pipes concedes that NSDD-75 was his "main contribution to the Reagan administration's foreign policy." Pipes, *Vixi*, 188. Roger Robinson contributed the sections focused on economic and financial warfare. Interviews with Roger W. Robinson, June 6 and 8, 2005.

6. Pipes stated this in an interview with Peter Schweizer. Schweizer, *Victory*, 131.

7. Clark in Schweizer, ed., *The Fall of the Berlin Wall*, 71–73.

8. NSDD-75, January 17, 1983, 1. NSDD is on file at Reagan Library.

9. Interview with Richard Pipes, September 9, 2002.

10. Interview with Richard Pipes, September 27, 2005.

11. Richard Pipes, *Survival Is Not Enough* (New York: Simon and Schuster, 1984), 262–64.

12. Clark in Schweizer, ed., *The Fall of the Berlin Wall*, 72.

13. Quote is taken from NSDD-75 third to last bullet on page 2.

14. McFarlane was interviewed by Peter Schweizer. Quoted in Schweizer, *Victory*, 132.

15. The two releases from the Moscow Domestic Service were released at 1940 and 2015

GMT on March 17, 1983. Transcripts are published as " 'Economic, Military Blackmail' " and "Directive 75 'Subversive' Anti-Soviet Plan" in *FBIS*, FBIS-SOV-18-MAR-83, March 18, 1983, A8–9.

16. G. Dadyants, "Pipes Threatens History," *Sotsialisticheskaya Industriya*, March 26, 1983, 3, published as "New Directive on USSR Trade 'Threatens History,' " in *FBIS*, FBIS-SOV-29-MAR-83, March 29, 1983, A6–7.

17. For the record, Pipes did not learn of the *Sotsialisticheskaya Industriya* article until over twenty years after its publication, and was perplexed how a Soviet journalist had learned of the highly classified NSDD-75. Pipes told me: "Are you sure of this? NSDD-75 was highly classified—how would a Soviet journalist hear about it?" Correspondence with Richard Pipes, December 24, 2005.

18. Reagan, "Remarks at the National Leadership Forum of the Center for Strategic and International Studies of Georgetown University," April 6, 1984.

19. He revisited this "economy first" point throughout 1984–85. Among others, see Reagan, "Address to the Nation and Other Countries on United States-Soviet Relations," January 16, 1984; "Interview With Lou Cannon, David Hoffman, and Juan Williams of the *Washington Post* on Foreign and Domestic Issues," January 16, 1984; "Remarks at the National Leadership Forum of the Center for Strategic and International Studies of Georgetown University," April 6, 1984; and "State of the Union Address," February 6, 1985.

20. Reagan, "Remarks at the Annual Washington Conference of the American Legion," February 22, 1983.

21. V. Bolshakov wrote: "When the present U.S. Administration took office in 1981, wholesale ideological sabotage against the socialist countries was openly elevated to the rank of U.S. state policy." The aim, said the *Pravda* analysis, was to "destabilize the existing system in the socialist community." V. Bolshakov, "Washington 'Crusaders' on the March," *Pravda*, January 31, 1983, 6, published as "Pravda Accuses U.S. of 'Ideological Aggression,' " in FBIS-SOV-9-FEB-83, February 9, 1983, A1–4.

22. Talk of an anti-Communist "crusade" was heard again at the time on Moscow's "Studio 9" program, a popular television broadcast blessed by the benefits of a monopoly on both opinion and the airwaves. Moscow TV's "Studio 9" program, February 26, 1983. Transcript published as "U.S. Anti-Communist 'Crusade,' " in FBIS-SOV-28-FEB-83, February 28, 1983, CC11.

23. February 19, 1983 TASS statement. Text is published as "Leader Speeches Show 'Crusade Against Communism,' " in FBIS-SOV-22-FEB-83, February 22, 1983, A5.

24. For a lengthy discussion, including sources on Reagan's thinking, see chapter 16 of *God and Ronald Reagan*.

25. Reagan, "Remarks at the Annual Convention of the National Association of Evangelicals," Orlando, Florida, March 8, 1983.

26. Clark in Schweizer, *Fall of the Berlin Wall*, 76.

27. Spencer shared this story with Tom Reed. Reed, *At the Abyss*, 240.

28. A marked-up draft of the speech on file at the Reagan Library, specifically in Folder 155, Box 9, of the Presidential Speeches section of the Presidential Handwriting File.

29. On this, Ed Meese is especially informative. Meese, *With Reagan*, 101–16. George Keyworth said that on the Saturday before the speech that Wednesday evening, "the only people who were knowledgeable, the only people who knew what was going on," were Reagan, McFarlane, Poindexter, Clark, and himself. That was it. Only those five. "A day or so later,"

Ray Pollock was "involved." Closer to the time of the speech, Gil Rye also became involved. Rye wasn't involved until the time came to write the necessary messages to various embassies. He confirmed that "it is true that not many people in the Defense Department had any idea that the President was going to make the speech." He said Cap Weinberger and Richard Perle were both in Portugal for a NATO conference. "They knew something was happening," said Keyworth, "but only on Sunday or Monday." "It was most definitely a surprise," summed Keyworth. Keyworth interviewed by Baucom, September 28, 1987, RRL, OHT, Folder 37, Box 8, 27–29.

30. Interview with Edward Teller, July 15, 2003. Also see transcript of Edward Teller interviewed by Donald Baucom, July 6, 1987, RRL, OHT, Folder 35, Box 8, 7–9.

31. I have discussed this with Bill Clark many times. Also see: Meese, *With Reagan*, 101–16.

32. Keyworth interviewed by Baucom, September 28, 1987, RRL, OHT, Folder 37, Box 8, 25.

33. Ibid., 25–26, 35.

34. Bill Clark, "President Reagan and the Wall," Address to the Council of National Policy, San Francisco, California, March 2000, 8–9.

35. George Shultz said, "SDI was entirely the president's idea." George Keyworth added: "There was never any single initiative by the Reagan administration that was so thoroughly created and invented in Ronald Reagan's own mind and experience. It was his decision. It was his creation." For quotes, see D'Souza, *Ronald Reagan*, 175 and Ken Adelman, *The Great Universal Embrace* (New York: Simon & Schuster, 1989), 298. That said, there were a handful of men who were influential in Reagan's thinking on missile defense, including Martin Anderson, Daniel O. Graham, Malcolm Wallop, Karl Bendetsen, James D. Watkins, Harrison H. Schmitt, Angelo Codevilla, and Edward Teller, among others. Especially important was a February 11, 1983 meeting between Reagan and his joint chiefs of staff. Among other sources, see William Broad, *Teller's War* (Simon & Schuster, 1992), 96–120, 136.

36. Reagan once remarked that "it kind of amuses me that everybody is so sure I must have heard about it [SDI], that I never thought about it myself. And the truth is, I did." Reagan, "Interview With Morton Kondracke and Richard H. Smith of *Newsweek* Magazine," March 4, 1985. George Keyworth stated: "I don't know of anything that was more clearly Ronald Reagan's than SDI was. There is absolutely no question that SDI originated with the president." George A. Keyworth interviewed by Donald Baucom, September 28, 1987, RRL, OHT, Folder 37, Box 8, 28.

37. See the important findings from Schweizer's *Reagan's War*, 84–85, where he talks of earlier Reagan meetings with another Californian involved in defense technology.

38. Interview with Edward Teller, July 15, 2003. See Edward Teller with Judith Shoolery, *Memoirs* (Cambridge, MA: Perseus, 2001), 508–9; and Edward Teller interviewed by Donald Baucom, July 6, 1987, RRL, OHT, Folder 35, Box 8, 1–2.

39. Edward Teller interviewed by Donald Baucom, July 6, 1987, RRL, OHT, Folder 35, Box 8, 1–2.

40. Interview with Edward Teller, July 15, 2003.

41. Among the examples of Reagan deploying Teller, in August 1982 he dispatched Teller to Sicily for the ERICE Conference on preventing nuclear war. Teller was thrilled with the assignment. "I was deeply honored by your asking me to present your message to the conference,"

Teller told Reagan, "and I did so with great pleasure and feeling of pride which I do not often experience." Teller wrote this in an August 25, 1982 letter, to which Reagan responded with an October 25 letter to Teller. The letters are filed at the Reagan Library in the presidential letters section, Folder 54, Box 4.

42. "Memorandum to the President, Subject: Letter from Edward Teller," from Jay Keyworth to President Reagan, July 29, 1982. The memo, printed on White House letterhead, is on file at the Reagan Library in the PHF:PR section, Folder 48, Box 4.

43. On this, see, among others, Anderson, *Revolution*, 80–85, 99; Shultz, *Turmoil and Triumph*, 261–63; and Lou Cannon testimony at University of Virginia's Miller Center is published in Thompson, ed., *Leadership in the Reagan Presidency, Pt II: Eleven Intimate Perspectives*, 61.

44. Reagan, *An American Life*, 550. See Reed, *At the Abyss*, 254–57.

45. At last, a solid work has been devoted to Reagan's hatred of nuclear weapons and desire to eliminate them. See Paul Lettow, *Ronald Reagan and His Quest to Abolish Nuclear Weapons* (New York: Random House, 2005); Oberdorfer, *The Turn*, 26; and Shultz, *Turmoil and Triumph*, 700.

46. Reagan, "Remarks at a Ceremony Marking the Annual Observance of Captive Nations Week," July 19, 1983.

47. Reagan, "Remarks to Private Sector Leaders During a White House Briefing on the MX Missile," March 6, 1985.

Reagan told Walter Cronkite on the fortieth anniversary of D-Day in Normandy, France on June 6, 1984: "Walter, I have said, and will continue to say, a nuclear war cannot be won. It must never be fought. And this is why the goal must be to rid the world once and for all of those weapons." Reagan, "Interview With Walter Cronkite of CBS News in Normandy, France," June 6, 1984.

48. Reagan, "Interview With Alastair Burnet of ITN Television of the United Kingdom," March 10, 1988.

49. Among many others, see Reagan, "Address Before a Joint Session of the Congress on the State of the Union," January 26, 1984; Reagan, "Debate Between the President and Former Vice President Walter F. Mondale," Kansas City, Missouri, October 21, 1984; Reagan, "Interview With Arrigo Levi of Canale 5 Television of Italy," March 10, 1988; Reagan, "Interview With Soviet Television Journalists Valentin Zorin and Boris Kalyagin," May 20, 1988; and Reagan, "Radio Address to the Nation on Soviet-United States Relations," December 3, 1988. Also see Reagan, *An American Life*, 288; and Kenneth W. Thompson, "The Reagan Presidency: Interview with Frank Carlucci," *Miller Center Journal*, 2 (Spring 1995): 43.

50. Quoted by William Pemberton, *Exit With Honor: The Life and Presidency of Ronald Reagan* (Armonk, NY: Sharpe, 1997), 131.

51. Interview with George P. Shultz, July 15, 2003, Stanford University, Palo Alto, California.

52. Reagan, "Remarks to a White House Briefing for Republican Student Interns on Soviet-US Relations," July 29, 1986.

53. Reagan, *An American Life*, 257. Also on MAD, see Reagan, "Interview With Morton Kondracke and Richard H. Smith of *Newsweek* Magazine," March 4, 1985; Reagan, "Interview With Foreign Journalists," April 25, 1985; Reagan, "Statement on the Fifth Anniversary of the Strategic Defense Initiative," March 23, 1988; and Joseph Coors, oral-history testimony, July 31, 1987, RRL, OHT, Folder 35, Box 8.

54. Reagan, "Remarks to the Institute for Foreign Policy Analysis at a Conference on the Strategic Defense Initiative," March 14, 1988; and Caspar W. Weinberger, *Fighting for Peace: Seven Critical Years in the Pentagon* (New York: Warner Books, 1991), 327.

55. Reagan, "Remarks to Citizens in Hambach, Federal Republic of Germany," May 6, 1985.

56. Reagan, "Remarks to the Institute for Foreign Policy Analysis at a Conference on the Strategic Defense Initiative," March 14, 1988.

57. Interview with Ed Meese, September 23, 1998.

58. Adelman quoted in Cannon, *Role of a Lifetime*, 278.

59. Reagan, "Remarks to the Institute for Foreign Policy Analysis at a Conference on the Strategic Defense Initiative," March 14, 1988.

60. Reagan, "Remarks to Administration Supporters at a White House Briefing on Arms Control, Central America, and the Supreme Court," November 23, 1987.

61. See Reagan, "Remarks at a Luncheon Hosted by the Heritage Foundation," November 30, 1987.

62. Reagan, *An American Life*, 571.

63. Reagan, "Address to the Conservative Political Action Conference," March 8, 1985.

64. Keyworth interviewed by Baucom, September 28, 1987, RRL, OHT, Folder 37, Box 8, 30.

65. Gergen interviewed on television series, "Television and the Presidency," Fox News Channel, December 25, 2000.

66. Cannon, *Role of a Lifetime*, 683. At one press conference in Moscow, for example, he spoke of SDI's potential to "make it impossible for missiles to get through the screen." Of course, it is possible that Reagan was willing to exaggerate SDI while in Moscow. Reagan, "President's News Conference Following the Soviet-United States Summit Meeting in Moscow," June 1, 1988.

67. Reagan, "Foreword Written for a Report on the Strategic Defense Initiative," December 28, 1984.

68. Later, he was also measured when negotiating directly with Gorbachev, a moment when bluster and exaggeration might be expected. In the third plenary session at the Geneva Summit on November 20, 1985, he told Gorbachev that "no one was sure whether SDI would work; the U.S. effort was designed only to find out if a defense was possible." It "would be years before this was known." See Details in "Geneva Meeting: Memcons," Box 92137, Folder 2, RRL.

69. Reagan, *An American Life*, 609.

70. Ibid., 608.

71. What made the idea promising, wrote Reagan, was that, if it worked, "and we then entered an era when the nations of the world agreed to eliminate nuclear weapons," it could serve as a "safety valve against cheating—or attacks by lunatics who managed to get their hands on a nuclear missile." This was precisely the thinking behind the U.S. Senate decision to vote 97 to 3 in January 1998 in support of President Clinton's proposal to pursue a missile-defense system. Reagan, *An American Life*, 608.

72. See Lou Cannon, "President Seeks Futuristic Defense Against Missiles," *Washington Post*, March 24, 1983, A1.

73. Moreover, the suggestion, advanced by some journalists, that Reagan got the SDI idea from B movies he once made in the 1950s, playing the role of Brass Bancroft, is ludicrous.

There is not a shred of evidence for this. The notion was advanced by people who preferred to ridicule rather than ascertain the truth.

74. Steven R. Weisman, "Now, Talk of New Strains Among the Top Aides," *New York Times*, March 31, 1983.

75. Reagan, "Interview With Morton Kondracke and Richard H. Smith of *Newsweek* Magazine," March 4, 1985.

76. Reagan quoted by Philip B. Kunhardt, Jr., Philip B. Kunhardt III, and Peter W. Kunhardt, *The American President* (New York: Riverhead Books, 1999), 296.

77. Reagan, "Interview With Morton Kondracke and Richard H. Smith of *Newsweek* Magazine," March 4, 1985.

78. Reagan wrote this in a February 4, 1985 letter to his friend V.H. Krulak. See Skinner, Anderson, and Anderson, eds., *Reagan: A Life in Letters*, 122.

79. Reagan wrote this in a November 1, 1984 letter to Drs. Ivy Mooring and John Shelton of Los Angeles. The letter is in Skinner, Anderson, and Anderson, eds., *Reagan: A Life in Letters*, 426. For a second example in the book, see the Reagan letter to Patrick Mulvey, 425.

80. Reagan, "The President's News Conference," January 9, 1985.

81. Ibid.

82. Ibid.

83. Moscow International Service commentary by Viktor Vasilyev, November 6, 1985, published as "Commentary on Reagan Interview with Journalists," in *FBIS*, FBIS-14-NOV-85, November 14, 1985, 1985, A9.

84. On another broadcast of "Studio 9," Valentin Zorin said: "You know that Reagan is now possessed of dreams of star wars and the militarization of space. This has been his fixation." See "Moscow TV's 30 June 'Studio 9' Program," transcript published in *FBIS*, FBIS-3-JUL-84, July 3, 1984, CC6. This take is seen in official TASS statement. To cite just two, one began and ended with the words "star wars," noting that Reagan's "plan" to build "space weapons" would jeopardize Soviet-American talks. The other statement quoted Soviet academician-scientist Horis Raushenbakh, who assured that Reagan's "militarization of space" would "put an end" to the hopes for future treaties between the two superpowers. The two TASS statements are published as "Concern Over 'Star Wars' Plans Expressed," in *FBIS*, FBIS-1-FEB-85, February 1, 1985, AA8–9.

85. Viktor Olin commentary for Moscow World Service, December 18, 1984, transcript published as "Olin Views U.S. Signing of 'Star Wars' Contracts," in *FBIS*, FBIS-19-DEC-84, December 19, 1984, AA4.

86. F. Aleksandrov, "How the 'Star Wars' Are Being Prepared," *Krasnaya Zvezda*, March 8, 1985, 5, published as "U.S. Preparation, Research for SDI Discussed," in *FBIS*, FBIS-12-MAR-85, March 12, 1985, AA5.

87. " 'Studio 9' Program Discusses 'Star Wars,' " May 25, 1985, transcript published in *FBIS*, FBIS-29-MAY-85, May 29, 1985, AA1–12.

88. See: Yuri Zhukov, "Choice," *Pravda*, December 17, 1984, 6, published as "Zhukov on Reagan Administration's USSR Policy," in *FBIS*, FBIS-19-DEC-84, December 19, 1984, AA1–3.

89. Valentin Falin, "Fact and Fancy," *Izvestia*, April 10, 1985, 5, published as "Izvestia's Falin Muses on SDI Issue, Part 1," in *FBIS*, FBIS-15-APR-85, April 15, 1985, AA3–5.

90. Yuri Gudkov, "False Promises and the Nuclear Reality," *New Times*, September 1985, 3–5, published as "U.S. Strategic Doctrine Encourages Arms Escalation," in *FBIS*, FBIS-7-OCT-85, October 7, 1985, AA5–10.

91. TASS statement by Vladimir Matyash, October 16, 1985, published as "Reagan Uses 'Clever Tricks' to Explain SDI," in *FBIS*, FBIS-16-OCT-85, October 16, 1985, AA5.

92. Moscow Domestic Service statement by Boris Adrianov, October 23, 1985, published as "Commentator Sees 'Nothing Defensive' About SDI," in *FBIS*, FBIS-28-OCT-85, October 28, 1985, AA1–2.

93. In one case before his first summit meeting with Gorbachev in Geneva, he sat for a testy interview with Soviet journalists. They asked the president about "star wars"—how and why he was seeking to deploy offensive missiles in space. Reagan struggled to explain the unfortunate origins of the term—not citing Kennedy by name—how "our press picked it up," and the subsequent "misconception" it conveyed. "We're not talking about star wars at all," he objected. "We're talking about seeing if there isn't a defensive weapon that does not kill people." The Soviet reporters were incredulous; after all, they had gotten the term from Reagan's own American media which, the Soviets surely surmised, was certainly more objective on the matter than Reagan. See Reagan, "Remarks in an Interview With Representatives of Soviet News Organizations, Together With Written Responses to Questions," October 31, 1985.

94. The minutes from this session of the May 31, 1983 Politburo meeting in Peter Schweizer, *Reagan's War*.

95. Quoted in "Inefficiencies hamper Soviet energy plans," *Baltimore Sun*, August 13, 1983.

96. Data provided by Roger Robinson. Interview with Roger Robinson, June 6, 2005. This is consistent with figures published by Peter Schweizer, who has estimated the loss at $7–15 billion per year, and the figures cited by Richard Pipes, who lists data projecting annual losses of $10 billion.

97. Pipes, *Vixi*, 158.

98. Among examples not quoted in this chapter, see Reagan, "Remarks at the Annual Washington Conference of the American Legion," February 22, 1983; and Reagan, "Remarks and a Question-and-Answer Session at a World Affairs Council Luncheon," Los Angeles, California, October 28, 1988.

99. This was much more frequent than references to most other countries.

100. Reagan, "Remarks to Polish Americans," Chicago, Illinois, June 23, 1983.

101. Reagan, "Statement on the Third Anniversary of Solidarity," June 23, 1983.

102. Reagan, "Remarks at a Ceremony Marking the Annual Observance of Captive Nations Week," July 19, 1983.

103. In August 1984, he found himself celebrating the annual Polish festival in Doylestown, Pennsylvania, where he quoted John Paul II: "Freedom is given to man by God as a measure of his dignity. . . . As children of God, we cannot be slaves"—a quote he repeated at a White House luncheon marking the fortieth anniversary of the Warsaw Uprising. Reagan, "Remarks at a White House Luncheon Marking the 40th Anniversary of the Warsaw Uprising," August 17, 1984; and Reagan, "Remarks at a Polish Festival," Doylestown, Pennsylvania, September 9, 1984.

104. Aleksandr Bovin, "A Face Not a Policy," *Izvestia*, January 10, 1982, 5, published as "Bovin on U.S. Poland Policy," in *FBIS*, FBIS-SOV-10-JAN-82, January 10, 1982, F5.

Additionally, Vitaly Korionov, in a January 1984 *Pravda* piece titled, "Production Line of Crimes and Hypocrisy," listed a cornucopia of Reagan efforts intended to sustain Solidarity and subvert Polish Communism. Korionov lamented: "As a result of the measures taken by Poland's leadership, the situation in the country began to stabilize, but people in Washington decided that this contradicted their 'vital interests' and gave additional instructions for launching a new campaign under the changed circumstances." He said that Washington's goal was to "push in every possible way the thesis about the desirability of the notorious Solidarity." Vitaly Korionov, "Production Line of Crimes and Hypocrisy," *Pravda*, January 10, 1984, published as " 'Unprecedented Wave' of Lies Seen in U.S.," in *FBIS*, FBIS-SOV-13-JAN-84, January 13, 1984, A4.

Reagan continued to express that desirability by both covert and overt means. On the overt side, he offered more statements of camaraderie with the Polish people.

105. Ibid.

106. Unfortunately for Bovin, his writings were widely judged by Pols as dishonest and shameless. Generally speaking, it was Soviet talk of human rights in Poland that was considered a hypocritical farce by the Polish people. Contrary to Bovin's writings, a majority of the Polish people felt that Ronald Reagan understood their human rights concerns far better than the Soviets did, and they recognized that the Soviets' so-called sympathy for human rights was nothing more than a cruel joke. As Krakow native Joseph Dudek put it, "I feel that few people understood the Communists with the exception of Reagan." (Source: Interview with Joseph Dudek.)

107. Ibid.

108. Interview with Wladyslaw Kaludzinski, October 31, 2005, translated by Tomasz Pompowski.

109. May 1983 poll in a chapter by Teresa Rakowska-Harmstone, "Communist Regimes' Psychological Warfare Against Their Socities: The Case of Poland," in Janos Radvanyi, ed., *Psychological Operations and Political Warfare in Long-Term Strategic Planning* (New York: Praeger, 1990), 103.

110. Edmund Morris, *Dutch*, 492; and William P. Clark, "President Reagan and the Wall," Address to the Council of National Policy, San Francisco, California, March 2000, 11.

111. John Barletta, *Riding With Reagan: From the White House to the Ranch* (New York: Citadel Press, 2005), 52.

112. Reagan's handwritten draft of the September 1, 1983 KAL statement is on file at the Reagan Library.

113. The last quote is cited in Hedrick Smith, "Reagan's Crucial Year," *New York Times*, October 16, 1983.

114. Reagan, "Remarks at the Annual Pulaski Day Banquet," New York, NY, September 25, 1983.

115. Ibid.

CHAPTER 13

1. These letters are quoted in my biography of Clark.

2. Bailey, *The Strategic Plan That Won the Cold War*, i.

3. This Marxist group was a renegade, competitor Marxist group. Bishop himself was a Marxist, albeit much less radical and violent. Bishop and his New Jewel Movement came to power in a 1979 coup.

4. Reagan, "Speech Announcing the Strategic Defense Initiative," March 23, 1983.

5. On this, and the entire operation in general, see Ed Meese's chapter in Edwin Meese, *With Reagan* (Washington, DC: Regnery, 1992), 213–27.

6. Reagan, "Remarks at a Summit Conference of Caribbean Heads of State," University of South Carolina, Columbia, South Carolina, July 19, 1984.

7. Quoted by Cannon, *The Role of a Lifetime*, 386–87.

8. Quoted in "D-Day in Grenada," *Time*, November 7, 1983; and Morris, *Dutch*, 501.

9. Figure is based on 58,000 dead from July 1959 through April 1975, which is a very broad time frame.

10. Quoted in Richard Harwood, "Tidy U.S. War Ends: 'We Blew Them Away,'" *Washington Post*, November 6, 1983.

11. Reagan, *An American Life*, 454.

12. Data provided in Meese, *With Reagan*, 219.

13. Ibid., 220–21. Numbers provided.

14. It is greatly unappreciated how seldom the tough-talking Reagan chose to use force. For instance, he deployed troops in combat far fewer times than Bill Clinton, not to mention other presidents. In the few cases where he used force, action was fast and decisive, largely successful, and usually a morale booster (another example was the bombing of Moammar Kaddafi in Libya in April 1986). Charles W. Dunn notes that this rapid, rare use of force allowed Reagan to avoid liberal criticism over the use of excessive force in the battle against Communism. He chose prudently, selecting spots that were eminently doable. See Charles W. Dunn, *The Scarlet Thread of Scandal* (Lanham, MD: Rowman-Littlefield, 2000), 150.

Judiciousness in the use of force had long been Reagan's inclination. In a May 1968 speech, he spoke of one of his favorite presidents: "Eisenhower understood the authority as well as the limitations of force in international politics; he was not afraid to make it count in a world where force still settles the fate of nations." "Speech to Republican State Central Committee Finance Dinner," Sheraton-Cleveland Hotel, Cleveland, May 22, 1968. Speech filed at Reagan Library, "RWR—Speeches and Articles (1968)," vertical files.

Reagan, too, felt force was necessary to settle scores—but not all scores at all times. He felt that one must recognize not only force's authority but also its limitations.

Agreeing that Reagan was "very reluctant to use power," contrary to his trigger-happy cowboy image, aide Ken Adelman offers an instructive example: Secretary of State Haig, according to Adelman, underwent "tremendous agony" the first two years of the administration because "he wanted to take it to the source and go after Cuba"; but Reagan vetoed the idea. Adelman explained: "Basically because Ronald Reagan, whenever he used force, wanted to make it pretty easy, pretty cheap, pretty quick, and pretty decisive. . . . [He was] quite moderate, and quite unwilling to really use much." See Kenneth Adelman in Hofstra conference (1993) proceedings, 240.

15. Quoted in Monica Crowley, *Nixon in Winter* (New York: Random House, 1998), 226.

16. Lawrence F. Kaplan, "We're All Cold Warriors Now," *Wall Street Journal*, January 18, 2000.

17. Reagan, "Keynoting the Conservative Decade," Remarks at the Heritage Foundation's 10th Anniversary Banquet, October 3, 1983.

18. Ibid.

19. Reagan, *An American Life*, 450–51.

20. Moscow TASS International Service, October 26, 1985, published as "Reagan Speaks on Grenada Invasion Anniversary," in FBIS-SOV-28-OCT-85, October 28, 1985, A8.

21. TASS commentary written by Konstantin Yuriyev, titled, "A Breakfast of the President (topical satire)," November 4, 1983, published as "Reagan 'Breakfast' Remark Prompts Satire," in FBIS-7-NOV-83, November 7, 1983, A2–3.

22. The Kaiser op-ed appeared in the *Washington Post*, October 30, 1983, C1.

23. TASS release, November 1, 1983, published as "Critical Review of Reagan Foreign Policy Cited," in FBIS-SOV-2-NOV-83, November 2, 1983, A4.

24. Vsevolod Ovchinnikov, "McFarlane's Revelations," *Pravda*, August 9, 1984, 4, published as "McFarlane Views on East-West Confrontation Hit," in FBIS-10-AUG-84, August 10, 1984, A3; TASS report, "Beneath the Veil of Hypocrisy," *Pravda*, July 28, 1984, 5, published as "Pravda Attacks Reagan Remarks on Grenada," in FBIS-1-AUG-84, July 28, 1984, A1. Also in *Pravda*, political observer Vsevolod Ovchinnikov angrily said of Reagan's action: "He demonstrated the desire to act as a world policeman, to commit violence with impunity, to guarantee the United States world supremacy." See Vsevolod Ovchinnikov, *Pravda*, October 29, 1983, published as "Reagan's Grenada, Lebanon Address Rebutted," in FBIS-1-NOV-83, November 1, 1983, A1.

25. Valentin Falin, "Interview? No, a Program," *Izvestia*, June 21, 1985, 4, published as "Falin Sees Reagan's Words, Deeds at Variance," in FBIS-27-JUN-85, June 27, 1985, A5–6.

26. Valentin Zorin speaking through Moscow Domestic Service, January 13, 1984, transcript published as "'International Situation: Questions & Answers,'" in FBIS-SOV-16-JAN-84, January 16, 1984, CC8–9.

27. Herb Meyer interviewed on "Reagan," *The American Experience*, PBS. On this campaign by Casey, see the work of Peter Schweizer, in particular, *Victory*, and *The Fall of the Berlin Wall*.

28. So, said Meyer, "We hit 'em high, we hit 'em low." Vulnerabilities were pinpointed. Stinger missiles were sent to Afghanistan; Bibles to Leningrad. "And we did a lot more," he adds. "We were throwing banana peels under their feet every day." This, said Meyer, was a by-product of "Reagan's historic decision" to start playing to win. Meyer in Hofstra conference (1993) proceedings, 126–27.

29. Peter Schweizer obtained a copy of the memo and cites it in *Reagan's War*, 231–32.

30. Reagan, "Informal Exchange With Representatives of *Le Figaro*," December 22, 1983.

CHAPTER 14

1. Reagan, "Interview With Lou Cannon, David Hoffman, and Juan Williams of the *Washington Post* on Foreign and Domestic Issues," January 16, 1984. Reagan did not let go: He heralded the operation on its second anniversary in October 1985, seizing the occasion as a continued shot in the arm to the American body politic.

2. Reagan, *An American Life*, 457–58.

3. He actually began seeing the changes as early as summer 1982, especially in the military. See Reagan, *An American Life*, 557–58.

4. Reagan, "State of the Union Address," January 26, 1984. For more such references later, see Reagan, "State of the Union Address," February 6, 1985; and Frederick J. Ryan Jr., ed., *Ronald Reagan: The Wisdom and Humor of the Great Communicator* (San Francisco: Collins, 1995), 11.

5. Reagan, "Remarks Upon Returning From the Soviet-United States Summit Meeting in Moscow," June 3, 1988.

6. He took the time to say so in his Farewell Address on January 11, 1989: "[O]ne of the things I'm proudest of in the last eight years [is] the resurgence of national pride . . . the recovery of our morale. America is respected again in the world and looked to for leadership."

He rightfully identified this as one of his "greatest triumphs." Reagan, "Farewell Address to the Nation," January 11, 1989.

7. The assessment of the left and academic community is powerful. The verdict from academe—itself overwhelmingly politically liberal—began surfacing early. It was evident in a 1985 poll by *National Journal*, published only months into Reagan's second term. The findings came from a questionnaire sent to members of the Presidency Research Group, an affiliate of the American Political Science Association. More than half of the 225 members replied, which, stated *National Journal*, provided a statistically representative sample. The group rated Reagan well, scoring him an A-minus/B-plus. Most of the assessments of Reagan's policies were negative, partly reflecting the policy biases of those doing the ratings. Where Reagan scored especially well was in the restoration of "morale and trust" to both the country and the office of the presidency. Those who gave Reagan high marks pointed to his "mastery of the presidential office," the restoration of national pride, and the "reviving of trust and confidence" in an institution "that in the post-Vietnam era had been perceived as being unworkable." See Dom Bonafede, "Presidential Scholars Expect History To Treat the Reagan Presidency Kindly," *National Journal*, 17, no. 14 (April 6, 1985): 743–47.

Harvard's renowned presidential scholar Richard Neustadt, who said Reagan gave Americans a sense that "all was well," recalled that watching Jimmy Carter in the presidency, one had to wonder if the presidency was "even possible." Neustadt speaking on documentary, "The American President," PBS, WNET-New York, 2000. Neustadt also said this in his classic work, *Presidential Power*.

Some political scientists refer to this rebirth of morale, for both the office and the country, as the "symbolic" aspect of Reagan's presidency. It was a key factor driving his surge upward in late 1990s rankings of presidents. Among those surveys was a major 2000 poll by the Federalist Society and the *Wall Street Journal*, where Reagan ranked as the eighth best president ever, placing him in the "near great" category. A key factor in Reagan's rising stock at the time was the 1990s erosion of the symbolic element of the presidency, due to the private behavior of President Bill Clinton. University of Louisville Political Scientist Gary Gregg spoke to this: "The symbolic aspects of the presidency are well under-rated. . . . This is one reason Reagan should be ranked higher than he generally is." See James Lindgren, "Ranking Our Presidents," 11.

There are countless academic authorities that credit Reagan with restoration to the nation and the presidency: Larry Berman, A. E. Jeffcoat, James T. Patterson, Robert Dallek, Christopher Layne, and others. See Larry Berman in Berman, ed., *Looking Back on the Reagan Presidency*, 5, 8–9. A. E. Jeffcoat, *From Covered Wagons to Clinton* (Bainbridge Island, WA: Winslow House Books). James T. Patterson in James M. McPherson, ed., "To the Best of My Ability," *The American Presidents* (New York: Dorling Kindersley, 2000), 288.

There are many more. Boston University's Robert Dallek says that Reagan "did more to restore a measure of confidence in the institution of the presidency than anything since the Kennedy administration." International relations scholar Christopher Layne, who was so critical of the Reagan Doctrine that he predicted it "would lead the administration into a political and moral morass," later conceded that, "The Reagan administration did restore the national pride and self-confidence that had been damaged by Vietnam and the 1979–80 Iran hostage crisis."

Dallek interviewed on "Reagan," *The American Experience*, PBS. See David Boaz, ed., *Assessing the Reagan Years* (Washington: Cato Institute, 1988), 95, 108.

Ed Harper of the Brookings Institution said that, "One of the most important things Ronald Reagan did was to erase the attitude of defeatism." This, he goes so far as to say, was a defeatism "not only about the manageability of government but also about the future of the planet." Harper points to the calamitous forecasts of the Club of Rome report. "The Carter administration had bought the Club of Rome Report hook, line, and sinker," he asserted. "The Club of Rome made all kinds of dire predictions, saying that by the 1990s we would be turning out lights all over the world. . . . There was a defeatism in the last years of the Carter administration that we haven't seen since." Ed Harper testimony before University of Virginia Miller Center is published in Thompson, ed., *Leadership in the Reagan Presidency, Pt II: Eleven Intimate Perspectives* (Landham, MD: University Press of America, 1993), 133.

Historian Alonzo Hamby agrees, writing that among Reagan's accomplishments, "He had brought back the presidency from its lowest ebb since Hoover." Commenting on Reagan's unimpressive showing in one of Arthur M. Schlesinger, Jr.'s presidential greatness polls, Hamby assures: "When passions cool after a generation or so, Ronald Reagan will be widely accepted by historians as a near-great chief executive." "Most of all," said Hamby, Reagan "restored the nation's confidence and had affirmed its faith in the future." He said that Reagan "excelled" in the "job" of the "restoration of confidence." Reagan was an "uplifter," an "outstanding national cheerleader." See Hamby in Adam Meyerson, ed., "How Great Was Ronald Reagan," *Policy Review*, no. 46 (Fall 1988), 32–33. From this symposium, also see the observations from George H. Nash and James Nuechterlein on pages 34–36. On Hamby, see Alonzo Hamby, *Liberalism and Its Challengers* (New York: Oxford University Press, 1992), 376, 388.

Strong words also come from Cold War expert John Lewis Gaddis, who was impressed at how Reagan after only his first few months in office had "managed to project—and therefore instill—a degree of self confidence that went well beyond anything his predecessor had achieved." Gaddis, "Hanging Tough Paid Off," 60–63.

8. Not surprisingly, Reaganites inside and outside the administration lauded the president for boosting spirits. These include Don Regan, Larry Speakes, Peter Rodman, and many more. See Donald Regan, *For the Record* (San Diego, CA: Harcourt Brace Jovanovich, 1988), 244; Larry Speakes, *Speaking Out* (New York: Scribner, 1988), 308; and Rodman, *More Precious Than Peace*, 542. Adam Meyerson of the Heritage Foundation cleverly referred to Reagan as the Great Rejuvenator. Burton Yale Pines concluded: "The mood in the 1970s was tragic . . . dreadful. Reagan reversed and healed this. See Burton Yale Pines, speech at the China Association for International Friendly Contact, the People's University of China, and the Shanghai Institute for International Studies, October 1988, published in *Policy Review*, Spring 1988.

9. Among them, journalist Michael Ruby, who wrote the lengthy *U.S. News & World Report* final assessment of the Reagan presidency, estimated that, "historians will conclude that it was during the Reagan decade that the U.S. finally overcame its Vietnam-induced loss of national nerve." Ruby finished with this: "[H]e has helped the nation regain its self-confidence. . . . Ronald Reagan convinced most of his countrymen that it was still morning in America. See Michael Ruby, "The lessons of the Reagan era," *U.S. News & World Report*, January 9, 1989, 18–26.

Over a full decade later, in February 2001, a vigorous debate over Reagan's legacy took place on the popular CNN political show "Capital Gang." A heated exchange ensued between conservative Robert Novak and liberals Mark Shields and Al Hunt. Shields and Hunt were critical of Reagan, finding much to still dislike. Both, however, agreed that he must be credited with "turning around" the presidency and the country at a time that seemed "ungovernable," as Hunt put it, with Shields voicing agreement. Mark Shields and Al Hunt on "The Capital Gang," CNN, February 10, 2001.

The examples in the days after Reagan's death could fill a book, and are not quoted here.

10. Mike Wallace speaking at conference, "The Reagan Legacy," Ronald Reagan Library, Simi Valley, CA, May 20, 1996; and Sidey speaking on documentary, "The American President," PBS, WNET-New York, 2000.

Reagan biographers agree. Lou Cannon has said so repeatedly. In his 1991 *Role of a Lifetime*, he wrote of how Reagan's speeches "did succeed in reviving national pride." See Cannon, *Role of a Lifetime*, 292. Also, in a 1999 C-SPAN interview, Cannon said that Reagan "restored America's confidence at a time when it was very low. He convinced the American people to believe in themselves again." In a 2001 CNN documentary, Cannon credited: "I think he'll be remembered as the person who restored hope and confidence to a country when it was really troubled." Cannon interviewed by Brian Lamb on C-SPAN's "American Presidents" series, December 6, 1999; and Cannon interviewed for CNN documentary, "The Reagan Years: The Great Communicator," Pt. II of series, CNN, February 2001.

11. In a way, "overnight" is not a stretch by Morris: On the day Reagan was inaugurated, those hostages in Iran were released—a huge shot in the arm for Uncle Sam, though it would be a big mistake to think the hostage release alone put the nation back on track. By 1984, however, it was certainly back. See Morris in Wilson, ed., *Power and the Presidency*, 125–26. Also, Edmund Morris interviewed by Leslie Stahl of "60 Minutes," September 26, 1999.

12. Doug Gamble, "On patriotism, the Gipper and Bill Clinton," *The Los Angeles Times*, December 17, 1999.

13. Later, from the apex of the Soviet leadership, Mikhail Gorbachev himself spoke to the morale issue and even Reaganomics. "It was widely accepted that the first four years of 'Reaganomics' had stabilized the US economy," acknowledged Gorbachev, "and that the Reagan administration had contributed to the strengthening of the 'American spirit.'" Gorbachev, *Memoirs*, 415.

14. Vladimir Simonov, "Political Portrait of Ronald Reagan," *Literaturnaya Gazeta*, May 25, 1988, 14, published as "Weekly Presents 'Political Portrait' of Reagan," in FBIS-SOV-88–102, May 26, 1988, 8, 10.

15. Michael Ellman and Vladimir Kontorovich, eds., *The Disintegration of the Soviet Economic System* (London and New York: Routledge, 1992), 18.

16. Schweizer, *Victory*, 166.

17. TASS release written by economic writer Vladimir Pirogov, December 29, 1983, published as "U.S. Seeks to 'Exhaust' USSR Economy," in *FBIS*, FBIS-SOV-29-DEC-83, December 29, 1983, A4.

18. These sources include Peter Schweizer, who wrote of the campaign in *Victory* and *Reagan's War*, as well as references made by Vasiliy Mitrokhin in his research and Anatoly Dobrynin in his memoirs. The term "active measures campaign" was a KGB intel term that referred to spreading propaganda, disinformation, and conducting operations against an individual, short of assassination.

19. On the campaign and this quote, see Schweizer, *Victory*, 171–73, 201.

20. Vitaliy Korionov, "Production Line of Crimes and Hypocrisy," *Pravda*, January 10, 1984, 4, published as "'Unprecedented Wave' of Lies Seen in U.S.," in FBIS-13-JAN-84, January 13, 1984, A1, A5.

21. Vladimir Lomeyko, "NATO's Procrustean Bed," *Literaturnaya Gazeta*, January 25, 1984, 14, published as "'Psychological War' Eyed as Cold War 'Crusade,'" in FBIS-3-FEB-84, February 3, 1984, CC5.

22. A. Gurov and V. Martynenko, "Imperialism's Economic Aggression," *Krasnaya Zvezda*, January 27, 1984, 3. Text is published as "U.S. Anti-Socialism Includes 'Economic Warfare,' " in *FBIS*, FBIS-SOV-2-FEB-84, February 2, 1984, CC1–4.

23. Specifically, they stated: "But what failed earlier, when the Soviet state did not possess such powerful scientific, technical, and production potential as it does now, is all the more doomed to failure under modern conditions. . . . The socialist countries counter imperialism's economic aggression with the further deepening of fraternal cooperation, economic integration, and unbreakable cohesion. . . . A resolution rebuff will continue to be dealt to the subversive actions of the 'economic NATO.' "

24. See Caldicott speaking on "Reagan," *The American Experience*, PBS.

25. To the best of my knowledge, this book is the first full examination and presentation of the Chebrikov document. The document came from the Central Committee Archives of the Communist Party of the former USSR. There are a number of different archives from the Soviet period, including, for example, the KGB Archives, the Comintern Archives, among others. The archives were opened in the early 1990s by the Boris Yeltsin government, whereupon scholars and journalists eagerly began digging into the documents. The Yeltsin government, for various reasons, closed many of the archives, including the Central Committee Archives, in the 1994–95 period.

The document that I cite here was apparently initially found by a *London Times* reporter in early 1992. The *London Sunday Times* subsequently published an article on the document, titled "Teddy, the KGB and the top secret file," in the February 2, 1992 edition. The *Times* article, however, was brief and included only a few quotes from the document; the newspaper published only a photo of a small section from the upper left corner of page one of the five-page document. Once the *Times* piece ran, a number of people scrambled to the archives to obtain their own copy of the document, which then began circulating in and around Moscow. Various individuals obtained copies. Shortly thereafter, the Russian government closed the file.

Among those who obtained a full copy was Herb Romerstein, author of the seminal work on the Venona papers, *The Venona Secrets*. Romerstein is a former staff member of the U.S. House Permanent Select Committee on Intelligence and a seasoned veteran of archival research from the Cold War period. Romerstein continues to do research from intelligence archives of former Communist countries. My copy was obtained through Marko Suprun, who got the copy from Herb Romerstein and Walter Zaryckyj. Romerstein provided details to me on the document's origin in a number of discussions in June 2005.

26. Tunney won his seat by defeating close Reagan friend George Murphy in the 1970 race. He lost his reelection bid in 1976 to Republican Samuel Hayakawa.

27. They were classmates at the University of Virginia Law School. See the November 16, 1970 profile of Tunney in *Time* magazine, and Charles T. Powers, "Can John Tunney Make It as a Heavyweight?" *West Magazine (Los Angeles Times)*, December 12, 1971. Both articles are posted on Tunney's Web site.

28. Ibid. Tunney quote is taken from *West Magazine* piece.

29. Ibid.

30. When told about the Tunney-Kennedy effort two decades later, Bill Clark said that he and others in the Reagan administration had heard that certain leftist politicians were reaching out to Moscow, though Clark could not recall specific names, including Tunney and Kennedy, and never pursued the information. Interview with Clark, July 7, 2005.

31. Tunney was tracked down by the *London Times*, which obtained a copy of the memo

in 1992. The *Times'* translation makes Kennedy look even worse on the matter of personal political motivations; it says that Tunney informed the Soviets that Kennedy was "directing his efforts at becoming president of the US in 1988." When asked about this specifically, Tunney dismissed Chebrikov's interpretation as "bullshit." See "Teddy, the KGB and the top secret file," *London Sunday Times*, February 2, 1992.

32. I suspect that some will also claim that Kennedy's overture was not completely unusual, since a number of liberal Democrats at the time (including the other Massachusetts senator, John F. Kerry) had gone so far as to openly meet with Communist enemies like Nicaragua's Daniel Ortega. The difference, however, is that Kennedy's overture was to Moscow, to Yuri Andropov himself. That was quite unprecedented.

33. Vasiliy Mitrokhin, "The KGB in Afghanistan," Working Paper No. 40, English edition, Christian F. Ostermann and Odd Arne Westad, eds., Cold War International History Project, Woodrow Wilson International Center for Scholars, Washington, DC, February 2002. A full copy of the working paper, in PDF format, is available at the Web site of the Wilson Center. See Herbert Romerstein, "Ted Kennedy, 'Collaborationist,'" *Human Events*, December 8, 2003.

34. Edward M. Kennedy, "A State of Disunion," *Rolling Stone* magazine, March 15, 1984, 11–12. It is not clear whether Kennedy meant a "wider war" in Central America, which he mentioned in the previous sentence, or a "winnable nuclear conflict" generally, which he referred to in the next sentence.

35. Tim Weiner, "Lies and Rigged 'Star Wars' Test Fooled the Kremlin, and Congress," *New York Times*, August 18, 1993, A1.

36. Interview with Peter Schweizer, September 19, 2005. In an e-mail exchange, Schweizer gave me the go-ahead to use this material and cite him as the source. Importantly, the *Times* article noted that the rigged test was intended to fool the Kremlin, and that it also (intentionally or not) fooled Congress. From what I could learn of the incident, the goal was not to mislead Congress. The goal was to mislead the Soviets.

37. The reporter added: "The total collapse of the European postwar order is being sought. To state it even more clearly: Socialism's right to exist as a world system is disrupted. He [Reagan] will 'not accept' Yalta, he will 'not accept' socialism." Hartmut Kohlmetz, "Attacks Against Yalta," East Berlin *Berliner Zeitung*, August 21, 1984, published as "Reagan View of Yalta Could Disrupt Postwar Order," FBIS, Eastern Europe, E1.

38. "Reagan Joke Echoes 'Crusade' Against Socialism," Sofia BTA, August 17, 1984, published in FBIS, Eastern Europe, August 20, 1984, C1; and Yoan Mateev, "The Mistake," *Sofia Rabotnichesko Deo*, August 18, 1984, published as "Reagan Joke Termed Technical, Political Mistake," FBIS, Eastern Europe, August 22, 1984, C1.

39. Genrikh Borovik, Moscow Television Service, August 31, 1984, transcript published as "Parallels Between Reagan, Hitler Policies Posed," in FBIS-4-SEP-84, September 4, 1984, DD4–5.

40. Reagan, "Remarks at a White House Luncheon Marking the 40th Anniversary of the Warsaw Uprising," August 17, 1984.

41. Moscow TASS International Service, September 10, 1984, published as "Reagan Again Questions Yalta Conference Decisions," in *FBIS*, FBIS-SOV-11-SEP-84, September 11, 1984, A4.

42. Valentin Zorin speaking on Moscow TV's "The World Today" program, September 7, 1984. Text is published as "U.S. Attempts to Destabilize USSR Backfire," in *FBIS*, FBIS-SOV-11-SEP-84, September 11, 1984, A3–4.

43. Andrei Gromyko, *Memoirs* (New York: Doubleday, 1989), 307.

44. Both letters are contained in a folder at the Reagan Library. See: PHF, PR, RRL, Box 10, Folder 147.

CHAPTER 15

1. Jack F. Matlock, Jr., *Reagan and Gorbachev* (New York: Random House, 2004), 106–7.

2. Reagan, "Remarks to Citizens in Hambach, Federal Republic of Germany," May 6, 1985.

3. The remainder of this statement said: "To win this struggle, to preserve our way of life, to maintain the peace, we must be strong and true to our ideals. And together we can meet the challenge." And then, "Future generations, not only in the United States but throughout the hemisphere, will be grateful for what we do today." Because, said the president, "We're passing to them the most precious gift of all—liberty." Reagan, "Remarks at a Fundraising Dinner for Senator Paula Hawkins," Miami, Florida, May 27, 1985.

4. Gorbachev's father was a nonbeliever; Reagan's father was Catholic but largely apathetic.

5. By October 1980, noted Archie Brown, Gorbachev already showed signs of being a "serious reformer." Now, it looked clear that the reformer was on his way to potentially becoming a future general secretary as well. See Archie Brown, *The Gorbachev Factor* (New York: Oxford University Press, 1997), ix–x.

6. Interview with Richard Pipes, April 7, 2005.

7. Brown, *The Gorbachev Factor*, 227–30.

8. Ibid, 4.

9. Said Yuzhin: "[T]he day could come when we might be asked to get rid of Reagan." See "Boris Yuzhin," *U.S. News & World Report*, October 18, 1999.

10. Andropov first met Gorbachev in a visit to the Stavropol territory in April 1969; they subsequently had regular meetings together throughout the 1970s. Brown says they were impressed by each other. As early as 1977, says Brown, Andropov saw Gorbachev as not merely a "brilliant man" but also as a future Soviet leader—four years before Reagan's presidency. See Brown, *The Gorbachev Factor*, 50.

11. Interview with Richard Pipes, April 7, 2005.

12. Brown, *The Gorbachev Factor*, 229–30.

13. Reagan, "Statement on the 40th Anniversary of the Yalta Conference," February 5, 1985.

14. Thatcher quoted in Ed Feulner, ed., *Leadership for America: The Principles of Conservatism* (Dallas, TX: Spence, 2000), 11.

15. Brock's testimony is published in Thompson, ed., *Leadership in the Reagan Presidency, Pt II: Eleven Intimate Perspectives*, 119.

16. Ambrose: "How Great Was Ronald Reagan? A Symposium," *Policy Review*, Fall 1988, 30.

17. Writing in *Presidential Studies Quarterly*, Andrew E. Busch of the University of Denver argued: "Reagan led to Gorbachev by fundamentally changing the international situation that the Soviet leadership faced." Busch, "Ronald Reagan and the Defeat of the Soviet Empire," 455–62. See Shattan, *Architects of Victory*, 270. Other respected academics, such as Sidney Milkis, Marc Landy, Vladimir Kontorovich, and Russell Bova, have acknowledged the influence of Reagan policy on Gorbachev, his rise, and *perestroika*. Bova cites Gorbachev's own words, including a May 1985 speech delivered two months after he came to power. "The necessity for accelerating

socioeconomic development is also caused by serious external factors," said Gorbachev, who thus pledged to hike Soviet defense spending in order to try to meet the heightened Reagan levels. He embarked on a new economic restructuring (*perestroika*) in part to provide the resources to handle that spending and to generally improve the Soviet economy. See Marc Landy and Sidney M. Milkis, *Presidential Greatness* (Lawrence, University Press of Kansas, 2000), 226; Vladimir Kontorovich, "The Economic Fallacy," *The National Interest*, Spring 1993, 44; and Russell Bova in Ellman and Kontorovich, eds., *The Disintegration of the Soviet Economic System*, 19, 43.

18. Gaddis, *The United States and the End of the Cold War*, 225n.

19. Leebaert, *The Fifty-Year Wound*, 512.

20. Morris interviewed on "Reagan," *The American Experience*, PBS.

21. Grinevsky speaking on May 8, 1998 at Brown University symposium, "Understanding the End of the Cold War."

22. Schweizer, *Reagan's War*, 241–45.

23. Gromyko said this to former U.S. Senator George McGovern. See Oberdorfer, *The Turn*, 89.

24. Ibid., 245.

25. Schweizer, *Victory*, 198. On Zaslavsky, see David Remnick, *Lenin's Tomb* (New York: Random House, 1994), 320–23.

26. Quoted in Schweizer, *Victory*, 198.

27. Falin said this in an interview with Marina Kalashnikova of *Vlast* magazine, published on the June 9, 2005 Web site of *Kommersant* (one of Russia's independent newspapers) under the headline, "Reagan Provoked Us to Slam the Door."

28. One visitor to Russia reported this display to me in 1999. Another Soviet who talks of Reagan's influence on the environment that created Gorbachev is Major-General Vladimir Slipchenko, a well-known Soviet academician who holds two doctorates, and worked in the central headquarters of the Soviet Ministry of Defense in the early 1980s. He spent much of his time traveling to Warsaw Pact countries visiting military institutions. In these countries, he said there was a consensus that the military burden had become "intolerable." As a result, said Slipchenko, "A Gorbachev had to appear in the Soviet Union. If it was not Gorbachev, it could have been any other person." Slipchenko speaking on May 8, 1998 at Brown University symposium, "Understanding the End of the Cold War."

29. All of these difficulties in measuring Gorbachev's sincere thinking are present when reading *Perestroika*. He produced it just two years after coming to power, at the height of the global excitement over perestroika. The book provides a clear delineation of the disparities between Gorbachev and Reagan. At the same time, it must also be read carefully with an understanding of the above caveats. I suggest reading it in tandem with Archie Brown's *The Gorbachev Factor*, or perhaps reading Brown first.

30. Mikhail Gorbachev, *Perestroika* (New York: Harper & Row, 1987), 24, 58, 62, 64, 161, 252–54.

31. Ibid., 128–30.

32. Ibid., 220.

33. Ibid., 66, 220.

34. Also see, among others: Ellman and Kontorovich, eds., *The Disintegration of the Soviet Economic System*, 20–27.

35. Gorbachev, *Perestroika*, 41, 106, 146.

36. In January 1987, the Party Plenum made official this thinking by formally defining

perestroika. "The main idea of our strategy," declared the Party Plenum, "is to unite the achievements of the scientific-technical revolution with a planned economy and to bring into action the entire potential of socialism."

37. The "potential of socialism" (note: Gorbachev used the word "socialism" interchangeably with "communism"), wrote Gorbachev, "had been underutilized. We realize this particularly clearly now in the days of the seventieth anniversary of our Revolution. We have a sound material foundation, a wealth of experience and a broad world outlook with which to perfect our society."

38. Gorbachev added: "The classics of Marxism-Leninism left us with a definition of the essential characteristics of socialism. They did not give us a detailed picture of socialism. They spoke of its theoretically predictable stages. It is our job to show what the present stage should be like." That next stage, perestroika, "revives" the "living spirit of Leninism," wrote Gorbachev.

39. Gorbachev, *Perestroika*, 10, 11, 17, 35, 45, 50, 66, 151, 253.

40. Ibid., 50.

41. Archie Brown also says, and this would at least partly explain Gorbachev's Leninist language in *Perestroika* in 1987, that "Until the [Soviet] system had been radically altered by the late 1980s, . . . most ideas for changing that system fundamentally had, indeed, to be presented as a return to Leninist first principles if they were to get off the ground in the real world of Soviet politics." Brown says that a "failure to appreciate" this has "led a number of Western authors wildly astray in their analysis of Gorbachev's speeches and public statements" and "public utterances."

In other words, Gorbachev had to cloak his actions in Leninist language. At the same time, as Brown himself notes, Gorbachev was a Leninist and long retained an "idealized" view of Lenin. Ultimately, he moved away from that idealized view, but certainly not totally. In short, the answer is much more complicated. The full picture is complex and requires nuance.

Brown adds that Gorbachev, being CPSU chief, was thus also "official guardian of Marxist-Leninist holy writ," and, further then, "as leader of a reform movement ready to question much of that doctrine, was in the uncomfortable position of being simultaneously both Pope and Luther."

Furthermore, Brown concedes that Gorbachev, like the vast majority of Soviet citizens, had an "idealized" view of Lenin, a consequence of seventy years of distortion and brainwashing. Brown, *The Gorbachev Factor*, 93.

42. Gorbachev, *Perestroika*, 25.

43. Gorbachev, *Perestroika*, 25, 45, 145.

44. Ibid., 145. Obviously, it is arrant nonsense to say that Vladimir Lenin rejected violence. No government in history, with the possible exception of Red China, which had many more victims at its disposal, killed as many innocents as the Soviet state that Lenin created. The killing spree began not under Joseph Stalin but by direct orders from Lenin, the man who instituted the gulag labor-camp system where tens of millions perished. Lenin established the secret police, first called the Cheka. Right after the October 1917 revolution, by 1918–19, the Cheka was averaging 1,000 executions per month for political offenses alone, without trial. See Alexander Solzhenitsyn, *Alexander Solzhenitsyn Speaks to the West* (London: The Bodley Head, 1978), 17. This number was proudly self-reported by the Cheka in its documents. The Cheka actually apologized that its data was incomplete, and boasted that the number was likely much higher.

W. H. Chamberlain, probably the first historian of the revolution, said that by 1920 the

Cheka had carried out 50,000 executions. See Douglas Brown, *Doomsday 1917: The Destruction of Russia's Ruling Class* (London: Sidgwick and Jackson, 1975), 174; and George Leggett, *The Cheka: Lenin's Political Police* (Oxford: Clarendon Press, 1981), 463–68.

Robert Conquest, drawing exclusively on Soviet sources, tallies 200,000 executions at the hands of the Bolsheviks under Lenin from 1917–23, and 500,000 when combining deaths from execution, imprisonment, and insurrection. See Robert Conquest's Congressional testimony, "The Human Cost of Soviet Communism" in Document No. 92–36, 92nd Congress, 1st session, U.S. Senate, Judiciary Committee, Subcommittee to Investigate the Administration of the Internal Security Act and Other Internal Security Laws, July 16, 1971, 5–33.

The essence of this early Red Terror was described by the ferocious Latvian, M. Y. Latsis, who Lenin appointed as chief of his killing machine: "We are exterminating the bourgeoisie as a class." Among other sources that cite this quote, see Brown, *Doomsday 1917*, 173. For a comparison of numbers, see Courtois, *Black Book*, 13; and Leggett, *The Cheka*, 463–68.

Page after page of this book could be filled with bloodcurdling directives written in longhand by Lenin, order after order whereby this political gangster requested that various groups and peoples, from kulaks to priests, be hanged or shot. For transcripts of Lenin directives, see Richard Pipes, ed., *The Unknown Lenin* (New Haven: Yale University Press, 1996), 1, 3, 8–11, 13–16, 46, 50, 55–56, 61, 63, 69, 71, 116–21, 127–29, and 150–55.

45. What Lenin actually said, in a famous 1920 remark before the All-Russian Congress, known universally and oft-repeated by Ronald Reagan, was the following: "We repudiate all morality that proceeds from supernatural ideas that are outside class conceptions. Morality is entirely subordinate to the interests of class war. Everything is moral that is necessary for the annihilation of the old exploiting social order and for uniting the proletariat." V. I. Lenin, *Collected Works, Vol. 31: April-December 1920* (Moscow: Progress Publishers, 1977), 291. See "On Soviet Morality," *Time*, February 16, 1981, 17.

46. Here is one example from pages 32 and 48–49: "In the West, Lenin is often portrayed as an advocate of authoritarian methods of administration. This is a sign of total ignorance of Lenin's ideas and of their deliberate distortion. In effect, according to Lenin, socialism and democracy are indivisible." It was Gorbachev who was totally ignorant of this repugnant, very violent man who created the gulag and gave the Communist Party a monopoly on all political power.

47. Ibid., 50–51.

48. Ibid., 10, 37, 42.

49. Ibid., 42.

50. Ibid., 31.

51. Gorbachev, *Perestroika*, 161–63.

52. Ibid., 165.

53. Ibid., 163–64.

54. Properly evaluating Mikhail Gorbachev is a daunting problem, for at least three reasons: first, the real man must be separated from the mythological figure canonized by Western hagiographers in academe, especially those eager to shift any credit for the Cold War's end away from Reagan, whose politics they despised. Yet, that record presents a second, sticky challenge: It is very difficult to assess what Gorbachev believed, especially regarding Communism, because as the reform-minded leader of the Soviet Marxist system he could not be forthcoming regarding his true goals and beliefs. His position depended on the support of hard-line Communists on the Central Committee and elsewhere. Prior to when Gorbachev at last firmly held

complete control of the General Secretaryship of the Central Committee in 1990, he likely would have been removed at a moment's notice if he had openly criticized Communism or spoke of abolishing the Soviet Communist system. Still, there is a third problem in gauging Gorbachev: He was on a lifelong journey from being a Marxist-Leninist who, as a young man, did not perceive Lenin's or even Stalin's culpability in Soviet crimes, to, at the end, a democratic socialist who favored a more Western European kind of socialism. He became, and remains, a staunch political pluralist. More confusing, he has long used Communist rhetoric and still seems to retain some Communist goals and sympathies, though he adamantly rejects totalitarianism. He never seemed to completely discard all Marxist-Leninist ideology.

55. Ibid., 175, 207, 215.

56. He pointed out this exception many times in his remarks. Reagan, "Remarks and a Question-and-Answer Session With Area High School Seniors," Jacksonville, Florida, December 1, 1987; Reagan, "Interview with Network TV Broadcasters," December 3, 1987.

57. Importantly, this seemed to contradict Gorbachev's understanding that socialism was approaching another stage in its continuing advancement along the dialectal plane, one that Marx and Lenin envisioned as eventually leading to world socialism; however, Gorbachev either convinced himself that such a global goal was never Lenin's or, as noted earlier, convinced himself that Lenin only advocated socialism's global advancement by peaceful means. Also, Gorbachev repeatedly spoke of some form of fuzzy quasi-democratic method of Lenin, of which Gorbachev was extremely vague. See: Gorbachev, *Perestroika*, 202, 210.

58. Ibid., 152.

CHAPTER 16

1. Sylvain Boulouque in Stephane Courtois et al., *The Black Book of Communism* (Cambridge: Harvard University Press, 1999), 713–25.

2. Ibid., 712.

3. These refugees began returning only after the post-September 11 U.S. invasion in late 2001, which removed the repressive Taliban government that took root after the Soviet chaos.

4. Selig S. Harrison, "Afghanistan," in Anthony Lake et al., *After the Wars* (New Brunswick, NJ: Transaction Publishers, 1990), 45–46; and Boulouque in Courtois et al., *The Black Book of Communism*, 713–25.

5. That said, an editor told me bitterly: "Reagan gave us the Taliban," referring to the terribly repressive Islamic regime that ran Afghanistan from 1996 to 2001 and harbored Osama Bin Laden and his September 11, 2001 suicide bombers. This criticism is not just unfair and unwarranted but very shortsighted. The Reagan administration goal was to defeat the Soviets in Afghanistan, not to create the Taliban. More pointedly, while it is true that many of those who made up the Taliban came from the Mujahedin, it is equally true that many of those who opposed and helped to remove the Taliban also came from the Mujahedin. Those anti-Taliban forces included the former Muj fighters who comprised the Northern Alliance, an Afghan coalition that fought alongside U.S. troops in the October–November 2001 overthrow of the Taliban. To charge that "Reagan gave us the Taliban" would be as unreasonable as asserting that Woodrow Wilson produced Hitler. It would be less of a stretch to blame the Soviets for giving the world the Taliban, via the ruination and chaos caused by the Red Army invasion. Ultimately, the emergence of the Taliban can be blamed only on the Taliban.

6. Reagan, "Statement on the Observance of the Afghan New Year," March 20, 1982.

7. Reagan, "Statement on the Situation in Afghanistan," December 27, 1981.

8. Reagan, "Remarks on Signing the Afghanistan Day Proclamation," March 10, 1982; Reagan, "Address at Commencement Exercises at Eureka College," Eureka, Illinois, May 9, 1982; and Reagan, "Informal Exchange With Representatives of *Le Figaro*," December 22, 1983.

9. Among other sources, see the January 1980 letter from Reagan to Edward Langley, in Skinner, Anderson, and Anderson, eds., *Reagan: A Life in Letters*, 433–34.

10. Schweizer, *Victory*, 25.

11. Melor Sturua, "Aggressors and Hypocrites," *Izvestia*, March 1, 1985, 5, published as "U.S. Conducting 'Dirty War' in Afghanistan," *FBIS*, FBIS-SOV-5-MAR-85, March 5, 1985, A7–9

12. Reagan, "Statement on the 6th Anniversary of the Soviet Invasion of Afghanistan," December 27, 1985.

13. On this, see Adelman, *The Great Universal Embrace*, 136–37, 140–41; D'Souza, *Ronald Reagan*, 188; Schweizer, *Victory*, 245–46; and Morris, *Dutch*, 561–62.

14. Adelman, *The Great Universal Embrace*, 136–37, 140–41.

15. Morris, *Dutch*, 561–62.

16. Reagan, "Proclamation 5621—Afghanistan Day, 1987," March 20, 1987.

17. Moscow radio statement in Persian, published as "Savchenko Scores U.S. Aid to Afghan 'Ruffians,'" in *FBIS*, FBIS-SOV-12-MAR-81, March 12, 1981, A1; TASS statement, March 26, 1985, published as "Reagan Remarks on Afghanistan Criticized," in *FBIS*, FBIS-SOV-27-MAR-85, March 27, 1985, D1–2; Melor Sturua, "Aggressors and Hypocrites," *Izvestia*, March 1, 1985, 5, published as "U.S. Conducting 'Dirty War' in Afghanistan," *FBIS*, FBIS-SOV-5-MAR-85, March 5, 1985, A7–9; and Melor Sturua, "Aggressors and Hypocrites," *Izvestia*, March 1, 1985, 5, published as "U.S. Conducting 'Dirty War' in Afghanistan," *FBIS*, FBIS-SOV-5-MAR-85, March 5, 1985, A7–9; and TASS statement, December 29, 1985, published as "Reagan 'Maliciously Distorted' Soviet Afghan Policy," *FBIS*, 30-DEC-85, December 30, 1985, A6.

18. Steve Coll, "Anatomy of a Victory: CIA's Covert Afghan War," *Washington Post*, July 19, 1992, A1.

19. Lyakhovskii's statement, published in Russian, is cited by William E. Odom, *The Collapse of the Soviet Military* (New Haven and London: Yale University Press, 1998), 103.

20. The precise date is still classified. Peter Rodman says that it was signed in March 1985. Rodman, *More Precious Than Peace*, 337.

21. NSDD-75, January 17, 1983, 4. NSDD is on file at Reagan Library.

22. I have filed a FOIA request with the federal government to try to obtain its release.

23. Schweizer, ed., *The Fall of the Berlin Wall*, 19.

24. Coll, "Anatomy of a Victory," A1.

25. Scott cites Gelb and other sources. James M. Scott, *Deciding to Intervene: The Reagan Doctrine and American Foreign Policy* (Durham, NC, and London: Duke University Press, 1996), 59.

26. Simpson, *NSDDs*, 446–47; and Jack Anderson and Dale Van Atta, "'Perverse Opposition' to Afghan Aid," *Washington Post*, April 26, 1987.

27. Rodman, *More Precious Than Peace*, 337.

28. Simpson, *NSDDs*, 446–7; and Jack Anderson and Dale Van Atta, "'Perverse Opposition' to Afghan Aid," *Washington Post*, April 26, 1987.

29. Rodman, *More Precious Than Peace*, 337; and Coll, "Anatomy of a Victory," A1.

30. Milt Bearden, *The Main Enemy*, 180, 210.

31. Scott, *Deciding to Intervene*, 59; Schweizer, *Reagan's War*, 299; and Coll, "Anatomy of a Victory," A1.

32. Coll, "Anatomy of a Victory," A1.

33. Quoted in Vernon Loeb, "Undercover to Hardcover; Warrior Milt Bearden Is Warming Up To a New Career," *Washington Post*, December 12, 1998.

34. Coll, "Anatomy of a Victory," A1; and Schweizer, *Victory*, 151–53, 242–54.

35. Ibid.

36. "Did he [Reagan] want to see the Soviets humiliated in Afghanistan?" asked Ed Meese, repeating the question that I posed to him. "Though I don't recall hearing such words from him, he may well have said that. That would not be out of character with his goals, considering the necessity he saw in stopping and rolling back Soviet aggression." Interview with Ed Meese, September 9, 2002.

37. Schweizer, *Victory*, 151–53.

38. Interview with Bill Clark, November 1, 2004.

39. Bearden, *The Main Enemy*, 210.

40. Schweizer, ed., *Fall of the Berlin Wall*, 19–21; Schweizer, *Reagan's War*, 234–35; and various references in Schweizer's *Victory*.

41. Interview with Ed Meese, September 9, 2002.

42. Steve Coll, "In CIA's Covert Afghan War, Where to Draw the Line Was Key," *Washington Post*, July 20, 1992, A1.

43. Coll, "In CIA's Covert Afghan War," A1.

44. Cannon, *Role of a Lifetime*, 323.

45. Coll, "Anatomy of a Victory," A1. This is also stated by George Crile in his compelling *Charlie Wilson's War* (New York: Atlantic Monthly Press, 2003), 504.

46. These figures are cited by Schweizer, *Victory*, 214; and Schweizer, *Reagan's War*, 236.

47. Such language was omnipresent in Soviet official statements and press reports in the 1980s. Among others, see Moscow TASS statement January 2, 1984, published as "DRA Spokesman Hits 'Adventurist' Reagan Policy," in FBIS, January 3, 1984, 3-JAN-84, D5; Alexander Timoshkin commentary in Arabic on Moscow radio, February 4, 1983, published as "Reagan Support for Zionists, Afghans Attacked," in FBIS, February 10, 1983, 10-FEB-83, A1; Melor Sturua, "Aggressors and Hypocrites," *Izvestia*, March 1, 1985, 5, published as "U.S. Conducting 'Dirty War' in Afghanistan," *FBIS*, FBIS-SOV-5-MAR-85, March 5, 1985, A7–9; and TASS statement, December 29, 1985, published as "Reagan 'Maliciously Distorted' Soviet Afghan Policy," in *FBIS*, 30-DEC-85, A6.

48. For references on Gorbachev's insistence in ending the arms race, see: Gorbachev, *Memoirs*, 136, 171, 415, 445–46.

49. Arbatov speaking on Studio 9 Program, June 18, 1987, transcript published as "Studio 9 Participants Discuss Reagan's Recent Speeches," in *FBIS*, FBIS-SOV-19-JUN-87, June 19, 1987, A9.

50. Reagan, *An American Life*, 237–38.

51. Interview with Cap Weinberger, October 10, 2002.

52. Weinberger speaking at the conference, "Reagan's War and the War on Terrorism," hosted by the American Enterprise Institute, Washington, DC, November 13, 2002. We can probably predict two partisan reactions to Weinberger's remarks: Liberals who for years

disregarded or attacked nearly everything Weinberger said, may suddenly take his word (or maybe *half* of his words) as the final statement on this issue; they will cite his remarks as proof that Reagan never pursued a buildup to bankrupt the USSR. On the other hand, conservatives, who have long championed Weinberger as a hero, will want to ignore some of these remarks.

53. Weinberger, *In the Arena*, 266, 270–1, 274–78, 280–1, 285–86, 329.

54. Richard Allen interviewed on "Reagan," *The American Experience*, PBS.

55. The Reagan people I spoke to confirmed this thinking. It is unnecessary to quote them all here, particularly since most of them used language almost identical to what Reagan himself said, which is quoted at length below. To cite just one example, in November 2002 Jeane Kirkpatrick said of Reagan: "He understood that the Soviet economy was in decline and would continue to decline . . . and that the Soviets could never win an arms race." Source: Kirkpatrick speaking at conference, "Reagan's War and the War on Terrorism," hosted by the American Enterprise Institute, Washington, DC, November 13, 2002.

56. A photograph of the handwritten speech is printed in *Reagan, In His Own Hand*, Skinner, Anderson, and Anderson, eds., 438–42.

57. "Reagan: 'It Isn't Only Washington . . .'" *National Journal*, March 8, 1980, 392.

58. Reagan, "Interview with the President," October 16, 1981.

59. Reagan, "Remarks and a Question-and-Answer Session With Reporters on the Second Anniversary of the Inauguration of the President," January 20, 1983.

60. Reagan, "Question-and-Answer Session With High School Students on Domestic and Foreign Policy Issues," January 21, 1983.

61. Ibid.

62. Ibid. Note Reagan's formulation: A) America needs to join the arms race against the Soviets, B) the Soviets will try to match the arms race, C) "hope" of "disarmament" will follow, and D) "this is what we're [Reagan administration] doing" because "we want peace."

63. Reagan, *An American Life*, 267.

64. Speech by Deputy R. N. Stakheyev, published in *Izvestia*, November 28, 1985, 3. Published in *The Current Digest of the Soviet Press*, 37, no. 48 (1985): 14.

65. "Chance Missed, Search Continues," *Izvestia*, October 17, 1986.

66. Boris Altshuler, a Soviet expert who for years estimated the annual portion of Soviet GDP that went to the military, claims that by 1989 that portion had risen to a staggering 73%, compared to a range of 40%–50% in the 1960s and 1970s. See Reed, *At the Abyss*, 213, 218–19.

67. One of the best authorities on this is Odom, *The Collapse of the Soviet Military*, 104, 119.

68. Statement by the Moscow World Service, February 4, 1984, published as "U.S., Soviet Defense Expenditure Contrasted," in *FBIS*, FBIS-6-FEB-84, February 6, 1984, A1.

69. Gorbachev, *Memoirs*, 215. Don Regan said that he took a close look at the Soviet budget during one of the arms summits. He gleaned that the Soviets dedicated 50% of their budget to defense alone, and 17% of GDP. Regan, *For the Record*, 296.

70. Reed, *At the Abyss*, 213, 218–19.

71. Shevardnadze November 1991 interview with *International Affairs* (Moscow). Reprinted in "Retrospective on the End of the Cold War," Woodrow Wilson School, Princeton University, February 25–27, 1993, 6.

72. His conversion is based on the "then-realistic black market exchange rate" of six rubles for one dollar.

73. Based on six credible sources, both American and Soviet, this was the number cited by

Princeton's Woodrow Wilson School at its February 1993 Cold War conference. "Retrospective on the End of the Cold War," Woodrow Wilson School, Princeton University, February 25–27, 1993, 7.

74. The height of Reagan defense spending as a percentage of GDP was 6.2% in 1986. In 1981 it was 5.1%; by 1988 it was 5.8%.

75. This was a 1980 figure. Trofimenko's conversion is based on the "then-realistic black market exchange rate" of six rubles for one dollar. Another source says that by 1992, when Russia was no longer Communist and thus more honest in its numbers, GDP was 15 trillion rubles. Reed, *At the Abyss*, 224.

76. Odom, *The Collapse of the Soviet Military*, 366.

77. The peak period for military spending (defense outlays) in the 1980s was $256.6 billion in 1989, up from $164.0 billion in 1980, measured in constant FY 1982 dollars. This $256.6 billion was 26.5% of total outlays for the year, or 5.9% of GDP for that year. The highest percent of GDP for any year in the 1980s was 6.5% in 1986. *Historical Tables, Budget of the United States Government, Fiscal Year 1992, Pt. 7* (Washington, DC: Government Printing Office, 1991), 66–70.

78. Genrikh Aleksandrovich (Henry) Trofimenko in Hofstra conference (1993) proceedings, 139.

79. Trofimenko in Hofstra conference (1993) proceedings, 139.

80. An excellent case in point is the two Koreas—the Communist North and non-Communist South. To cite a typical year, in 1995 North Korea's military spending was roughly 23% of GDP, compared to a mere 3% for South Korea. Yet, because South Korea is vastly more prosperous than the North, despite the fact that the North holds the vast majority of the peninsula's land and resources, North Korea's total military spending was almost three times lower than South Korea's ($5 billion vs. $14 billion).

81. There were other contributory factors to the deficit, some Reagan's doing and some not.

82. Weinberger said that Reagan told him this "frequently," a point that Weinberger himself has made frequently. Among other examples, see Weinberger, *In the Arena*, 275.

83. Gorbachev's remarks as recorded by Morris, *Dutch*, 561.

84. Alexander Bessmyrtnykh interviewed on "Reagan," *The American Experience*, PBS. This book could be filled with such testimonies, from the Russian in the street, to military officials, to apparatchiks, to the Soviet ministry, to Gorbachev. I cut twenty pages of testimonies from this chapter.

85. Princeton University, Woodrow Wilson School of Public and International Affairs, "Retrospective on the End of the Cold War," A Conference Sponsored by the John Foster Dulles Program for the Study of Leadership in International Affairs, Princeton, NJ, February 25–27, 1993; and Bruce Olson, "SDI, Chernobyl Said to Break Cycle of Nuclear Buildup," *Executive News Service*, February 26, 1993. His remarks are also published in Wohlforth, ed., *Witnesses to the End of the Cold War*, 31–32.

86. The Soviet press articles cited throughout this chapter offer examples of both perspectives.

87. Still, as Soviet officials and Gorbachev aides and confidantes attest—such as Anatoly Chernyaev and Nikolai Detinov—Gorbachev was at the very least fearful of a workable limited antimissile system. One of the better presentations of this Soviet thinking, provided by high-level Soviet officials themselves, is a May 8, 1998 discussion held at a Brown University symposium, "Understanding the End of the Cold War," a transcript of which is provided by Brown University.

88. Gorbachev, *Memoirs*, 407.

89. Chernenko reportedly said of SDI: "The goal is . . . to deprive [the USSR] of the possibility of a reciprocal strike in case of nuclear aggression against it." Atanas Atanasov, "In Moscow, in March," *Rabotnichesko Delo* (Sofia, Bulgaria), March 8, 1985, 6, published as "Marshal Tolubko Interviewed on Geneva, SDI," in *FBIS*, FBIS-12-MAR-85, March 12, 1985, AA4.

On the Soviet TV program, "Top Priority," Sergey Plekhanov of the Institute for USA and Canada Studies said that a "shield" or "astrodome which would protect the United States from Soviet missile[s] is a pie in the sky, there's no possibility of building that." He maintained that the real purpose of SDI was to ensure the United States a "first-strike capability." " 'Top Priority' Program Addresses U.S. Policies," transcript published in *FBIS*, FBIS-30-JAN-87, January 30, 1987, A2.

90. These Gorbachev remarks were made at the 2:30–3:40 PM session on November 19. "Geneva Meeting: Memcons of Plenary Sessions and Tete-A-Tete," November 19–21, 1985, declassified May 2000, RRL, Box 92137, Folder 2. Gorbachev told Reagan in a December 1985 letter that, "Viewing the SDI program from such a position the Soviet leadership inevitably arrives at one conclusion: in the current actual conditions, the 'space shield' is needed only by the side which is preparing for a first (preemptive) strike." See Reagan, *An American Life*, 646.

91. Grinevsky speaking on May 8, 1998 at Brown University symposium, "Understanding the End of the Cold War."

92. Slipchenko continued: "Marshal Sergeev eventually shut it [the program] down as ineffective. Why did the money get invested there? The investment began early, of course, and once it got started, it continued." Slipchenko speaking on May 8, 1998 at Brown University symposium, "Understanding the End of the Cold War."

93. Tim Weiner, "Lies and Rigged 'Star Wars' Test Fooled the Kremlin, and Congress," *New York Times*, August 18, 1993, A1.

94. Quoted in Morris, *Dutch*, 544. In an April 28, 1987 interview Reagan said it was "very obvious" that Gorbachev "is faced with a tremendous economic problem, and a great deal of that problem has been aggravated, made worse, by their military buildup." Reagan, "Interview with White House Newspaper Correspondents," April 28, 1987.

95. A tense Gorbachev reported to his Central Committee that the White House "clearly wants to pull us into a second scenario of the arms race. They are counting on our military exhaustion. And they are trying to pull us in on the SDI." Also at this session, a pained Gorbachev lamented that his government was "stealing everything from the people" in trying to keep up with the Reagan spending. The Soviet attempt to match the United States was "turning the country into a military camp." These notes from the May 8, 1987 Communist Party of the Soviet Union Central Committee Session are found in Anatoly Chernyaev's diary, held at the Gorbachev Foundation. Credit goes to Peter Schweizer for finding these notes. Schweizer, *Reagan's War*, 258.

96. Gorbachev said this at the third plenary meeting, held from 11:30 AM to 12:40 PM. "Geneva Meeting: Memcons of Plenary Sessions and Tete-A-Tete," November 19–21, 1985, declassified May 2000, RRL, Box 92137, Folder 2.

97. Likewise, Gorbachev said this at the third plenary meeting. "Geneva Meeting: Memcons of Plenary Sessions and Tete-A-Tete," November 19–21, 1985, declassified May 2000, RRL, Box 92137, Folder 1, 2–12.

98. Anatoly Dobrynin, *In Confidence* (New York: Random House, 1995), 591.

99. Regan, *For the Record*, 297.

100. "Let there be no doubt," Reagan said in February 1987 in reference to SDI and the technical challenge it presented to the Kremlin, "we have no intention of being held back because our adversary cannot keep up." Reagan, "Address to the Conservative Political Action Conference," February 19, 1987.

101. Shultz, *Turmoil and Triumph*, 690–710.

102. These were the words of Reagan Secretary of Defense Frank Carlucci, referring to one such episode during the 1987 INF negotiations. Thompson, "Interview with Frank Carlucci," 52.

103. Kenneth Adelman in Hannaford's *Recollections of Reagan*, 5.

104. Reagan, "The President's News Conference," January 9, 1985. On January 14, 1988, he said flatly: "SDI is not a bargaining chip." Reagan, "Statement on the Soviet-U.S. Nuclear and Space Arms Negotiations," January 14, 1988.

105. Reagan, *An American Life*, 548, 608.

106. Reagan, "Statement on SDI," March 23, 1987.

107. Reagan, "Interview With Foreign Broadcasters on the Upcoming Soviet-United States Summit Meeting in Geneva," November 12, 1985.

108. Reagan, "The President's News Conference," October 22, 1987.

109. Ed Harper testimony is published in Thompson, ed., *Leadership in the Reagan Presidency, Pt II*, 125.

110. Thatcher recalls a discussion she had with Reagan on SDI: He admitted to her that he was unsure whether it could work, though he felt it warranted investigation. Yet, said Thatcher, Reagan told her that even if SDI was not developed it would be an enormous economic burden on the USSR: "[The president] argued that there had to be a practical limit as to how far the Soviet government could push their people down the road of austerity." Margaret Thatcher, *The Downing Street Years* (New York: HarperCollins, 1993), 467. Richard Allen spoke to this: "I don't believe Reagan saw SDI as a shield you could put over the United States, and it would be invulnerable. I believe that he thought that this was far out enough that if we put some big bucks behind it, the Soviets would have to fall for it." Schweizer, *Victory*, 136, 277–78. Reagan's new national security adviser, John Poindexter, said that Reagan "understood just what he had in the SDI system, and that it posed a grave challenge to the Kremlin." For additional illuminating statements on SDI and the Reagan buildup as forces for change in the Soviet Union, see remarks by Soviet officials on the public television documentary, "Messengers from Moscow," WNET-New York, PBS, 1995.

111. NSDD-153, January 1, 1985. NSDD on file at RRL.

112. James A. Nathan, "Decisions in the Land of Pretend: U.S. Foreign Policy in the Reagan Years," *Virginia Quarterly Review*, Winter 1989, 13. Nathan says he was told this by a senior NSC aide.

113. Edward Wright testimony before University of Virginia's Miller Center in Thompson, ed., *Leadership in the Reagan Presidency, Pt II: Eleven Intimate Perspectives*, 182.

114. Transcript of Nitze's remarks in Wohlforth, ed., *Witnesses to the End of the Cold War*, 40.

115. Ken Adelman said that Reagan kept the SDI program from being hashed and rehashed by his staff and "gummed to death" by the bureaucracy. "No other president," wrote Ken Adelman, "would have been as romantic, even visionary, as to launch and push the concept the way Reagan did." Ken Adelman, *The Great Universal Embrace* (New York: Simon & Schuster, 1989), 298.

116. "Geneva Meeting: Memcons of Plenary Sessions and Tete-A-Tete," November 19–21, 1985, declassified May 2000, RRL, Box 92137, Folder 2.

117. Reagan, *An American Life*, 14–15, 637.

118. Quoted in Morris, *Dutch*, 570.

119. Reagan, *An American Life*, 708.

120. Reagan, "Address to a Special Session of the European Parliament," Strasbourg, France, May 8, 1985.

121. Reagan, "Remarks at the Annual Convention of the American Legion," Seattle, Washington, August 23, 1983.

122. Reagan, "Interview With Morton Kondracke and Richard H. Smith of *Newsweek* Magazine," March 4, 1985.

123. Reagan, "Remarks and Q & A With Reporters on Domestic and Foreign Policy Issues," March 25, 1983.

124. Interview with Bill Clark, August 24, 2001.

125. "Reagan's Foreign Policy—His No. 1 Aide Speaks Out," *U.S. News & World Report*, May 9, 1983.

126. See Schweizer's works, including his *The Fall of the Berlin Wall*, 40; Also see Pipes, *Vixi*, 158.

CHAPTER 17

1. Data cited by Reed, *At the Abyss*, 215. At the time, the United States was second and Saudi Arabia was third.

2. See Schweizer's works, including his *The Fall of the Berlin Wall*, 45–46.

3. Paul Taylor, "Furor Over Remarks Fails to Dismay Bush," *Washington Post*, April 14, 1986; Schweizer, *Victory*, 242–43; and Richard Alm, "Alarm bells over cheaper oil; while consumers cheer plummeting prices, Vice President Bush and others are warning enough is enough," *U.S. News & World Report*, April 14, 1986, 24. A read of the *U.S. News & World Report* article demonstrates the extent to which the media was unaware of the true reason for the increase in Saudi oil production and low prices.

4. Interview with Roger Robinson, June 6 and 8, 2005. See Richard V. Allen, "The Man Who Changed the Game Plan," *The National Interest*, Summer 1996, 64; Schweizer, *Victory*, 140–44, 261; and Schweizer, *Reagan's War*, 239.

5. CIA report is cited in Schweizer, *Victory*, 261–63.

6. Cited in Schweizer, *Victory*, 261–63.

7. On this, Peter Schweizer gives a number of examples. Schweizer, *Victory*, 263.

8. Ibid., 166.

9. See analysis by Grigorii Khanin in Ellman and Kontorovich, eds., *The Disintegration of the Soviet Economic System*, 77.

10. Richard V. Allen, "The Man Who Changed the Game Plan;" and Schweizer, *Victory*, 262.

11. Gorbachev, *Memoirs*, 216.

12. Mikhail Gorbachev, *On My Country and the World* (New York: Columbia University Press, 2000), 26.

13. Gorbachev, *Memoirs*, 177.

14. Ellman and Kontorovich, eds., *The Disintegration of the Soviet Economic System*, 2–3, 26.

NOTES 383

15. Schweizer, *Victory*, 255–56, 261.

16. Prince Turki al-Faisal, "Allied Against Terrorism," *The Washington Post*, September 17, 2002, A21.

17. Bush's role in this was very interesting. For a discussion, see Paul Kengor, *Wreath Layer or Policy Player? The Vice President's Role in Foreign Policy* (Lanham, MD: Rowman-Littlefield, 2000).

18. Reagan, "The President's News Conference," October 1, 1981.

19. Weinberger told this to Peter Schweizer. Quote is in Schweizer, ed., *The Fall of the Berlin Wall*, 44.

20. Reagan, "Remarks and an Informal Exchange With Reporters on the U.S. Arms Sale to Saudi Arabia," June 3, 1986.

21. The one obstacle was probably the ambivalence and personal dealings of Vice President Bush.

22. The date of this was September 26, 1986. This information was taken from a variety of sources, including: Rodman, *More Precious Than Peace*, 339–40; Fred Barnes "Victory in Afghanistan: The Inside Story," *Reader's Digest*, December 1988, 87–93; and interviews.

23. Barnes, "Victory in Afghanistan," 87.

24. Rodman, *More Precious Than Peace*, 336–37.

25. Barnes, "Victory in Afghanistan," 91.

26. Cited in Rodman, *More Precious Than Peace*, 339. Schweizer says that the Stingers downed 75% of their targets in the first year. Schweizer, *Victory*, 269–70. Steve Coll says that estimates vary widely, from 30% to 75%, though he cited no source. Coll, "In CIA's Covert Afghan War," A1.

27. Martin Schram, "Reagan Urges U.S. Mideast Presence," *Washington Post*, January 10, 1980, A3.

28. Other contemporary sources from 1980, including Lou Cannon in the *Washington Post*, reported that Reagan supported sending arms (in general) to the Mujahedin. Lou Cannon, "A Vision of America Frozen in Time," *Washington Post*, April 24, 1980, A2.

Surely, someone must have advised Reagan on the Stingers; nonetheless, the earliest on-the-record account of this idea (as far as I could find) was from the mouth of Ronald Reagan.

29. Statement from the Moscow Domestic Service, published as "Afghan 'Basmachs' Visit Washington Seeking Missiles," in *FBIS*, FBIS-SOV-26-FEB-81, February 26, 1981, A1.

30. On this, see especially Rodman, *More Precious Than Peace*, 337–38.

31. Ibid., 338.

32. Barnes, "Victory in Afghanistan," 88–89.

33. Simpson, *NSDDs*, 446–47.

34. Rodman, *More Precious Than Peace*, 337–38.

35. Barnes, "Victory in Afghanistan," 90.

36. Rodman, *More Precious Than Peace*, 339; and Barnes, "Victory in Afghanistan," 90.

37. McFarlane was interviewed by Schweizer in Schweizer, *Reagan's War*, 255–56.

38. I was unable to ascertain whether these were the same group of individuals, nor the precise day of the visit.

39. Coll, "Anatomy of a Victory," A1.

40. Rodman, *More Precious Than Peace*, 339.

41. Jack F. Matlock, *Autopsy of an Empire* (New York: Random House, 1995), 747–48.

42. Reagan, *An American Life*, 660. Later recalling Gorbachev's "strong motives for wanting to end the arms race," Reagan asserted: "The Soviet economy was a basket case, in part because of enormous expenditures on arms. He had to know that the quality of American military technology, after reasserting itself beginning in 1981, was now overwhelmingly superior to his. He had to know we could outspend the Soviets on weapons as long as we wanted to" (14–15).

43. Reagan to William F. Buckley, Jr., May 5, 1987 in Skinner, Anderson, and Anderson, eds., *Reagan: A Life in Letters*, 418–19.

44. Gorbachev speaking on Soviet television, October 14, 1986. Remarks provided by Peter Schweizer.

45. Shevardnadze November 1991 interview with *International Affairs* (Moscow), in "Retrospective on the End of the Cold War," Woodrow Wilson School, Princeton University, February 25–27, 1993, 6.

46. Bessmertnykh speaking during February 25–27, 1993 conference at Princeton University. Remarks published in Wohlforth, ed., *Witnesses to the End of the Cold War*, 31–32.

47. Ibid., pp. 47–48.

48. Estimate is provided by Peter Schweizer in *Reagan's War*.

49. Ellman and Kontorovich, eds., *The Disintegration of the Soviet Economic System*, 13–16, 19. Michael Ellman and Vladimir Kontorovich, economics professors with expertise on the Soviet economy, said that it was the "unprecedented peacetime escalation of military expenditures and the other costs of being a superpower" that sapped the USSR. They note that while the USSR was doomed in the long run, it was certainly not predestined to disintegrate in the late 1980s. That demise, Ellman and Kontorovich found, "could have been alleviated by appropriate policies." Even Yuri Andropov had notable success by applying strict policies, particularly by replacing ineffective managers and corrupt officials and ministers; he also benefited from a series of better harvests in the agricultural sector. In all, the Soviet economy had recovered from its 1979–82 decline.

50. Akhromeev quoted by Ellman and Kontorovich, "The collapse of the Soviet system and the memoir literature," *Europe-Asia Studies*, 49, no. 2 (March 1997): 259.

CHAPTER 18

1. Reagan, "Remarks on East-West Relations at the Brandenburg Gate in West Berlin," June 12, 1987.

2. Ibid.

3. Interview with Peter Robinson, October 8, 2004.

4. Studio 9 Program, June 18, 1987, published as "Studio 9 Participants Discuss Reagan's Recent Speeches," in *Foreign Broadcast Information Service*, FBIS-SOV-19-JUN-87, June 19, 1987, A1–10.

5. Gorbachev, *Perestroika*, 199.

6. This is almost verbatim what he had written in *Perestroika*. See Gorbachev, *Perestroika*, 200; and Brown, *The Gorbachev Factor*, 244–45. According to Brown, not until February 1990 did Gorbachev tell West German Chancellor Helmut Kohl that it was up to the Germans to determine the kind of state they would have and on what timetable. At this point, of course, the matter was well out of Gorbachev's hands. To his credit, Gorbachev kept Soviet troops out of the process, though, again, such a solution at that point would have been almost impossible.

7. Brown, *The Gorbachev Factor*, 249.

8. Ibid. To be sure, Gorbachev later moved away from this sentiment. Yet, this language, especially shared privately, in no way helped those hoping for democracy in East Germany and Eastern Europe as a whole. To the contrary, it very much hurt their efforts.

9. The first of these was in an August 7 interview with the West German publication *Bild Zeitung*. After describing the wall as an "affront to the human spirit" and a symbol of the "failings of totalitarian regimes," he was asked: "When do you believe the wall can be torn down?" He replied: "I would like to see the wall come down today, and I call upon those responsible to dismantle it." He urged the wall come down again in an August 12 news conference. The skeptical reporter retorted: "How realistic is it, though? Some critics have suggested it raises false hopes for those beyond the wall." Reagan said he did not think so. He assessed again on August 13—the wall's twenty-fifth anniversary—stating that the wall "cannot be" a permanent structure and affirming that it "will come down." See, respectively, Reagan, "Written Responses to Questions Submitted by Bild-Zeitung of the Federal Republic of Germany," August 7, 1986; Reagan, "The President's News Conference," August 12, 1986; and Reagan, "Statement on the 25th Anniversary of the Berlin Wall," August 13, 1986.

10. Reagan, "Written Responses to Questions Submitted by Deutsche Presse-Agentur of the Federal Republic of Germany," June 2, 1987.

11. Reagan, "Written Responses to Questions Submitted by Die Welt of the Federal Republic of Germany," June 12, 1987.

12. Reagan, "Statement on the Fifth Anniversary of the Founding of the Solidarity Movement in Poland," August 31, 1985.

13. Reagan, "Statement on the Lifting of Economic Sanctions Against Poland," February 19, 1987.

14. Robinson, *How Ronald Reagan Changed My Life*, 99–100, 192.

15. Reagan, "Address to the Conservative Political Action Conference," Washington, DC, February 19, 1987, in Roberts, ed., *A City Upon a Hill*, 120.

16. In May, he made a number of freedom statements in West Germany. On May 5, 1985 in Bitburg, he claimed: "[W]e can see a new dawn of freedom sweeping the globe. And we can see in the new democracies of Latin America, in the new economic freedoms and prosperity in Asia . . . that the light from that dawn is growing stronger." Reagan, "Remarks at a Joint German-American Military Ceremony at Bitburg Air Base in the Federal Republic of Germany," May 5, 1985. Traveling west to Oklahoma for another political fundraiser on June 5, he told his fellow Americans: "Like it or not, freedom depends on us." Reagan, "Remarks at a Fundraising Luncheon for Senator Don Nickles," Oklahoma City, Oklahoma, June 5, 1985.

In a July 18, 1986 speech to, of all groups, the members of the American Legion Auxiliary's Girls Nation, Reagan said that the "challenges America must face in the world" were "twofold." "The first is expanding the boundaries of democracy and freedom by curbing, in the face of totalitarian expansion, that urge on the part of some governments to seek domination of even more territory and peoples." The second was to reduce the threat of nuclear war. He said he was "confident" that both of the goals would be achieved. Reagan, "Remarks to Members of the American Legion Auxiliary's Girls Nation," July 18, 1986.

In his January 27, 1987 State of the Union speech, Reagan said it was crucial to prevent Communism's spread into the Western Hemisphere. Reagan, "State of the Union Address," January 27, 1987.

On January 24, 1985, Reagan declared that "free and democratic government is the birthright of every citizen of this hemisphere." Reagan, "Remarks at the Western Hemisphere Legislative Leaders Forum," January 24, 1985.

17. Reagan, "Remarks at the Western Hemisphere Legislative Leaders Forum," January 24, 1985. See Reagan, "Remarks to the International Forum of the Chamber of Commerce of the United States," April 23, 1986; Reagan, "State of the Union Address," January 27, 1987; Reagan, "Address to the Permanent Council of the OAS," October 7, 1987.

18. Reagan, "Remarks at the Western Hemisphere Legislative Leaders Forum," January 24, 1985.

19. This was November 25, 1986.

20. Reagan, "Remarks to Civic Leaders at a White House Briefing on Aid to the Nicaraguan Democratic Resistance," January 20, 1988.

21. The recurring Chautauqua conference was long an interactive dialogue on relations between the two nations.

22. Reagan, "Remarks on Soviet-US Relations at the Town Hall of California Meeting," Los Angeles, August 26, 1987.

23. Reagan, "Remarks at the 40th Anniversary Conference of the United States Advisory Commission on Public Diplomacy," September 16, 1987.

CHAPTER 19

1. Barnes, "Victory in Afghanistan," 91.

2. Tarasenko is quoted in Wohlforth, ed., *Witnesses to the End of the Cold War*, 141.

3. Barnes, "Victory in Afghanistan," 87.

4. Crile, *Charlie Wilson's War*.

5. John Lewis Gaddis, *U.S. News & World Report*, October 18, 1999, 43.

6. I'm indebted to the research of Stephen F. Knott, who identified these examples in his "Reagan's Critics," *The National Interest*, Summer 1996, 72.

7. Russell Watson and others, "Inside Afghanistan," *Newsweek*, June 11, 1984, 55.

8. Russell Watson and John Barry, "Insurgencies: Two of a Kind," *Newsweek*, March 23, 1987, 32.

9. Richard Cohen, "Why Aid Afghanistan?" *Washington Post*, January 2, 1985, A15; and Richard Cohen, "The Soviets' Vietnam," *Washington Post*, April 22, 1988, A23.

10. Gorbachev, *Memoirs*, 171, 197, 365, 138.

11. Adelman, *The Great Universal Embrace*, 228.

12. I'm indebted to military historian Tod Reiser for this insight.

13. These are the words of Congressman Charlie Wilson. Crile, *Charlie Wilson's War*, 522–23.

14. I found at least a half dozen clear *wall calls* before Reagan's June 12, 1987 speech, including the three occasions prior to his presidency. Further, I found an additional nine wall calls or affirmations made publicly by Reagan between July 24, 1987 and August 12, 1988, and there may well have been more. Here are examples of Reagan calling on Gorbachev or the USSR to tear down the wall (this is not a complete list): Reagan, "Remarks to the Captive Nations Conference," July 24, 1987; Reagan, "Remarks on Soviet-US Relations at the Town Hall of California Meeting," Los Angeles, August 26, 1987; Reagan, "Remarks at the 40th Anniversary Conference of the United States Advisory Commission on Public Diplomacy," September 16, 1987; Reagan, "Radio Address to the Nation Following the North Atlantic Treaty

Organization Summit Meeting," Brussels, Belgium, March 5, 1988; Reagan, "Address to the Citizens of Western Europe," May 23, 1988; Reagan, "Remarks to the Paasikivi Society and the League of Finnish-American Societies," Helsinki, Finland, May 27, 1988; Reagan, "Remarks on Signing the Captive Nations Week Proclamation," July 13, 1988; Reagan, "Statement on the 27th Anniversary of the Berlin Wall," August 12, 1988; and Reagan, "Radio Address to the Nation on Soviet-United States Relations," December 3, 1988. At Moscow State University on May 31, 1988, Reagan called the Berlin Wall "one sad reminder of a divided world" and then said, "It's time to remove the barriers that keep people apart." Reagan, "Remarks and a Question-and-Answer Session With the Students and Faculty at Moscow State University," May 31, 1988.

15. Reagan, "Address to the Citizens of Western Europe," February 23, 1988.

16. Also see: Reagan, *An American Life*, 705–7.

17. Igor Korchilov, *Translating History: Thirty Years on the Front Lines of Diplomacy With a Top Russian Interpreter* (New York: Scribner/Drew Books, 1997), 156.

18. Reagan, "Remarks to Members of the Royal Institute of International Affairs," London, June 3, 1988.

19. Reagan, "Remarks to Members of the Royal Institute of International Affairs in London," June 3, 1988.

20. Reagan, "Remarks on Signing the Captive Nations Week Proclamation," July 13, 1988.

21. For yet another important speech in the months ahead, see Reagan, "Remarks at Georgetown University's Bicentennial Convocation," October 1, 1988.

22. Reagan, "Remarks at the Republican National Convention," New Orleans, Louisiana, August 15, 1988. He updated the tally in an October 1, 1988 speech at Georgetown.

23. Reagan, "Proclamation 5869—Polish American Heritage Month," September 28, 1988.

24. NSDD-320, November 20, 1988, 1–5; and Simpson, *NSDDs of Reagan and Bush*, 58.

25. Writing in *Pravda* on March 27, 1986, Kolesnichenko had no illusions about what Reagan was up to: He was "trying" to " 'roll back communism,' to exhaust the USSR's economy and undermine its political system." This "effort," which the Reagan administration was seeking through "methods that they employ over and over," which was tantamount to "reversing the wheels of history," would be an "unrealizable dream," assured the confident comrade Kolesnichenko. T. Kolesnichenko, "In a Web of Stereotypes—Washington's Imperial Ambitions," *Pravda*, March 27, 1986, 4, published as "Has Washington Lost 'Spirit of Geneva,' " in *The Current Digest of the Soviet Press*, 38, no. 13 (1986): 1–2. Also see Tomas Kolesnichenka, from the "Vremya" newscast, Moscow Television Service, July 15, 1988, published as "Editor Raps 'Hostile' Reagan Speech," in FBIS-SOV-88-137, July 18, 1988, 29.

26. In November 1986, in what can indeed be best characterized as a raving diatribe, Arbatov analyzed a revealing Reagan speech to the Ethics and Public Policy Center in Washington. The occasion was the thirtieth anniversary of the Soviet crushing of the 1956 Hungarian uprising, in which tens of thousands were killed by Red tanks. President Dwight Eisenhower had agonized that the United States could do nothing to support those Hungarians literally dying to be free. He decided not to intervene. Reagan, who admired Ike, lamented that America had "stood by, hands folded." In a calmer moment, Arbatov said it was clear that if Reagan had been president in 1956, the United States would not have done nothing. He went further: "No matter how you look at it there is only one way to interpret all this," insisted Arbatov. "U.S. policy is reverting not to 1956 but to 1918, when American troops participated in the ignominiously failed intervention in Soviet Russia." Back then, Woodrow Wilson sent U.S. troops

to try to unseat the Bolsheviks during the Russian civil war. That, assured Arbatov, was Reagan's goal—to dislodge the Bolsheviks from power. The goal, he said, was to "conquer" the "empire of evil." G. Arbatov, "Not Just for the Sake of It; On R. Reagan's Speech," *Pravda*, November 21, 1986, 4, published as "Arbatov Rebuttal to 18 Nov Reagan Speech," in *FBIS*, FBIS, November 21, 1986, A4–5.

27. Reagan, said Ponomarev, would "wreck them by means of an arms race." His "chief aim" against the Soviet Union and other socialist states remained to seek "every way possible to destroy socialism and exact social revenge in the world arena." The Crusader, who did childish things like proclaim an official "Year of the Bible," retained a religious devotion to "fighting communism on a 'world scale.'" Manki Ponomarev, "The United States: Policy With No Future," *Krasnaya Zvezda*, March 8, 1987, 3, published as "Army Paper on Links Between Reagan, 'Truman Doctrine,'" in *FBIS*, March 13, 1987, A6–8.

CHAPTER 20

1. Reagan, "Remarks and a Question-and-Answer Session at the University of Virginia," Charlottesville, Virginia, December 1988.

2. Reagan, "Remarks to the National Chamber Foundation," November 17, 1988.

3. Reagan, "Farewell Address to the Nation," January 11, 1989.

4. Ibid.

5. Andrew Nagorski, "Reagan Had It Right," *Newsweek*, October 21, 2002, 68.

6. Gennady Vasilyev, "A Political Portrait: Reagan's Best Role," *Pravda*, January 20, 1989, 6. Published in *The Current Digest of the Soviet Press*, 41, no. 3 (1989): 1.

7. This information was shared with me by Zawitkowski. Interview with Chris M. Zawitkowski, November 9, 2005.

8. Reagan gave the crucifix to his friend Bill Clark, who donated it to the Villanova Prep School (Order of Saint Augustine) in Ojai, California in June 1984. Chris Woodka, "Secretary of Interior Clark honored," *San Francisco Chronicle*, July 7, 1984.

9. Daniel Yergin and Joseph Stanislaw, *The Commanding Heights: The Battle for the World Economy* (New York: Simon and Schuster, 1998), 273–74.

10. Gorbachev, *Memoirs*, 478–79. In a 1993 conference at Princeton, Sergei Tarasenko spoke of how the nation elections in Poland in June 1989—in which Solidarity emerged the big winner—convinced him and the Soviet leadership, particularly the foreign ministry, that the Soviet system would break up. Wohlforth, ed., *Witnesses to the End of the Cold War*, 112–13.

11. Bernstein, "The Holy Alliance," 28–35.

12. Carl Bernstein and Marco Politi say $50 million in total. Peter Schweizer says $8 million annually. Other estimates seem close to these. Bernstein and Politi, *His Holiness*, 357; and Schweizer, *Victory*, 75–76.

13. So was Voice of America (VOA). Along with sources in the Western Hemisphere like Radio Marti, which penetrated Cuba, RFE and VOA were among Reagan's favorite projects. These forces of freedom were put under the control of Reagan's close friend, Charles Wick, who headed USIA. Some of them—Radio Marti in particular—were considered way too provocative by the State Department. It was a continuing battle for Wick to get them up and running; only because of Reagan's full support did Wick succeed.

14. Interview with Joseph Dudek, Radek Sikorski, March 3, 2003. "This was the only

source of media we had of what was going on in the free world. We learned who was being imprisoned and it helped in keeping the spirits of Poles up. I listened to them all the time."

15. *In the Face of Evil: Reagan's War in Word and Deed* (American Vantage Films and Capital Films I, LLC, 2005).

16. The Walesa quote is from the proceedings of "The Failure of Communism: The Western Response," a conference sponsored by Radio Free Europe / Radio Liberty, November 15, 1989, in Munich Germany.

17. Interview with Lech Walesa, April 25, 2005.

18. Interview with Jan Winiecki, March 11, 2003.

19. Pipes, *Vixi*, 167, 183. Secretary of State Shultz in particular did not see the Poland situation through the same ideological lens as Reagan, Pipes, Clark, and others. Pipes notes that Shultz pushed to lift the sanctions on Poland and prodded Reagan to a more conciliatory position.

20. Marek Jan Chodakiewicz, "Miracle of Solidarity Ended Communism," *Human Events*, September 26, 2005, 9.

21. Bernstein and Politi, *His Holiness*, 260.

22. "President Reagan's political support helped it survive martial law to become the decisive catalyst in the eventual chain reaction of communist collapse at the end of the 1980s." James H. Billington, "The Foreign Policy of President Ronald Reagan," Address to the International Republican Institute Freedom Dinner, Washington, DC, September 25, 1997, 2.

23. Arch Puddington, "Voices in the Wilderness: The Western Heroes of Eastern Europe," *Policy Review*, Summer 1990, 34–35.

24. Piecuch was interviewed by Schweizer. Schweizer, *Reagan's War*, 236.

25. Arch Puddington found that Poles liked Reagan's vitality, sense of humor, and willingness to call a spade a spade. They admired his anti-Soviet rhetoric, particularly the "Evil Empire" speech. That explosive remark, as well as the president's prediction that Communism would end up on the "ash heap of history," thrilled Poles. Reagan had spoken their language. Bartak Kaminski, a Polish émigré teaching at the University of Maryland, explained that Reagan was the first world leader in the post-détente era who was willing to express ideas about the Soviets that were shared by most Poles. Kaminski pointed to the strong policy response to the declaration of martial law. He said that this was imperative in undermining the legitimacy of the Jaruzelski government. It stood in marked contrast to the shameful friendliness that the West and Nixon administration extended to the odious regime of Nicolai Ceausescu in Romania. In Romania, notes Kaminski, such accommodation had the effect of undercutting the opposition. Jerzy Warman, a student activist in Polish politics, agreed. Warman pointed to the Reagan defense buildup and aid to anti-Communist forces in the Third World, which he believed sent a signal to the entire Soviet bloc that the Communists "simply could not hope to win." Puddington, 34–35.

Poles applauded Reagan's verbal cruise missiles launched at Soviet Communism. "He was right, 100% right," said Polish citizen Joseph Dudek of Reagan's labeling the USSR an "Evil Empire." "For the oppressed in Poland it was relief that someone had the courage to stand up for the right thing and name the Communist system as it really was. . . . Most of the people in Poland were hoping he'd do something to expose Russia and the Communists and end it [Soviet domination of Poland]. They wanted him to fix the mistakes Churchill and Roosevelt made after World War II." Dudek also pointed to Reagan's "Tear-Down-This-Wall" speech.

For Dudek and other Poles, Reagan's Cold War candor was both a symbol and a weapon, and his choice phrases gave them hope. A native of Krakow who is careful to also emphasize the role of the Pope, Dudek says that Reagan was among the few outside of Poland who understood the Communists, and that his "confident speaking" gave the Solidarity leaders support. Dudek assures: "He was a part of keeping alive the hopes of Polish freedom." He said Poles viewed Reagan as someone who would say almost anything in standing up to the Communists. Interview with Joseph Dudek.

Credit to Reagan also came from Boguslaw W. Winid. Winid has a doctorate in history from Warsaw University and served at the Polish Embassy in Washington, DC. He was a researcher at Warsaw University's American Studies Center. Asked what role, if any, the United States played in Solidarity's victory over Communism, Winid replied: "I think it played quite an important role. This was particularly true with President Ronald Reagan . . . who clearly defined the Soviet Union as an Evil Empire. And though many arguments have been made about the value of his approach, I believe it was important that he portrayed situations as right or wrong and not as a gray area. Moral issues were important in dealing with a country like the Soviet Union. Reagan's hard approach toward the Soviet Union was very helpful from the Polish point of view and made him very popular in Poland." Winid's talk February 24, 1994, in Kenneth W. Thompson, ed., *The Presidency and Governance in Poland: Yesterday and Today* (Lanham, MD: University Press of America, 1997), 115.

From inside the Polish Communist government that tried to smash Solidarity, General Kiszczak, who was Poland's Interior Minister in the 1980s, later said: "The assistance from [the] American government for Solidarity was essential." Kiszczak interviewed by Schweizer, Schweizer, *Reagan's War*, 276.

Vladimir Bukovsky agrees. For twelve years Bukovsky languished in the Communist gulag, where he was shifted back and forth among prisons, work camps, and, as was the cruel Communist custom, even lunatic asylums. He now lives in Cambridge, England. By the time of Solidarity's struggles, he was a free man living in the West, albeit one intimately informed of the Poland situation. Asked about the Reagan administration's importance to the survival of Solidarity, he remarked succinctly: "It was crucial." Interview with Vladimir Bukovsky, March 8, 2003.

From within Solidarity, a telling testimony comes from Jan Winiecki, a Warsaw native who earned doctorates in economics and public administration at the University of Warsaw and was an economic adviser to the Solidarity underground in the 1980s. "Of all the things most key to Solidarity's survival, most important was the sheer will, sheer desire, of the Polish people," Winiecki stresses. Yet, he says the Reagan impact on Solidarity was significant: "It's very important for those underground to know they'll have support diplomatically if they're repressed. They knew they could count on Reagan and his administration for this rhetorical, moral, public support—this political support. It raised their spirits that they could survive."

Winiecki believes that the three main external impetuses to Solidarity were, in order of happenstance, the pope's 1979 visit, 1981 martial law, and Reagan's advocacy from 1981–89. He believes that Reagan's role was fundamental to the Soviet collapse, and that the Reagan administration accelerated the implosion by at least a decade. He points to not only Reagan's support of Solidarity but also to SDI and the arms race. "This was because the USSR tried to meet the Reagan challenge through the arms race, the technological competition of SDI," said Winiecki, who also cites aid to Solidarity among causal factors. Interviews with Jan Winiecki, March 17, 1998, March 4, 1999, and March 7, 2000.

26. Walesa spoke at a conference on "The Reagan Legacy," held at the Reagan Library on

May 20, 1996. Mack Reed, "Walesa Hails Reagan at Daylong Seminar," *Los Angeles Times*, May 21, 1996, A1, A18.

27. For his part, Reagan called Walesa a heroic figure. Reagan, "Remarks on Signing the Human Rights Day, Bill of Rights Day, and Human Rights Week Proclamation," December 8, 1988, in John O'Sullivan, "Friends at Court," *National Review*, May 27, 1991, 4.

28. In April 2005, Walesa told me that Reagan had "emboldened" and "encouraged" him in the 1980s. He spoke of Reagan's "testimony to the truth and liberty" and "his understanding of life as living in light without a lie," which "was always an inspiration for me." Walesa's words were translated by Tomasz Pompowski, a senior editor and reporter at *Fakt*, the largest newspaper in Poland. As he translated, Pompowski could not help but add that Reagan had meant a "great deal" to him as well. Interview with Lech Walesa, April 25, 2005.

29. Walesa spoke at a conference on "The Reagan Legacy," held at the Reagan Library on May 20, 1996. Mack Reed, "Walesa Hails Reagan at Daylong Seminar," *Los Angeles Times*, May 21, 1996, A1, A18.

30. Lech Walesa, speech at the conference, "The Reagan Legacy," Ronald Reagan Library, Simi Valley, California, May 20, 1996.

31. Reagan, *An American Life*, 303.

32. Ibid., 301.

33. Reagan, "Address at Commencement Exercises at Eureka College," Eureka, Illinois, May 9, 1982.

34. A reader can literally count on one hand the total number of references to Solidarity in Edmund Morris' *Dutch* and Lou Cannon's *Role of a Lifetime* and still have two fingers two spare. Incredibly, Lech Walesa is not mentioned even once in either 870-plus-page work.

35. Clark said, "In Afghanistan, the Soviets lost face; in Poland, they lost an empire." He said this twice. See "The Pope and the President: A key adviser reflects on the Reagan Administration," interview with Bill Clark, *Catholic World Reporter*, November 1999; and Bill Clark, "President Reagan and the Wall," Address to the Council of National Policy, San Francisco, California, March 2000, 8.

36. See Radek Sikorski, "Christmas Day in Romania," *National Review*, January 22, 1990, 23–24.

37. Maciek Gajewski, "In Solidarity's Cradle, Poles Applaud Reagan," *United Press International*, September 16, 1990; and "Poles give Reagan a hero's welcome," *Reuters*, September 16, 1990.

38. This, of course, was a play on the face of Helen of Troy that launched a thousand ships. Gajewski, "In Solidarity's Cradle, Poles Applaud Reagan."

39. Interviews with Radek Sikorski, February 28 and March 3, 2003.

CHAPTER 21

1. Dobrynin, *In Confidence*, 632.

2. In the mid-1990s the Heritage Foundation began publishing a popular Index of Economic Freedom, which measures degrees of economic freedom. Former Communist countries like the Czech Republic and Estonia ranked as high as eleven, far freer than Western European economies like France, Italy, and Spain (Gorbachev's favored model).

3. Brown, *The Gorbachev Factor*, 306.

4. Ibid., 91, 130, 252, 267.

5. For a superb source on this episode, which includes interviews with all of the participants, see the PBS documentary, "Yeltsin," produced by Pacem Productions and First Circle Films, 2000. The documentary was broadcast nationally in August 2000. Shevardnadze resigned on December 20, 1990. In his speech to the Fourth Congress of People's Deputies to the USSR, he said he was resigning as Foreign Minister and warned against "the onset of dictatorship" in the USSR.

6. Brown, *The Gorbachev Factor*, 280.

7. Kryuchkov speaking on "Yeltsin" documentary.

8. Brown, *The Gorbachev Factor*, 280–82, 383–84, n145. It can also be said that even if Gorbachev did not order the violence, he was responsible for hiring those who did, and, worse, for failing to subsequently remove them from their posts.

9. Here, Gorbachev strongly and quickly condemned the shootings and said firmly, "I resolutely reject all speculation and suspicion" that he was responsible. Brown, *The Gorbachev Factor*, 282.

10. Gorbachev speaking on "Yeltsin" documentary.

11. Editorial, "Gorbachev's Tanks," *The New Republic*, February 4, 1991, 9–10.

12. The Tbilisi, Georgia incident took place in April 1989, a few weeks prior to the Tiananmen Square massacre in China. As in Beijing, thousands of young people participated in peaceful demonstrations over several days, seeking democracy and independence from Gorbachev's USSR, until the night of April 8–9 when Soviet troops responded brutally. They used asphyxiation agents—poisonous gas—fired from canisters. The majority of those who choked to death were young women. Rather than hold the USSR together, the incident further propelled the independence movement throughout the various republics. Soviet Communist authorities were ripping the union apart by the very policies and actions intended to keep it together. Archie Brown says that Gorbachev did not authorize this use of force by Communist authorities and had in fact urged that the demonstration be resolved peacefully through dialogue. Brown, *The Gorbachev Factor*, 265–66.

13. One of the better testimonies to how Gorbachev neither planned nor desired the upheaval in Eastern Europe was a December 1992 review essay by Michael Cox in the scholarly journal *Soviet Studies*. Michael Cox, "Beyond the Cold War in Europe: A Review Article," *Soviet Studies*, 44, no. 6 (December 1992): 1099–1103.

14. Brown, *The Gorbachev Factor*, 268.

15. In his superb work published by Yale University Press, William E. Odom cautiously discusses the situation in December 1991 in which Gorbachev summoned Marshal Yevgenny Shaposhnikov to the Kremlin and, according to Shaposhnikov, raised the prospect of using the military to preserve Gorbachev's hold on power. As Odom notes, this account can only be confirmed by Gorbachev, "who is hardly going to do so." William E. Odom, *The Collapse of the Soviet Military* (New Haven and London: Yale University Press, 1998), 352–54.

16. Matlock, *Reagan and Gorbachev*, 302.

17. Ibid., 304–6.

18. The headline read: "Soviet Leaders Agree to Surrender Communist Party Monopoly on Power." The accompanying news article was by Francis X. Clines.

19. Jack Matlock agrees with me on these points. See Matlock, *Reagan and Gorbachev*, 317–18.

20. Brown, *The Gorbachev Factor*, 105–6, 193–94.

21. Ibid., 96.

22. Reagan, "Address to the Cambridge Union Society," Cambridge, England, December 5, 1990, in Ryan, ed., *Ronald Reagan: The Wisdom and Humor of the Great Communicator.*

23. This is shown nicely, but not intentionally, in an excellent, detailed chronology of Soviet events listed in the index of Jack Matlock's *Autopsy of an Empire*, 749–59.

24. Valery Boldin, *Ten Years That Shook the World* (New York: Basic Books, 1994), 336. Ambassador Jack Matlock, a keen observer as close to the scene as anyone, put it well when he summarized: "When he came to power, Gorbachev still believed in the Soviet system. . . . He was convinced that the instrument to achieve . . . changes would be the Communist Party itself. . . . Gorbachev's policies eventually . . . destroy[ed] the system they were intended to save." Matlock, *Reagan and Gorbachev*, 110.

25. Reed, *At the Abyss*, 225.

26. Gorbachev gave this speech on December 25, 1991.

27. Princeton University, Woodrow Wilson School of Public and International Affairs, "Retrospective on the End of the Cold War," A Conference Sponsored by the John Foster Dulles Program for the Study of Leadership in International Affairs, Princeton, NJ, February 25–27, 1993; and Bruce Olson, "SDI, Chernobyl Said to Break Cycle of Nuclear Build up," *Executive News Service*, February 26, 1993. His remarks are also in Wohlforth, ed., *Witnesses to the End of the Cold War*, 31–32.

28. Interview with Alexander Donskiy, October 2, 2000. Donskiy now travels with the Kiev symphony.

29. Lukhim quoted by Robert McFarlane in his memoir, *Special Trust* (New York: Cadell & Davies, 1994), 235.

30. Trofimenko in Hofstra conference (1993) proceedings, 136–37.

31. Ibid., 138–40. *The Economist* of London agreed: "He [Reagan] said that he would spend the Soviet Union into submission; he did." *The Economist*, June 5, 1993.

32. Trofimenko in Hofstra conference (1993) proceedings, 134–45, particularly 138.

33. Trofimenko stated: "[T]he greatest flimflam man of all time, Mikhail Gorbachev, was made the Nobel Peace Prize laureate." Trofimenko in Hofstra conference (1993) proceedings, 138–45; and Andrew E. Busch, "Ronald Reagan and the Defeat of the Soviet Empire," *Presidential Studies Quarterly*, Summer 1997, 455–62.

34. A surprising source of agreement is Strobe Talbott, *Time* magazine's chief Reagan critic and later the Clinton administration's top man on Russia. Talbott conceded that SDI had "given the Soviets an incentive to return to the bargaining table and offer serious proposals in the hope of tightening the bonds of arms control around SDI itself." He agreed that after the Soviets had walked out of arms control talks in 1984, "SDI was a factor in luring them back to the bargaining table." "For that Reagan deserves credit," conceded Talbott. "SDI helped elicit from the Soviets a dizzying barrage of proposals." Strobe Talbott, "Grand Compromise: SDI could end the arms-control stalemate," *Time*, June 23, 1986, 22, 25.

35. Morris in Wilson, ed., *Power and the Presidency*, 131.

36. Cannon testimony at University of Virginia's Miller Center in Thompson, ed., *Leadership in the Reagan Presidency, Pt II: Eleven Intimate Perspectives*, 61–62.

37. Shultz, *Turmoil and Triumph*, 264, 690, 699, 709.

38. Interview with Edward Teller, July 15, 2003.

39. Thatcher said that SDI proved "central to the West's victory in the Cold War." "Looking back," she said, "it is now clear to me that Ronald Reagan's original decision on SDI was the single most important of his presidency." Thatcher, *Downing Street Years*, 463.

40. Gaddis, *The United States and the End of the Cold War*, 130–31; and Gaddis, "Hanging Tough Paid Off," 11–14.

41. Garthoff's exact words were that the Reagan military buildup and SDI "posed a military challenge that the Soviet Union was economically and technologically hard pressed to meet. . . . Gorbachev understood that the Soviet Union could not afford to match or overmatch the United States militarily." Garthoff maintains that Gorbachev understood that he did not need to try to match or overmatch the United States. Gorbachev, he argues, consequently felt that the world would be better off without military competition: "He did not lose the arms race, he called it off." Garthoff, *The Great Transition*, 775.

42. Ibid., 614, 605–6, 635, 720. See Reagan, "The President's News Conference," December 8, 1988; and "Remarks Following the Soviet-United States Summit Meeting in Moscow," June 2, 1988.

43. Reagan, "The President's News Conference," December 4, 1985; and "Remarks and a Question-and-Answer Session With Broadcast Journalists on the Meetings in Iceland With Soviet General Secretary Gorbachev," October 14, 1986.

44. Ronald Reagan, *An American Life*, 508.

45. Letter located in ES, NSC, HSF: Records, USSR, RRL, Box 41, Folder 8890009-8890639.

46. Gorbachev, *Memoirs*, 457. He has made identical or near identical remarks in Gorbachev speaking on "Reagan," *The American Experience*, PBS; and Gorbachev, July 1997 statement, in *A Shining City: The Legacy of Ronald Reagan* (New York: Simon and Schuster, 1998), 222.

47. Gorbachev speaking on "Reagan," *The American Experience*, PBS.

48. "Ron and Gorby Remember the Good Old Days," *Reuters*, September 18, 1990.

49. Jack Matlock, who attended the dinner, relays the exchange in *Reagan and Gorbachev*, 326.

50. Gorbachev speaking on "Ronald Reagan: A Legacy Remembered," History Channel productions, 2002.

CHAPTER 22

1. Morris, *Dutch*, 667.

2. Deaver also makes the point that Reagan himself saw his lifeguarding as a parable of his larger life. Deaver, *A Different Drummer*, 14–15.

3. Interview with Olive and Savila Palmer and Marion Emmert Foster, Heritage Square nursing home, Dixon, Illinois, June 22, 2001.

4. The authoritative Harvard University Press study, *The Black Book of Communism*, estimates 100 million dead. The book was published prior to the two to three million deaths that occurred in North Korea in the late 1990s. Also, it uses a figure on the USSR that is conservative and, judging from recent seminal works on Mao's China, may have underestimated the death toll in Red China by ten million. No doubt, at least 100 million were killed under Communism, and likely many more.

5. Data was provided by Freedom House staff, April 8, 2002.

6. Thomas Mann, director of the Governmental Studies program at the Brookings Institution, agrees that a "success of the Reagan administration was the process of democratization—the march toward free governments around the world." Thomas Mann in Berman, ed., *Looking Back on the Reagan Presidency*, 26. Professor Robert A. Pastor, a Latin America expert at Emory University and the Carter Center, who is stingy in his praise of Reagan administration policy, concedes that, "the 1980s saw an unprecedented wave of democratization in Latin America" in particular. Robert A. Pastor in Berman, ed., *Looking Back on the Reagan Presidency*, 33.

7. Gorbachev letter was posted at the Web site of the Reagan Library and Museum.

8. Vladimir Isachenkov and Jim Heintz, "Reagan Mourned in Former 'Evil Empire,'" *Moscow Times*, June 7, 2004.

9. David E. Hoffman, "Hastening an End to the Cold War," *Washington Post*, June 6, 2004.

10. Russert interviewed by Chris Jansing, MSNBC, June 6, 2004.

11. Ted Anthony, "U.S. and the World Mourn Reagan's Death," Associated Press, June 5, 2004. Kennedy's statement was also posted on the Web site of the Reagan Library and Foundation.

12. Dan Pavel, "Superpower United States: From Reagan to Bush Jr.," *Bucharest Ziua*, June 14, 2004. Published by FBIS as "Romanian Daily Praises Reagan Role in Ending Cold War, Rejects Anti-US Attitude."

13. Alexei Pankin, "Revising Reagan's Role in Soviet History," *The Moscow Times*, June 15, 2004.

14. Statement from the Polish news agency PAP, "Former Polish President Says Reagan Helped to Overthrow Communism," June 5, 2004. Published by FBIS as "Former Polish President Says Reagan Helped to Overthrow Communism."

15. "A Salute to a Great President—And A Modest Proposal," *Budapest Business Journal*, June 14, 2004.

16. Vladimir Isachenkov and Jim Heintz, "Reagan Mourned in Former 'Evil Empire,'" *Moscow Times*, June 7, 2004.

17. Gillian Flaccus, "Immigrants from Former Soviet Union Mourn Reagan," *Associated Press*, June 9, 2004.

18. Quoted in "Asia Remembers Former US President Reagan as Friend, Vanquisher of Communism," *Paris AFP (North European Service)*, June 6, 2004.

19. Mohammad Ashraf Azeem, "Ronald Reagan's America," *Islamabad Khabrain*, June 9, 2004. Translated and published by FBIS as "Pakistan Columnist Lauds Ronald Reagan's Policies, Period," June 9, 2004.

20. Those future generations can look back with thanks to a number of Cold War leaders: John F. Kennedy and Nikita Khrushchev stepped back from the abyss, and many more were dedicated to avoiding that fate, from Truman to Carter. But it was Reagan and Gorbachev who were there at the end—an end completely unforeseen when Reagan arrived at 1600 Pennsylvania Avenue in January 1981. Those two leaders deserve special gratitude for ending the Cold War peacefully, with no nuclear weapons launched.

21. In mid-September 1961, Reagan spoke to the Press Club of Orange County, California. "There can be only one end of the war we are in," said Reagan firmly: "wars end in victory or defeat." Schweizer, *Reagan's War*, 35, in which he cites "Warns of Red Menace: Film Star Ronald Reagan to Speak," *Bakersfield Californian*, September 16, 1961, 19–20; and Ronald Reagan, "Encroaching Control," *Vital Speeches of the Day*, September 1961.

EPILOGUE

1. John Lewis Gaddis, "The Long Peace," *International Security*, 10, no. 4 (Spring 1986). Reprinted in Sean M. Lynn-Jones and Steven E. Miller, eds., *The Cold War and After: Prospects for Peace* (Cambridge and London: The MIT Press, 1993), 33.

2. Clark shared this during a February 22, 1999 presentation in Washington, DC. For a transcript, see: Clark in Schweizer, ed., *The Fall of the Berlin Wall*, 75.

3. Marc Fisher, "The Old Warrior at the Wall," *Washington Post*, September 13, 1990, D1–2.

4. Ibid.

5. Lawrence K. Altman, "Reagan's Twilight—A special report," *New York Times*, October 5, 1997.

ACKNOWLEDGMENTS

It is impossible to thank all of those who played a role in this book. Their contribution is partly reflected in the 1,200 endnotes, which comprise over a quarter of the overall text. I'm often asked how many people I interviewed in the course of researching and writing this book over the last roughly ten years. My best estimate is that I've done hundreds of interviews.

Those that gave their time were Reagan officials as diverse in their views as George Shultz and Cap Weinberger, both of whom frequently disagreed in the 1980s but, fortunately for me, agreed to talk with me. Weinberger shared his insights several times. He died a week before I handed in the final draft of this manuscript.

I appreciate the time of men as well-known as Lech Walesa (my special thanks to his translater, Tomasz Pompowski) as well as people as little known as Dixonites Marion Emmert Foster, the Palmer sisters, Ron Marlow (Dixon's best kept secret), and a host of others. The latter individuals reaffirm my continuing advice to Reagan researchers: You can't know Ronald Reagan without going to Dixon, Illinois.

I spoke to a number of Reagan speechwriters: Ben Elliott, Peter Robinson, Ken Khachigian. I talked to many Reagan friends, from Charlie Wick to the extraordinary Bill Clark, the subject of my next book. Most of those I spoke to gave more than one interview. This is especially true of Clark and is also true of Richard V. Allen, Ed Meese, Richard Pipes, Michael Reagan, and others. I commend Bill and Michael Reagan for their commitment to telling the truth about Ronald Reagan's dedication to human life, from the gulag to the womb.

I was especially appreciative to Edward Teller. To the best of my knowledge, I was the last person to interview Dr. Teller before he died. As I walked into his modest home in Palo Alto, California on July 15, 2003, and caught my first glance at him sitting horizontally in a reclining chair, I was shocked. This man of such renown and prestige was on his deathbed. I had hoped for fifteen minutes with Teller. After my first question, he paused and murmured: "We have no time." I told him that was fine and I would go. I asked if he wanted me to get his nurse. He said yes. When I returned, he was ready to talk. We proceeded to talk for almost an hour and a half. He became increasingly alert. Still, he struggled greatly, and his words were often indiscernible.

I felt compelled to ask Teller a question I had not intended: *If he didn't mind*, I asked, *could he tell me about his spiritual views? Did he believe in God?* The situation seemed to beg the question. He told me emphatically: "I strongly believe that I should not talk about things I don't understand." The scientist within came rushing forth: Teller said he did not have adequate "information" to make a determination about God's existence. When I asked again later, when the matter presented itself naturally, he retreated to his earlier position: "You ask me about God, and I will say, *'I don't know.'* And if I don't know, I won't talk about it." I smiled and told him I would not try again.

When I insisted (for his sake) that it was indeed time for me to go, he finally relented to let me leave. He bid me farewell by saying simply, "Go write a good book." He managed to sign a copy of his memoirs for me, without ever getting up or lifting his head. I told him I would return the gesture by sending a signed copy of my book once it was published. As I knew, I would never get the chance: shortly after I returned to Grove City, Edward Teller died on September 9, 2003 at the age of 95.

I believe I also may have done the final interview with a less-known but likewise significant individual named Gus Weiss, a Reagan NSC staff member whose story was so intriguing that snippets of it—though certainly not the entire story—made the front page of the *Washington Post* on February 27, 2004.

Other than Ronald Reagan himself, I interviewed every person that I wanted to speak to for this book. All were very cooperative and eager that this story be told. I remember that Professor Fred Greenstein, in his seminal work on Eisenhower, said that he had conversations with associates of all modern presidents since FDR, and that none were more unified than Eisenhower's in admiration of their leader. I found the same with Reagan's staff, especially those who knew him since the California days, men like Clark, Meese, and Weinberger.

In regard to research, I'm the only Reagan researcher who had access to the private collection of letters sent by Reagan to Mary Joan Roll-Seiffert, who in the 1940s was the president of the Pittsburgh chapter of Reagan's Hollywood fan club. I was also told that I was among the first to read the Lorraine Wagner letter collection held by the Young America's Foundation (YAF). The Wagner-Reagan collection of correspondence is rich, and reaches from the 1940s to the 1990s. Also with the help of YAF, and specifically Floyd Brown and Andrew Coffin, I was given full access to the bookshelves at the Reagan Ranch in the Santa Ynez Mountains of California. I was permitted to spend four hours pulling each book off the shelf and searching it for personal annotations from Reagan. I was able to document not only the types of books that Reagan read but also what he remarked upon and thought about these books as he read them—meaning, I read his marginalia. The subject matter in the books ranges from economic theory to treaty agreements with the USSR to biographies to westerns and cowboy novels.

Likewise, the documents at the Reagan Library were obviously very illuminating. In the summers of 2001, 2003, and 2005, I benefited from the diligence of archivists like Greg Cumming, Cate Sewall, and Ben Pezzillo, as well as Kirby Hanson, Holly Bauer, Duke Blackwood, and other Reagan Library staff—a solid group of people. I read a large number of documents only recently declassified. This book placed its heaviest emphasis on primary sources. All declassified NSDDs were carefully read. I tried to read every speech, letter, or Cold War–related document in the Presidential Handwriting File (PHF) at the Reagan Library. In each case, I ascertained Reagan's exact input in each. The PHF is a file of every declassified document featuring any Reagan handwriting. From this, I was able to determine *precisely* Reagan's input. I read literally thousands of letters to and from Reagan during his presidency (and long before then). The correspondence is as diverse as Soviet general secretaries like Yuri Andropov to men like Bob Hope, Richard Nixon, Malcolm Muggeridge, Billy Graham, and men and women from Everytown, America. I read the Memoranda of Conversations from Geneva, Reykjavik, and the other summits, most declassified only as recently as 2000–2001.

There were a number of Grove City College students who were especially helpful. Among my excellent researchers were John McCay, Jen Velencia, Hans Yehnert, Jeff Chidester, Betsy Christian, Allan Edwards, Leah Ayers, Jennifer Moyer, and Melissa Harvey. John vetted all of the publicly available Reagan *Presidential Documents*, which cover 10,000 pages of text. This is a massive, official record of every presidential statement. Any document that featured key words like "USSR," "Communism," "Poland," "Brezhnev," and so forth was flagged by John—*before* the collection was transferred online! Jen Velencia searched various newspaper archives for all pre-presidential references to Reagan from the 1940s to the 1970s; this was done for the *New York Times, Washington Post,* and *Wall Street Journal,* as well as a number of select California papers and conservative publications. Jennifer Moyer stepped up and dug out articles buried even deeper from the 1940s. Betsy Christian (with the help of Henry Johnston) read issues of *Pravda* (in Russian) from October through November 1947.

Melissa Harvey's work was particularly impressive. She completed a comprehensive search of Reagan material from Soviet newspaper and media archives—a volume of incredible material that, to my knowledge, only I possess. Melissa flagged all Reagan references from the 1950s through 2000 in the archives of FBIS and the *Current Digest of the Soviet Press.* FBIS is the Foreign Broadcast Information Service, which on a daily basis has long translated all foreign media—print, TV, radio, etc.—and published it in the United States. Through FBIS, I obtained Soviet and Eastern bloc opinions on Reagan. I was able to peruse innumerable analyses of Reagan by the likes of *Pravda, Izvestia,* TASS, the Moscow Television Service, and untold others. This involved thousands of articles. It was because of Melissa's research that this book is titled *The Crusader.* Indeed, I had once read in a 1996 book by Norman Wymbs that "inside the Kremlin," Reagan was referred to as "Crusader." This was news to me. I searched but couldn't confirm the reference. I called Wymbs in September 2001. He couldn't remember his source for the reference. I soon forgot about it. Yet, once I found countless examples of numerous Soviets calling Reagan "the Crusader" in media archives, I quickly sensed I had an unexpected title for this book.

Also among Grove City College students, Elaine Rodemoyer meticulously incorporated my handwritten edits from many drafts. When she graduated, Rachel Bovard completed the final round. I'm very fortunate to teach at a college filled with extremely intelligent, honest, hard-working, unselfish, and reliable students. Likewise helpful was Grove City College's splendid library staff, particularly Joyce Kebert.

I must also express appreciation for the spectacular research of Martin and Annelise Anderson and Kiron Skinner, who have uncovered thousands of Reagan handwritten documents, from letters and speeches to those eye-opening radio broadcasts. In so doing, they've produced a treasure trove for researchers like myself to ascertain the real Reagan. Their good work will live on for many years to come. All future Reagan scholars owe them.

Special appreciation goes to Marko Suprun, who brought to my attention the Chebrikov–Andropov letter on Ted Kennedy. Marko knows the horrors of Soviet Communism; his father survived the 1930s Ukrainian genocide. Herb Romerstein, the authority on the Venona papers and Soviet archives, first provided the document to Walter Zaryckyj, who turned it over to Marko for translation. They were all very gracious. Walter is correct in referring to Herb as a "national treasure."

I should note here that at least a half-dozen independent sources, scholars with rich expertise and experience in working with the Soviet archives—all of whose names appear in this Acknowledgments section—carefully read the Chebrikov–Andropov letter and judged it authentic. None had any doubts. Richard Pipes, the Harvard professor of Sovietology, who has been a member of the Harvard faculty since 1950 (he is now professor emeritus), told me that he believes he had read the document before, probably in Moscow in 1992 when he was doing archive work on the Lenin papers. This was the period when the *London Times* reported on the document and on John Tunney's travels to Moscow on behalf of certain U.S. senators. The *Times* featured a photograph of the upper left corner of the document, which listed the person to whom the letter was directed, plus some added information. Very careful, diligent research left no other conclusion than that the document is legitimate.

I also thank my agent, Leona Schecter, for her hard work and persistence, as well as her husband, Jerry, for his excellent advice. I cherish their friendship.

I also cherish my friendships with Peter Schweizer and Chuck Dunn. When I first suggested the concept for this book, I suggested it to Peter, with the intent of prodding *Peter* (not myself) to write the book. Peter responded: "Maybe *you're* the guy to write that one." About the same time, Chuck Dunn, a Grove City College dean, impressed upon me the importance of book writing generally. They created a monster.

Looking back, the actual writing of the book began even before Peter and Chuck as an independent study supervised in 1995 (and the research started well before then) by my doctoral advisor at the University of Pittsburgh, Paul Y. Hammond, a gentleman, a scholar, and a mentor. Pitt has been lucky to have him all these years.

Another early influence was Lee Edwards, who has now been publishing material on Ronald Reagan for five decades.

I also appreciate the support of Grove City College colleagues, especially Michael Coulter, Marv Folkertsma, Lee Wishing, John Sparks, Bill Anderson, John Moore, and Dick Jewell.

I'd like to reiterate my love for my wonderful family, my wife Susan and four children, Paul, Mitch, Amanda, and Abigail, and my parents, Paul and Gloria. To my father: thanks for working so hard to turn around that clueless teenager. To Susan: thanks for helping me block out the time to write, not to mention so much more than that. I've been blessed with much more than I deserve.

Finally, I've benefited mightily from the publishing savvy of Judith Regan and the editorial skills of the superb Cal Morgan and Matt Harper. If not for Matt's fantastic guidance, this book would not read the way it now does (that's an understatement). Matt did a truly magnificent job. And I greatly appreciate the support of a number of foundations: grants for this book were provided by the Smith Richardson Foundation, the Earhart Foundation, and the Historical Research Foundation. Earhart was kind enough to provide two grants, including a second grant for the summer 2003 trip to the Reagan Library.

Most notably, the first person to support this work was an unheralded entrepreneur from Pittsburgh named B. Kenneth Simon, who became a friend. I was deeply saddened to learn upon my return from a summer of research at the Reagan Library in 2003 that Ken had unexpectedly passed away. Ken so much wanted to see this book completed. This book is dedicated to him.

PAUL KENGOR
Grove City, Pennsylvania
April 17, 2006

INDEX